Acclaim for Robert Wright's

NONZERO:

The Logic of Human Destiny

"An original, accessible and thought-provoking view of history . . . full of rich detail, ingenious insight and bold argument." —*The Economist*

"Wright carries his learning lightly, and his bold attempt to uncover parallels between organic evolution and the development of human cultures makes for a compelling synthesis. . . . Wright is right about so many things: evolution is seeded with inevitabilities, cultures have common trajectories, and human history has seen great hopes and terrible crimes but is capable of achieving a final destiny."
—Simon Conway Morris, *The New York Times Book Review*

"An extraordinarily insightful and thought-provoking book. . . . Wright does an astonishingly effective job of finding directionality in history, not just over the past few thousand years but over the almost four billion years since the beginning of life on earth."
—Francis Fukuyama, *The Wilson Quarterly*

"A dazzling tour of world history. . . . Although he takes into account the tooth-and-claw battles of nations, the vanished empires, social violence and chaos, the shocks and changes of technology, Mr. Wright finds pattern and meaning in history. We are moving toward connectedness, toward one world. . . . Does that mean we can rest on our laurels and simply let the game go on? No; Mr. Wright wants us consciously to take charge." —*The Ottawa Citizen*

"A must for history buffs . . . highly readable. . . . Wright's chatty, informal style makes all the difference." —*The Singapore Straits Times*

"At long last, here is a 'millennium book' that is definitely worth reading. . . . An enormously skillful summary of everything you always wanted to know about history and science but were afraid to ask." —David Davidar, *The Hindu* (India)

ROBERT WRIGHT

NONZERO

Robert Wright is the author of *Three Scientists and Their Gods* and *The Moral Animal,* which was named by *The New York Times Book Review* as one of the twelve best books of the year and has been published in nine languages. A recipient of the National Magazine Award for Essay and Criticism, Wright has published in *The Atlantic Monthly, The New Yorker, The New York Times Magazine, Time,* and *Slate.* He was previously a senior editor at *The New Republic* and *The Sciences.* He lives in Washington, D.C., with his wife and two daughters.

ALSO BY ROBERT WRIGHT

The Moral Animal

Three Scientists and Their Gods

NONZERO

THE LOGIC OF HUMAN DESTINY

ROBERT WRIGHT

VINTAGE BOOKS
A DIVISION OF RANDOM HOUSE, INC.
NEW YORK

FIRST VINTAGE BOOKS EDITION, JANUARY 2001

The Library of Congress has cataloged the Pantheon edition as follows:
Wright, Robert, 1957–
Nonzero : the logic of human destiny / Robert Wright.
p.   cm.
Includes bibliographical references and index.
ISBN 0-679-44252-9
1. Social evolution.   2. Human evolution.   3. History.   I. Title.
GN360.W75   2000
303.4'21—dc21   99-040859

Vintage ISBN: 0-679-75894-1

Book design by JoAnne Metsch

www.vintagebooks.com

Printed in the United States of America
10 9 8 7 6 5 4 3 2 1

For
Eleanor
and
Margaret

As man advances in civilization, and small tribes are united into larger communities, the simplest reason would tell each individual that he ought to extend his social instincts and sympathies to all the members of the same nation, though personally unknown to him. This point being once reached, there is only an artificial barrier to prevent his sympathies extending to the men of all nations and races.

—*Charles Darwin*

CONTENTS

Contents

PART III   FROM HERE TO ETERNITY

NONZERO

THE STORM BEFORE THE CALM

A great many internal and external portents (political and social
upheaval, moral and religious unease) have caused us all to feel,
more or less confusedly, that something tremendous is at present
taking place in the world. But what is it?

—*Pierre Teilhard de Chardin*

The Nobel laureate Steven Weinberg once ended a book on this note:
"The more the universe seems comprehensible, the more it also seems
pointless." Far be it from me to argue with a great physicist about how
depressing physics is. For all I know, Weinberg's realm of expertise,
the realm of inanimate matter, really does offer no evidence of overarching purpose. But when we move into the realm of animate
matter—bacteria, cellular slime molds, and, most notably, human
beings—the situation strikes me as different. The more closely we
examine the drift of biological evolution and, especially, the drift of
human history, the more there seems to be a point to it all. Because in
neither case is "drift" really the right word. Both of these processes
have a direction, an arrow. At least, that is the thesis of this book.

People who see a direction in human history, or in biological evolution, or both, have often been dismissed as mystics or flakes. In some
ways, it's hard to argue that they deserve better treatment. The
philosopher Henri Bergson believed that organic evolution is driven
forward by a mysterious "élan vital," a vital force. But why posit something so ethereal when we can explain evolution's workings in the
wholly physical terms of natural selection? Pierre Teilhard de Chardin,
the Jesuit theologian, saw human history moving toward "Point
Omega." But how seriously could he expect historians to take him,
given that Point Omega is "outside Time and Space"?

On the other hand, you have to give Bergson and Teilhard de Chardin some credit. Both saw that organic evolution has a tendency to create forms of life featuring greater and greater complexity. And Teilhard de Chardin, in particular, stressed a comparable tendency in human history: the evolution, over the millennia, of ever more vast and complex social structures. His extrapolations from this trend were prescient. Writing at the middle of this century, he dwelt on telecommunications, and the globalization it abets, before these subjects were all the rage. (Marshall McLuhan, coiner of "global village," had read Teilhard.) With his concept of the "noosphere," the "thinking envelope of the Earth," Teilhard even anticipated in a vague way the Internet—more than a decade before the invention of the microchip.

 Can the trends rightly noted by Bergson and Teilhard—basic tendencies in biological evolution and in the technological and social evolution of the human species—be explained in scientific, physical terms? I think so; that is largely what this book is about. But the concreteness of the explanation needn't, I believe, wholly drain these patterns of the spiritual content that Bergson and Teilhard imputed to them. If directionality is built into life—if life naturally moves toward a particular end—then this movement legitimately invites speculation about what did the building. And the invitation is especially strong, I'll argue, in light of the phase of human history that seems to lie immediately ahead—a social, political, and even moral culmination of sorts.

As readers not drawn to theological questions will be delighted to hear, such speculation constitutes a small portion of this book: a few cosmic thoughts toward the end, necessarily tentative. Mostly this book is about how we got where we are today, and what this tells us about where we're heading next.

THE SECRET OF LIFE

On the day James Watson and Francis Crick discovered the structure of DNA, Crick, as Watson later recalled it, walked into their regular lunch place and announced that they had "found the secret of life." With all due respect for DNA, I would like to nominate another candidate for the secret of life. Unlike Francis Crick, I can't claim to have discovered the secret I'm touting. It was discovered—or, if you prefer, invented—about half a century ago by the founders of game theory, John von Neumann and Oskar Morgenstern.

They made a basic distinction between "zero-sum" games and "non-zero-sum" games. In zero-sum games, the fortunes of the players are inversely related. In tennis, in chess, in boxing, one contestant's gain is the other's loss. In non-zero-sum games, one player's gain needn't be bad news for the other(s). Indeed, in highly non-zero-sum games the players' interests overlap entirely. In 1970, when the three *Apollo 13* astronauts were trying to figure out how to get their stranded spaceship back to earth, they were playing an utterly non-zero-sum game, because the outcome would be either equally good for all of them or equally bad. (It was equally good.)

Back in the real world, things are usually not so clear-cut. A merchant and a customer, two members of a legislature, two childhood friends sometimes—but not always—find their interests overlapping. To the extent that their interests do overlap, their relationship is non-zero-sum; the outcome can be win-win or lose-lose, depending on how they play the game. (For elaboration on non-zero-sum logic, and discussion of the classic non-zero-sum game "the prisoner's dilemma," see appendix 1.)

Sometimes political scientists or economists break human interaction down into zero-sum and non-zero-sum components. Occasionally, evolutionary biologists do the same in looking at the way various living systems work. My contention is that, if we want to see what drives the direction of both human history and organic evolution, we should apply this perspective more systematically. Interaction among individual genes, or cells, or animals, among interest groups, or nations, or corporations, can be viewed through the lenses of game theory. What follows is a survey of human history, and of organic history, with those lenses in place. My hope is to illuminate a kind of force—the non-zero-sum dynamic—that has crucially shaped the unfolding of life on earth so far.

The survey of organic history is brief, and the survey of human history not so brief. Human history, after all, is notoriously messy. But I don't think it's nearly as messy as it's often made out to be. Indeed, even if you start the survey back when the most complex society on earth was a hunter-gatherer village, and follow it up to the present, you can capture history's basic trajectory by reference to a core pattern: New technologies arise that permit or encourage new, richer forms of non-zero-sum interaction; then (for intelligible reasons grounded ultimately in human nature) social structures evolve that realize this rich

potential—that convert non-zero-sum situations into positive sums.[†]
Thus does social complexity grow in scope and depth.

This isn't to say that non-zero-sum games always have win-win
outcomes and never have lose-lose outcomes. (Badly governed societies are littered with losses, and history is littered with the remains of
badly governed societies.) Nor is it to say that the powerful and
the treacherous never exploit the weak and the naïve; exploitation—
ranging from clear-cut parasitism to subtler inequity—is often possible
in non-zero-sum games, and history offers plenty of examples. Still,
on balance, over the long run, non-zero-sum situations produce more
positive sums than negative sums, and more mutual benefit than parasitism. As a result, people become embedded in larger and richer webs
of interdependence.

This basic sequence—the conversion of non-zero-sum situations
into mostly positive sums—had started happening at least as early as
15,000 years ago. Then it happened again. And again. And again.
Until—voilà!—here we are, riding in airplanes, sending e-mail, living
in a global village.

I don't mean to minimize the interesting details that populate most
history books: Sumerian kings, barbarian hordes, medieval knights, the
Protestant Reformation, nascent nationalism, and so on. In fact, I try to
give all of these their due (along with such too-often-neglected exemplars of the human experience as native American hunter-gatherers,
Polynesian chiefdoms, Islamic commercial innovations, African kingdoms, Aztec justice, and precocious Chinese technology). But I do
intend to show how these details, though important in their own right,
are ultimately part of a larger story—to show how they fit into a framework that makes thinking about human history easier.

After surveying human history, I will briefly apply to organic history the same organizing principle. Through natural selection, there
arise new "technologies" that permit richer forms of non-zero-sum
interaction among biological entities: among genes, or cells, or animals, or whatever. And the rest, as they say, is organic history.

In short, both organic and human history involve the playing of ever-
more-numerous, ever-larger, and ever-more-elaborate non-zero-sum
games. It is the accumulation of these games—game upon game upon

† This and all subsequent daggers refer to elaborative notes that can be found in the Notes
section at the end of the book.

game—that constitutes the growth in biological and social complexity that people like Bergson and Teilhard de Chardin have talked about. I like to refer to this accumulation as an accumulation of "non-zero-sumness." Non-zero-sumness is a kind of potential—a potential for overall gain, or for overall loss, depending on how the game is played. The concept may sound ethereal in the abstract, but I hope it will feel concrete by the end of this book. Non-zero-sumness, I'll argue, is something whose ongoing growth and ongoing fulfillment define the arrow of the history of life, from the primordial soup to the World Wide Web.

You might even say that non-zero-sumness is a nuts-and-bolts, materialist version of Bergson's immaterial élan vital; it gives a certain momentum to the basic direction of life on this planet. It explains why biological evolution, given enough time, was very likely to create highly intelligent life—life smart enough to generate technology and other forms of culture. It also explains why the ensuing evolution of technology, and of culture more broadly, was very likely to enrich and expand the social structure of that intelligent species, carrying social organization to planetary breadth. Globalization, it seems to me, has been in the cards not just since the invention of the telegraph or the steamship, or even the written word or the wheel, but since the invention of life. The current age, in which relations among nations grow more non-zero-sum year by year, is the natural outgrowth of several billion years of unfolding non-zero-sum logic.

YOU CALL THAT DESTINY?

Any book with a subtitle as grandiose as "The Logic of Human Destiny" is bound to have some mealy-mouthed qualification somewhere along the way. We might as well get it over with.

How literally do I mean the word "destiny"? Do I mean that the exact state of the world today has been inevitable for eons? No, on two counts.

First, I'm talking not about the world's exact, detailed state, but about its broad contours, such as the scope and nature of its political and economic structures. Second, I'm not talking about something that was literally inevitable—100 percent guaranteed ever since the dawn of history, or the dawn of life, or whatever. Still, I am talking about something whose chances of transpiring were very high—something that was "in the cards" in the sense that the deck was stacked heavily in its favor.

Some people may consider it cheating to use the word "destiny" when you mean not "inevitable" but "exceedingly likely." Would you consider it cheating to say that the destiny of a poppy seed is to become a poppy? Obviously, a given poppy seed may *not* become a poppy. Indeed, the destiny of some poppy seeds seems—in retrospect, at least—to have been getting baked onto a bagel. And even poppy seeds that have escaped this fate, and landed on soil, may still get eaten (though not at brunch) and thus never become flowers.

Still, there are at least three reasons that it seems defensible to say that the "destiny" of a poppy seed is to become a poppy. First, this is very likely to happen under broadly definable circumstances. Second, from the seed's point of view, the only alternative to this happening is catastrophe—death, to put a finer point on it. Third, if we inspect the essence of a poppy seed—the DNA it contains—we find it hard to escape the conclusion that the poppy seed is programmed to become a poppy. Indeed, you might say the seed is *designed* to become a poppy, even though it was "designed" not by a human designer, but by natural selection. For anything other than full-fledged poppyhood to happen to a poppy seed—for it to get baked onto a bagel or eaten by a bird—is for the seed's true expression to be stifled, its naturally imbued purpose to go unrealized.

It is for reasons roughly analogous to these that I will make an argument for human destiny. Of course, the human-poppy analogy gets most contentious when we ponder the third reason: Is it fair to say that our species has some larger "purpose"? Is there some grand goal that life on earth was "designed" to realize? I think the reasons for answering yes are stronger than many people—especially many scientists and social scientists—realize. Still, as I've already suggested, this question is slippery, and answers to it must be speculative. In contrast, destiny in the more modest sense of the word—a likely outcome, an outcome that history naturally moves toward—is a fairly concrete proposition, more clearly open to empirical appraisal. This book is a full-throated argument for destiny in this sense of "direction," along with a more reticent argument for destiny in the sense of "purpose."

THE CURRENT CHAOS

Neither biological evolution nor human history is a smooth, steady process. Both pass through thresholds; they can leap from one equilibrium to a new, higher-level equilibrium. To some people, the current era has the aura of a threshold; it has that unsettling, out-of-control feeling that can portend a major shift. Technological, geopolitical, and economic change seem ominously fast, and the fabric of society seems somehow tenuous.

For instance: World currency markets are rocked by the turbulent force of electronically lubricated financial speculation. Weapons of mass destruction are cultivated by rogue regimes and New Age cults. Nations seem less cohesive than before, afflicted by ethnic or religious or cultural faction. Health officials seriously discuss the prospect of a worldwide plague—the unspeakably gruesome Ebola virus, perhaps, or some microbe we don't yet know about, spread around the world by jet-propelled travelers. Even tropical storms seem to have grown more intense in recent decades, arguably a result of global warming.

It sounds apocalyptic, and some religiously minded people think it literally is. They have trouble imagining that this rash of new threats could be mere coincidence—especially coming, as it has, at the end of a millennium. Some fundamentalist Christians cite growing global chaos as evidence that Judgment Day is around the corner. A whole genre of best-selling novels envisions "the Rapture," the day when true believers, on the way to heaven, meet Christ in midair, while others, down below, find a less glamorous fate.

In a sense, these fundamentalists are right. No, I don't mean about the Rapture. I just mean that growing turmoil does signify, by my lights, a distinct step in the unfolding of what you could call the world's destiny. We are indeed approaching a culmination of sorts; our species seems to face a kind of test toward which basic forces of history have been moving us for millennia. It is a test of political imagi-nation—of our ability to accept basic, necessary changes in structures of governance—but also a test of moral imagination.

So how will we do on this test? Judging by history, the current turbulence will eventually yield to an era of relative stability, an era when global political, economic, and social structures have largely tamed the new forms of chaos. The world will reach a new equilibrium, at a level of organization higher than any past equilibrium. And the period we are now entering will, in retrospect, look like the storm before the calm.

Or, on the other hand, we could blow up the world. Remember, even poppy seeds don't always manage to flower.

Indeed, at this moment in history, when social organization approaches the global level, and technologies of destruction reach commensurate scope, the arrow of history starts to quiver. Though getting to this threshold has long been so probable as to border on the inevitable, whether we successfully pass through it is another question altogether. And, while I'm basically optimistic, an extremely bleak outcome is obviously possible.

For that matter, even if we avoid blowing up the world, several moderately bleak outcomes are possible. One can imagine, within the bounds of possibility suggested by the trajectory of the past, future political structures that grant more freedom or less, more privacy or less, that foster more order or less, more wealth or less. One purpose of this book is to aid in exploring this "wiggle room"—in choosing among such alternative futures and in realizing the choice.

At least as big as the question of where exactly our future path leads is how chaotically it leads there. History, even if its basic direction is good, can proceed at massive, wrenching human cost. Or it can proceed more smoothly—with costs, to be sure, but with more tolerable costs.

All told, our menu of options is rich, ranging from self-annihilation to graceful adaptation, and emphatically including the middle prospect of a long and turbulent adjustment full of strife and suffering. It is the destiny of our species—and this time I mean the inescapable destiny, not just the high likelihood—to choose.

PART
I

A

BRIEF

HISTORY

OF

HUMANKIND

Chapter One

THE LADDER OF
CULTURAL EVOLUTION

An idea isn't responsible for the people who believe in it.
—*Don Marquis*

In the early twentieth century, anthropologists commonly referred to particular groups of people as "savages." Technically speaking, this was not an insult (though it seldom came off as a compliment). "Savagery" was just a stage in the orderly history of human cultures. There had been a time when all human beings were savages, but then some of them got a cultural promotion—to "barbarians." Or, at least, to "low" barbarians. Barbarism had three subdivisions—lower, middle, and upper—and a culture, after passing through them, could cross the threshold into civilization. At that point its people could start writing books in which other cultures were called savage.

The leading proponent of this savagery-barbarism-civilization scale was Lewis Henry Morgan, who unveiled it in his 1877 book *Ancient Society.* Writing less than two decades after Darwin published *On the Origin of Species,* Morgan depicted human cultures as things that evolve. "It can now be asserted upon convincing evidence that savagery preceded barbarism in all the tribes of mankind, as barbarism is known to have preceded civilization," Morgan wrote. These "three distinct conditions are connected with each other in a natural as well as necessary sequence of progress."

Morgan was one of the world's first anthropologists. He was an aficionado of American Indian societies, whose depredation by white men he deplored. After Sitting Bull massacred Custer's men at the Little Bighorn, Morgan came out in defense of Sitting Bull.

That's not to say Morgan was a radical. In addition to being a self-trained scholar, he was a well-to-do lawyer. Still, his book was embraced by Karl Marx and Friedrich Engels, who found it consistent with their own teleological view of history. Like them, Morgan traced history's direction to material factors, including technology. And like them, he stressed changing notions of ownership. They loved his idea that man's initial, "savage," condition was communal, with no private property. Engels warmly quoted Morgan's prediction that in the end cultural evolution would restore some of this primal egalitarianism. "Democracy in government, brotherhood in society, equality in rights and privileges, and universal education," Morgan had written, "foreshadow the next higher plane of society to which experience, intelligence and knowledge are steadily tending."

The view that inexorable forces of history had created civilization, that cultures "evolve" in broadly predictable fashion and would keep doing so, was also held by the sociologist Herbert Spencer, who loathed Marxism. Another proponent was John Stuart Mill, whose politics were somewhere between Spencer's and Marx's. Back then, the idea of directionality in history was almost conventional wisdom. Ideology entered the picture only when you discussed the mechanism behind history's obviously patterned course, and extrapolated into the future.

TRENDS BECOME UNTRENDY

During the early twentieth century, the conventional wisdom changed. The ranking of some societies as "higher" than others seemed increasingly unsavory, especially to scholars on the left. In anthropology, the eminent Franz Boas led an assault on the idea that human cultures tend to move in any particular direction. His most famous student, Margaret Mead, would later summarize the Boasian credo: "We have stood out against any grading of cultures in hierarchical systems which would place our own culture at the top and place the other cultures of the world in a descending scale according to the extent that they differ from ours. . . . We have stood out for a sort of democracy of cultures, a concept which would naturally take its place beside the other great democratic beliefs."

Support for the Boasian perspective was intense and eventually overwhelming. In 1918, an essay in *American Anthropologist* attacked the idea of cultural evolution as "the most inane, sterile, and pernicious theory ever conceived in the history of science"—and, moreover—"a cheap toy for the amusement of big children." By 1939, another anthropologist could report that "[cultural] evolutionism can muster hardly a single adherent." Meanwhile, the idea of directional history wasn't faring well among historians, either. During the nineteenth century many of them had seen history as progress fueled by reason; the conscious, rational pursuit of the good would bring ever-expanding freedom and political equality. But after two world wars in which clever technologies had killed millions, the words "rational" and "good" didn't seem hallmarks of humankind; and, with fascism a recent memory and totalitarian communism still strong, "freedom" didn't seem to be history's goal.

Further, hadn't the enemies of freedom, Hitler and Stalin, believed that history was on their side? Maybe, then, theories of historical directionality weren't just wrong, but dangerous! After the Second World War, two of the most famous thinkers of this century—Isaiah Berlin and Karl Popper—took up arms against such theories. In the slim volume *Historical Inevitability*, Berlin attacked the notion "that the world has a direction and is governed by laws, and that the direction and the laws can in some degree be discovered by employing the proper techniques of investigation." Popper, in *The Poverty of Historicism*, announced that he had proved—literally *proved*—that predicting the future is flat-out impossible.

After Berlin and Popper wrote, the kind of "bigthink" they opposed—"speculative history," or "metahistory"—became an endangered species. In the 1960s, one philosopher of history observed that historians "tend to use the term 'metahistorian' to mark deviations from normal professional activity in either the law-seeking or the pattern-seeking direction." Not much has changed since then. The one pattern-seeking work of history to make a big splash over the past two decades—*The End of History*—was written not by a historian but by a political scientist, Francis Fukuyama. Oddly, pondering laws of history is less deviant behavior for a political scientist than for a historian.

Opponents of "metahistory" have often been candid about their motivations. The dedication to Popper's book reads, "In memory of the countless men and women of all creeds or nations or races who fell victims to the fascist and communist belief in Inexorable Laws of Historical Destiny." I'll argue that Popper's analysis—and Berlin's analysis, and Boas's analysis—was doubly wrong: wrong not just about whether directional views of the past can be valid, but about whether they are especially dangerous.

A SHORT-LIVED INSURRECTION

The war on directionality was not as successful within anthropology as within history. At mid-century, small pockets of resistance began to develop, notably at the University of Michigan. There a firebrand named Leslie White rebelled against Boasian anti-evolutionism and devoted his life to resuscitating and refining Morgan's ideas. This revival spawned some advances in the field (including labels less offensive than "savage" and "barbarian"). Indeed, as we'll see, White's students and allied colleagues laid a still-usable foundation for reassessing world history. Nonetheless, in the 1970s, as multiculturalism gained popularity, theories that seemed to rank the world's cultures lost popularity. By the time of White's death in 1975, cultural evolutionism was falling into neglect, if not disrepute.

Today the one part of anthropology that still harbors much sympathy for evolutionism is not White's field—cultural anthropology—but archaeology. To be sure, most archaeologists don't espouse as strong a version of cultural evolutionism as will be espoused in this book; they don't believe that the progression toward more complex society was essentially inevitable, from the Stone Age right up to globalization. Still, archaeologists can't help but notice that, as a rule, the deeper you dig, the simpler the society whose remains you find. Plainly, change in the structure of societies tends to happen sooner or later, and is more likely to raise complexity than to lower it.

In a way, it's odd that the greatest sympathy for evolutionism is found among scholars who study the distant past. For events of this century, and especially of the last few decades, suggest that the arrow of history identified by some social scientists of the nineteenth century is roughly on target. Lewis Morgan's essential point was right: the endless impetus of cultural evolution has pushed society through several thresholds over the past 20,000 years. And now it is pushing society through another one. A magnificent new social structure—our future home—is being built before our eyes.

To say that history has a direction is not to embrace all the ideas associated with early cultural evolutionism or nineteenth-century pro-gressivist history. It is not, for example, to say that history is a process of general improvement; or to blithely predict the triumph of free-dom and equality in all their dimensions. Indeed, though I think his-tory is on the side of human freedom in one sense, there is another sense in which freedom is shrinking. If there is something magnifi-cent about the social structure that now seems to be emerging—the social structure that history has long been moving us toward—there is something terrifying about it, too. Fortunately, this structure, even if hard to escape in the long run (and unwise to escape in the long run) is by no means inevitable in all its apsects.

Anyway, the question of whether history's basic arrow will on bal-ance make us freer or less free, will make our lives better or worse, is one I'll defer for now. I do think that in some respects history's basic direction makes human beings morally better, and will continue to do so. But that isn't the immediate point. The immediate point, to be made over the next thirteen chapters, is that if we leave morality aside and talk about the objectively observable features of social reality, the direction of history is unmistakable. When you look beneath the roiled surface of human events, beyond the comings and goings of particular regimes, beyond the lives and deaths of the "great men" who have strutted on the stage of history, you see an arrow beginning tens of thousands of years ago and continuing to the present. And, looking ahead, you see where it is pointing.

Chapter Two

THE WAY WE WERE

A common principle of intelligence meets us in the savage, in
the barbarian, and in civilized man.

—*Lewis Henry Morgan*

Mark Twain considered the Shoshone Indians of western North
America "the wretchedest type of mankind I have ever seen up to this
writing." They "have no villages, and no gathering together into
strictly defined tribal communities." A young Charles Darwin, observing the Fuegian Indians of South America, reported that their
dwellings were "like what children make in summer, with boughs of
trees." After ticking off examples of Fuegian uncouthness and inhu-
manity, Darwin wrote in a letter, "I feel quite a disgust at the very
sound of the voices of these miserable savages."

Such dismay was often expressed when nineteenth-century white
men encountered native American cultures. That is one thing that
bothered anthropologists such as Franz Boas about theories of cultural
evolution. In placing western cultures atop a universal ladder of
progress, they seemed merely the academic expression of an already
too-common European supremacism. Boas wrote: "The tendency to
value our own form of civilization as higher, not as dearer to our
hearts, than that of the whole rest of mankind is the same as that
which prompts the actions of primitive man who considers every
stranger his enemy, and who is not satisfied until the enemy is killed."

There is arguably something paradoxical about criticizing the deni-
gration of the primitive on grounds that it is primitive. Still, Boas's
heart was in the right place. He had a genuine and courageous concern for the downtrodden. He worried especially that notions of cul-

tural superiority would get conflated with notions of biological superiority, reinforcing racism. His book *The Mind of Primitive Man,* which traced the psychology of hunter-gatherers to a distinctive culture, not distinctive genes, was burned by the Nazis.

Boas's fear that racists would gleefully seize on notions of cultural evolution was not misplaced. Generally speaking, racism will use any tool that is handy. And if cultural evolutionism seems to imply that Europeans are biologically superior to native Americans, then it is a handy tool. Still, it is important to understand that, logically, cultural evolutionism implies no such thing.

One premise of cultural evolutionism is "the psychic unity of humankind"—the idea that people everywhere are genetically endowed with the same mental equipment, that there is a universal human nature. The psychic unity of humankind is the reason that around the world, on every continent, cultural evolution has moved in the same direction. The arrow of human history begins with the biology of human nature.

That arrow, as I've noted, points toward larger quantities of non-zero-sumness. As history progresses, human beings find themselves playing non-zero-sum games with more and more other human beings. Interdependence expands, and social complexity grows in scope and depth.

One way to start seeing this link between human nature and human history is to take a look at the "wretched" Shoshone and other basic hunter-gatherer societies—or, in the technical terminology of the nineteenth century, other "savages." They demonstrate how even the simplest societies are congenitally prone to increasing complexity; and how, nonetheless, quirks of environmental circumstance can slow the rate of increase.

There is one other reason to inspect these societies: to help us reconstruct the distant past. The ancestral cultures of all modern societies were hunter-gatherer cultures. Archaeologists have found their remnants—their spearheads and stone knives, the fireside bones of their prey—across Africa, Europe, Asia, the Americas. But archaeologists can't reconstruct the social lives of these peoples in much detail. The closest we can come to that is studying the few existing hunter-gatherer societies and reading accounts of how other hunter-gatherers lived before industrial society changed them.

Over the past two centuries, anthropologists and other travelers have documented hunter-gatherer life on all continents, ranging from the Chenchu of India to the Chukchi of Siberia, from the !Kung San of southern Africa to the Ainu of Japan, from the aborigines of Australia to the Eskimo of the Arctic, from the Fuegians of South America to the Shoshone of North America. To study these vanishing—mostly vanished—ways of life is to dimly glimpse the early stages of our own cultural evolution. The Shoshone and Fuegians observed by Twain and Darwin weren't "living fossils"—they were anatomically modern human beings, just like you or me—but their *cultures* were living fossils.

THE BARE MINIMUM

Mark Twain is not the only person to have commented on the rudimentary social structure of the Shoshone, who inhabited the Great Basin of North America, around present-day Nevada. One book on native American cultures discusses them under the heading "The Irreducible Minimum of Human Society." The largest stable unit of social organization was the family, and the male head of the family was the "entire political organization and its whole legal system." The Shoshone did spend part of the year in multifamily "camps." But the camps were less cohesive than, say, those of the !Kung San, the much-studied hunter-gatherers of the Kalahari desert in Africa. For months at a time Shoshone families would go it alone, roaming the desert with a bag and a digging stick, searching for roots and seeds.

What might account for the small gradation of complexity that separates the !Kung and the Shoshone? One good candidate is the fact that the !Kung lived amidst giraffes. The tracking and killing of giraffes, and the retrieval of meat before scavengers get it, calls for cooperation. Perhaps more important, a giraffe is more than one family can eat before the meat spoils. So for giraffe hunters to live in family-sized groups would be to waste meat—and to waste a chance to collect the IOUs that come from sharing it.

Such IOUs are a classic expression of non-zero-sumness. You give someone food when his cupboard is bare and yours is overflowing, he reciprocates down the road when your cupboard is bare, and you both profit, because food is more valuable when you're hungry than when you're full. Hunter-gatherers everywhere act in accordance with this logic. One chronicler of Eskimo life has observed, "the best place for [an Eskimo] to store his surplus is in someone else's stomach."

Hunting big animals encourages sharing not just because leftover meat can spoil but also because hunting is a chancier endeavor than gathering—so using current surplus to insure against future shortage pays especially big non-zero-sum dividends. All told, then, it is not surprising that social complexity tends to be higher among hunter-gatherers who rely heavily on big game. The more important big game is, the more non-zero-sumness there is, the more society organizes to harness that non-zero-sumness—to turn it into positive sums.

So, to the cultural evolutionist, the explanation for the Shoshone's having the "irreducible minimum" in social complexity is not (as Boas might fear) that the Shoshone are stupid. It's that their surroundings were—in this respect, and in other respects that we'll come to—less conducive to rapid cultural evolution than some other surroundings.

How can we be sure that the Shoshone would indeed have the wherewithal to evolve culturally if given the chance? Funny you should ask. Although the Shoshone had no big game to hunt, jackrabbits were afoot, sometimes in abundance. To harvest them, the Shoshone employed a tool too large for one family to handle—a net hundreds of feet long into which rabbits were herded before being clubbed to death. On such occasions, the requisite social structure would materialize. More than a dozen normally autonomous families would come together briefly and cooperate under a "rabbit boss." Though the Shoshone spent most of their time with the "irreducible minimum" of organization, the sudden appearance of non-zero-sumness brought latent social skills to the fore, and social complexity grew.

To say that reaping non-zero-sum benefits elevates social complexity borders on the redundant. The successful playing of a non-zero-sum game typically *amounts to* a growth of social complexity. The players must coordinate their behavior, so people who might otherwise be off in their own orbits come together and form a single solar system, a larger synchronized whole. And typically there is division of labor within the whole. (Some people make the nets, some people man the nets, some people chase the rabbits.) One minute you're a bunch of independent foragers, and the next minute you're a single, integrated rabbit-catching team, differentiated yet united. Complex coherence has materialized.

Note that this particular bit of social self-organization was mediated by a technology, the rabbit net. The invention of such technologies—technologies that facilitate or encourage non-zero-sum interaction—is a reliable feature of cultural evolution everywhere. New technologies create new chances for positive sums, and people maneuver to seize those sums, and social structure changes as a result.

A successful Shoshone rabbit hunt would culminate in a "fandango." Sounds like a spontaneous and carefree celebration—and indeed, fandangos featured, as one anthropologist put it, "gambling, dancing . . . philandering." Still, as scholars have noted, the fandango was eminently utilitarian. First, it distributed fresh meat among the rabbit hunt's various kinds of workers. Second, it was an occasion for trading such valuables as volcanic glass. Third, it was a chance to build up a network of friends. (Even the ritual exchange of knickknacks, though economically trivial, can be a way to bond, forming the conduits for future favor-swapping of greater moment.) Fourth, the fandango was an opportunity to trade information about, say, the location of other food.

All of these are non-zero-sum functions, and the last is especially so. Giving people data, unlike giving them food or tools, has no inherent cost. If you know of a place where the supply of pine nuts far exceeds your own family's needs, it costs nothing to share the information with a friend. So too if you know the location of a den of poisonous snakes. Sometimes, of course, surrendering information *is* costly (as when the supply of nuts *doesn't* exceed your family's needs). Still, data are often of little or no cost and great benefit; swapping them is one of the oldest forms of non-zero-sum interaction. People by their nature come together to constitute a social information processing system and thus reap positive sums. The fandango, the academic conference, and the Internet are superficially different expressions of the same deep force.

IN THE GENES

Though Shoshone life, like life everywhere, seems to have been filled with non-zero-sum calculation, "calculation" isn't quite the right word. When people interact with each other in mutually profitable fashion, they don't necessarily realize exactly what they're doing. Evolutionary psychologists have made a strong—in my view, compelling—case that this unconscious savviness is a part of human nature, rooted ultimately in the genes; that natural selection, via the evolution of "reciprocal altruism," has built into us various impulses which, however warm and mushy they may feel, are designed for the cool, practical purpose of bringing beneficial exchange.★

Among these impulses: generosity (if selective and sometimes wary); gratitude, and an attendant sense of obligation; a growing empathy for, and trust of, those who prove reliable reciprocators (also known as "friends"). These feelings, and the behaviors they fruitfully sponsor, are found in all cultures. And the reason, it appears, is that natural selection "recognized" non-zero-sum logic before people recognized it. (Even chimpanzees and bonobos, our nearest relatives, are naturally disposed to reciprocal altruism, and neither species has yet demonstrated a firm grasp of game theory.) Some degree of social structure is thus built into our genes.

Actually, the genetic basis of social structure goes beyond reciprocal altruism. Love of kin is human nature. In every hunter-gatherer society the family is the basic molecule of social organization. But the genetic logic behind families is another story, best saved for later. For now the point is that human nature itself, unadorned by technology, carries mutual benefit, and thus social structure, beyond the confines of family. The arrow of human history, the arrow that heads toward more non-zero-sumness and deeper and vaster social complexity, doesn't begin at zero. A universal feature of hunter-gatherer societies, some anthropologists say, is "generalized reciprocity"—not just within families but between them.

It is important to be clear on what "generalized" does and doesn't mean. Ever since Morgan wrote *Ancient Society* and Marx and Engels read it, people have been straining to portray our ancestral hunter-gatherer cultures as dens of communal bliss. Richard Lee, a pioneering observer of the !Kung San, contends that hunter-gatherer societies lend "strong support" to the idea that "a stage of primitive communism prevailed before the rise of the state and the breakup of society into classes." He writes that "the giving of something without an expectation of equivalent return" is "almost universal among foraging

---

★Theories from evolutionary psychology, including the theory of reciprocal altruism, rest on eclectic argument. Doing justice to these theories is itself a book-length endeavor. So this book, while grounded in evolutionary psychology's view of human nature, won't try to justify that view in detail. Readers can turn to my previous book, *The Moral Animal,* for elaboration.

peoples," and he cites in particular the !Kung's habit of sharing food "in a generalized familistic way."

Sounds great. Yet Lee also notes that the !Kung often argue, and that "accusations of improper meat distribution, improper gift exchange, laziness, and stinginess are the most common topics of these disputes." However "generalized" the giving among the !Kung, it is expected to be ultimately symmetrical.

Here again, though, the expectation isn't merely a matter of conscious calculation. When we accuse others of laziness or stinginess, we are driven by something deeper and hotter than sheer reason—by a feeling of moral indignation, of just grievance. And that feeling—found in cultures everywhere, and expressed in predictable circumstances—seems to be grounded in our genes. According to evolutionary psychologists, it is part of the emotional equipment designed by natural selection to govern reciprocal altruism, to help us play non-zero-sum games profitably.

THE TROUBLE WITH NON-ZERO-SUMNESS

Why would this sort of vigilance have been so crucial to our ancestors' prospects that genes conducive to it would flourish? There are two properties of non-zero-sum games—two kinds of pitfalls—that make instinctive wariness vital.

One is the problem of cheating, or parasitism. People may accept your generosity and never repay it. Or they may sit around getting a suntan while everyone else is rabbit hunting and then expect to have charbroiled rabbit for dinner. (Game theorists call this "free riding"—contributing nothing to the pie of positive sums created by collective action, yet cheerfully eating a piece.)

In the intimate context of hunter-gatherer life, moral indignation works well as an anti-cheating technology. It leads you to withhold generosity from past non-reciprocators, thus insulating yourself from future exploitation; and all the grumbling you and others do about these cheaters leads people in general to give them the cold shoulder, so chronic cheating becomes a tough way to make a living. But as societies grow more complex, so that people exchange goods and services with people they don't see on a regular basis (if at all), this sort of *mano-a-mano* indignation won't suffice; new anti-cheating technologies are needed. And, as we'll see, they have materialized again and again—via cultural, not genetic, evolution.[†]

The other principle of game theory that makes wariness adaptive is subtler than cheating. Within almost any real-life non-zero-sum game lies a zero-sum dimension. When you buy a car, the transaction is, broadly speaking, non-zero-sum: you and the dealer both profit, which is why you both agree to the deal. But there is more than one price at which you both profit—the whole range between the highest you would rationally pay and the lowest the dealer would rationally accept. And *within* that range, you and the dealer are playing a zero-sum game: your gain is the dealer's loss. That's the reason bargaining takes place at car dealerships.

Oddly, such zero-sum games are ultimately a tribute to the magic of non-zero-sumness. Watch me pull thirty rabbits out of my hat: Just take twenty Shoshone who, if hunting individually, would be lucky to snag one rabbit each, and turn them into a team. Presto!—fifty rabbits instead of twenty! It is the question of how to divvy up this magical thirty-rabbit surplus that the zero-sum game, the bargaining, is about.

Of course, you could just divide the rabbits equally. But haven't some people worked harder than others, or brought rarer talents to the project? Besides, if you do divide the spoils equally, the bargaining just gets subtler: some people may try to do slightly less work than average for their 1.5 rabbits—not so little work that their contribution isn't a net plus, but less work than you're doing. This zero-sum tension, this implicit bargaining, is the reason that hunter-gatherer societies feature gripes about laziness and stinginess. At least, it is the second reason, the first being out-and-out cheating.

These two reasons, applied over millions of years of biological evolution, have given people everywhere an innate tendency to monitor the contributions of others, whether consciously or unconsciously. In all cultures friendships have underlying tension. In all cultures workplaces feature gossip about who is a slouch and who is a team player. In all cultures people scan the landscape for the lazy and the ungrateful, and rein in their generosity accordingly. In all cultures, people try to get the best deal possible.

I stress the natural dearth of selfless giving—true, pure altruism, indifferent to ultimate payoff—not to show that truly communist economies aren't practical. The twentieth century has already made that point. I'm just trying to get clear on the parts of human nature that, in conjunction with technological evolution, give history its basic shape. And one of those parts is firm self-interest.

This fact may be disappointing from some moral standpoint. But if you are a fan of complex social organization, it is a godsend. Human nature's laser-like focus on ultimate payoff is a prime mover of cultural evolution. Instinctively enlightened self-interest is the seed that has grown into modern society. At the heart of every modern capitalist economy—as at the heart of the hunter-gatherer economies from which they evolved—is the principle of exchange. One hand washes the other, and both are better off than they would be alone—the very definition of a well-played non-zero-sum game. The difference between the two economies lies in the number of hands involved and the intricacy of their interdependence, two quantities that cultural evolution has a stubborn tendency to raise.

SOCIAL STATUS

The bent for reciprocal exchange is not the only aspect of human nature that helps propel society toward complexity. Evolutionary psychologists—and, for that matter, non-evolutionary psychologists—have shown that human beings naturally pursue social status with a certain ferocity. We all relentlessly, if often unconsciously, try to raise our standing by impressing peers. And we naturally, if unconsciously, evaluate other people in terms of their standing. We especially value the friendship of high-status people (since alliance with the powerful tends to come in handy), and we especially fear their disfavor (since the enmity of the powerful tends not to). Human beings evolved amid social hierarchies, and our minds are designed to negotiate them.

This gives social complexity a head start in at least two ways. First, deference to the high and mighty (though the deference is far from reflexive, as we'll see) paves the way for complex hierarchical organization. Hunter-gatherer peoples may or may not have a formally designated "headman," but they recognize a leader when the occasion demands. Just look at how readily a Shoshone "rabbit boss" pops up when the rabbits do.

Second, the ongoing quest for social status is a great spur to cultural innovation. We don't know who invented the rabbit net, but we can safely assume it didn't hurt his or her popularity. And so it is in all societies: one sure way to elevate your standing is to create something that is widely adopted and praised.

There is an irony here. To compete for high-status positions is to play a zero-sum game, since they are by definition a scarce resource.

Yet one way to compete successfully is to invent technologies that create new non-zero-sum games. This is one of various senses in which the impetus behind cultural evolution, behind social complexification, lies in a paradox of human nature: we are deeply gregarious, and deeply cooperative, yet deeply competitive. We instinctively play both non-zero-sum and zero-sum games. The interplay of these two dynamics throughout history is a story that takes some time to tell. For now I'll just say that, though they have been responsible for much suffering, the tension between them is, in the end, creative.

One point of dwelling on human nature is to stress that the arrow of human history, though awe-inspiring, is not mystical or uncanny. Technological evolution, and cultural evolution more broadly, are not alien forces, visited upon the human species from the great beyond, magically imbued with non-zero-sumness. Technology and other forms of culture come from within. The directionality of culture, of history, is an expression of our species, of human nature.

Indeed, cultural evolution is *doubly* reflective of human nature. Humans not only generate cultural innovations; they pass judgment on them. You can write any song you want, but other people will have to find it appealing if it is to spread. Your brain may give birth to any technology, but other brains will decide whether the technology thrives. The number of possible technologies is infinite, and only a few pass this test of affinity with human nature. One could invent, say, a battery-powered, helmet-mounted device that at random intervals jabs a sharp stick into the face of the helmet wearer. But a robust market for such a device is unlikely to materialize. So a battery-powered face jabber is unlikely to affect human history as profoundly as, say, the telephone. Or even the rabbit net.

NATURE'S SECRET PLAN

The mixture of cooperative and competitive human instincts, the subtle but potent quest for status, the ingenuity it fuels—these were evident long before Darwinian theory came along to explain their reason for being. In the eighteenth century, for example, Immanuel Kant noted the "unsocial sociability" of man, with special emphasis on the "unsocial" part and its ironic consequences. "Through the desire for honor, power or property, it drives him to seek status among his fellows, whom he cannot *bear* yet cannot *bear to leave*." Via this quest for status, "the first true steps are taken from barbarism to culture, which in fact consists in the social worthiness of man." Thus commences "a continued process of enlightenment" as "all man's talents are now gradually developed, his taste cultivated."

Without these "asocial qualities (far from admirable in themselves)," human beings "would live an Arcadian, pastoral existence of perfect concord, self-sufficiency and mutual love. But all human talents would remain hidden forever in a dormant state, and men, as good-natured as the sheep they tended, would scarcely render their existence more valuable than that of their animals." In that event, "the end for which they were created, their rational nature, would be an unfilled void. Nature should thus be thanked for fostering social incompatibility, enviously competitive vanity, and insatiable desires for possession or even power."

Kant made those remarks in an essay called "Idea for a Universal History with a Cosmopolitan Purpose," which suggested that human history embodied a "hidden plan of nature." Perhaps as history unfolds, he wrote, we will see "how the human race eventually works its way upward to a situation in which all the germs implanted by nature can be developed fully, and in which man's destiny can be fulfilled here on earth." Kant imagined that this destiny would include enduring peace among nations, ensured by a kind of world governance—a final, ironic payoff for millennia of antagonism and "unsocial" striving.

This was all conjecture, Kant stressed. Writing in 1784, before the harnessing of electricity, before the telegraph or typewriter or computer, he admitted that so far there was just "a little" evidence of such a "purposeful natural process." Only time would tell. "For this cycle of events seems to take so long a time to complete, that the small part of it traversed by mankind up till now does not allow us to determine with certainty the shape of the whole cycle, and the relation of its parts to the whole." Well, that was then.

Chapter Three

ADD TECHNOLOGY AND
BAKE FOR FIVE MILLENNIA

The propensity to truck, barter, and exchange one thing for another . . . is common to all men.

—Adam Smith

What is society, whatever its form may be? The product of men's reciprocal action. . . . Assume a particular state of development in the productive faculties of man and you will get a particular form of commerce and consumption.

—Karl Marx

When Europeans, beginning with Columbus, entered the New World in the fifteenth and sixteenth centuries, there were a number of things they didn't pause to appreciate before commencing with the pillaging. One is that they had happened upon a rare and precious natural experiment. The ancestors of native Americans had migrated from northeast Asia during the late Stone Age, the "Upper Paleolithic." Then, around 10,000 B.C., with the climate warming, the land they'd walked across was deluged by the Bering Sea. The Old World and the New World were now two distinct petri dishes for cultural evolution. Any basic trends inherent in the process should be evident in both.

The experiment wasn't perfect. Certainly by 2000 B.C., and possibly earlier, the Eskimos (also known as the Inuit) had boats. Though paddling across the Bering Sea wasn't the kind of thing you would do for weekend recreation, and travel from one Alaskan village to another was often arduous, there now existed the theoretical possibility for innovations to move glacially from Asia into North America. Still, for most of prehistory, cultural change in the New World appears to have been indigenous, and even during the last few thousand years contact with the Old World was tenuous. The two hemispheres, west and east, are the closest things to huge, independent examples of ongoing cultural change that this planet has to offer.[†]

There is one other reason that primitive American cultures are so enlightening. As of Columbus's voyage, they had an advantage over primitive Eurasian societies as objects of study. Namely, they still existed; they had not been steamrolled by the expansion of Old World civilizations. And, though Columbus and other Europeans tried to make up for lost time with their own steamrolling, they were not wholly effective. Observed and recorded in the New World was an unprecedented array of cultures, with diverse technologies and social structures.[†] From this diversity a few basic patterns emerge, patterns that turn out to be consistent with the archaeological remains of those steamrolled Old World cultures. Native American cultures thus offer unique evidence of the universal impetus toward cultural complexity.

Indeed, they virtually show that impetus in action. Snapshots of the different American "cultural fossils" amount to a kind of time-lapse sequence in which cultural evolution pushes social complexity beyond the level of the Shoshone, toward the modern world.

TWO KINDS OF ESKIMO

Consider two kinds of Eskimo. The Nunamiut Eskimos are comparable to the Shoshone—a basically family-level social organization that occasionally reaches a higher level. (During seasonal caribou migrations, renowned hunters lead big hunts.) The Nunamiut's neighbors, the Tareumiut, are closely related, and speak the same language. But there's one difference: whereas the Nunamiut live inland, the Tareumiut live on the coast, and hunt whales. And, as the cultural anthropologist Allen W. Johnson and the archaeologist Timothy Earle have noted, this whale-hunting technology seems to have propelled Tareumiut social organization up the ladder of complexity.

Each whaling boat is run by an *umealiq*—a "boat owner"—who recruits a crew that includes such specialists as a helmsman and a harpooner. There is no better metaphor for a non-zero-sum relationship than being "in the same boat." (And that's especially true when capsizing means death by freezing.) But in this case the non-zero-sumness extends well beyond one boat.

It often takes several boats to kill a whale, and boat owners must coordinate both the hunt and the food distribution. Perhaps building on this interdependence, the boat owners within a village have created a kind of joint insurance policy. If any one owner has fallen on hard times, he and his crew can draw food from other owners, with the promise of future reciprocation. The Tareumiut say proudly, "We don't let people starve," and indeed the long winters—the season of scarcity—are less precarious for them than for the Nunamiut.

This spreading of risk doesn't end at the village's bounds. An *umealiq* who has had a banner whaling season invites boat owners from other villages to a "Messenger Feast," where he lards them with surplus blubber and meat. This may seem magnanimous, but, as with the smaller-scale "altruism" among the !Kung, generosity is a veneer; future reciprocation is de rigueur. Like insurance policyholders, the region's boat owners are playing a non-zero-sum game, finding in large numbers security against misfortune. Long before economists were drawing graphs showing how diversified stock portfolios could serve the human aversion to risk, cultures were evolving by the same logic.

The Tareumiut are more socially complex than the Shoshone, the !Kung, or the Nunamiut. Their villages, with 100 to 200 people, comprise numerous families, living interdependently year round. These villages are truly "suprafamilial," whereas members of the smaller !Kung camps are often so closely related by blood or marriage as to be more like a big extended family.

THE NORTHWEST COAST INDIANS

The Tareumiut, with their entrepreneurial boat owners and their intricate whale hunts, belie the standard image of the simple hunter-gatherer society. But not nearly so much as the natives of the Northwest Coast of North America—the Salish, the Haida, the Kwakiutl, the Nootka, the Chilkat, and others. These peoples, arrayed north and south of the present-day border between Canada and the United States, had taken yet another step up the ladder of complexity.

In the popular mind these peoples are best known for the "Potlatch," the famously ridiculous ritual in which local chiefs indulged in fierce duels of generosity. It got to the point, sometimes, where they would prove their wealth by heaping prized possessions not just on one another but on bonfires. But the culture of the northwestern native Americans illustrates more than the human penchant for showing off. Namely: the ongoing conversion of non-zero-sumness into positive sums, and the resultant growth in social complexity.

The Northwest Coast Indians were blessed with mindboggling natural bounty. The salmon in their rivers may not have been so dense that, as one explorer claimed, "you could walk across their backs," but they were dense. There were also halibut, cod, and herring, and the sea was rich with shellfish, sea otters, seals, and whales. And then there was the incomparable candlefish—so oily that supposedly you can stick a wick in it and use it to light a room.

Diverse game called for diverse technology. The Nootka had an array of fishhooks ranging from a heat-treated spruce hook for halibut to a bone hook for cod. They made harpoons and tied them to inflated sealskin floats, to sap the energy of struggling whales. Boats ranged from one-man canoes to eight-man whalers to sixty-foot cargo boats. The Nootka had traps for bear, for deer, for elk. They had four kinds of salmon traps, ranging from cubic- to cone-shaped, some as big as a small house. (The actual houses, suburban ranch-styles, were routinely larger than 2,000 square feet and sometimes as large as 4,000.) There were smokehouses for curing fish, cellars for storing cured fish, and watertight cedar boxes for storing berries.

Not all the technology was so utilitarian. Luxury goods ranged from ornate copper shields to decorative robes whose creation was an exercise in economic interdependence. Chilkat women spun the yarn from the wool of mountain goats and made twine out of cedar bark imported from Indians to the south. The yarn was dyed one of four colors, including a true blue (rare among hunter-gatherers) that was made by importing copper from the north and soaking it in urine. On a loom, the women wove intricate patterns—animals or abstractions. The finished product was exported to various Northwest Coast Indians whose aspiration was to someday be buried in an attractive robe.

Much of this technology involved that classic non-zero-sum game, division of labor—through which, as Adam Smith noted, a group of people can expand overall output. Though all Northwest Coast Indian families would hunt and gather, many also had a sideline craft—carpentry, say—that was handed down through the family.

These native Americans also played the non-zero-sum game played by the Tareumiut and the Shoshone: collective hunting, as reflected in their whaling fleets and the huge fish traps they affixed to the river floor with massive posts. These things were major capital investments. To build a salmon trap or a whaling boat took weeks. The workers had to be paid for their labor, if only in the sense of being fed. So before building began, resources had to be saved and committed to the project.

Capital investment and division of labor are things we take for granted. They happen naturally in an economy with a currency, a stock exchange, and a bond market. The Northwest Coast Indians didn't have a capitalist economy, or even a currency, yet they managed to play the same basic non-zero-sum games capitalists play. How? Through the great enemy of Adam Smith aficionados: centralized planning.

THE BIG MAN GOES TO MARKET

The chief planner was the political leader, the "Big Man." He held the allegiance of a clan, maybe a village. He orchestrated the building of salmon traps or fish cellars, and he made sure that some villagers specialized in, say, making canoes that other villagers could then use. To pay for all of this he would take one-fifth, or even half, of a hunter's kill. Some of this revenue would be returned to the people in the form of chief-sponsored feasts, but for the most part this was simple taxation, used for public goods.† Only, in this case, many of the "public goods" were things that a modern capitalist society might deem private goods. (The U.S. government doesn't have a Bureau of Canoe-making.)

Needless to say, the Big Man skimmed a little off the top. He lived in a nicer-than-average house and owned a nicer-than-average wardrobe. Whether he skimmed off more than he "deserved" is a complex question that gets at an unresolved academic dispute about how exploitive ruling classes are. We'll get to this debate later. For now I'll just note that skimming a little off the top isn't exactly unheard of in a modern economy. Investment banking isn't charity work.

The Northwest Coast Indians' rudimentary "government" wasn't only a stand-in for the market. It did things that governments do even in capitalist societies. For example, if fishermen were allowed to compete without restraint, they could deplete the salmon stock, hurting everyone. This is an instance of what the biologist Garrett Hardin famously called the "tragedy of the commons"—a textbook non-zero-sum problem, in which overgrazing of public land by privately owned herds would be ruinous, so all herd owners can benefit by mutual restraint. The Northwest Coast Indians solved the problem by deciding when fishing would begin and end, much as governments today enforce a hunting season so that deer and ducks will live to die another day. There was even a specialist—a kind of "fishing warden"—who would go around from trap to trap, inspecting the haul to decide when the fishing must end.

Northwest government also blunted misfortune. Goods that Big Men gathered as tax—blankets, sea otter furs, hammered copper—were in times of scarcity traded for food with another region's Big Man, and the food then divvied up among followers. Here Big Men were together tapping one of the most hallowed forms of socially integrating non-zero-sumness: the diffusion of risk, as practiced by the !Kung with their giraffe-meat dinner parties and by Eskimo boat owners with their inter-village feasts. The more widely this risk is spread, the better for all concerned. And the Northwest Coast Indians spread it as widely as any known hunter-gatherer people, "even across 'tribal' and linguistic boundaries," as Johnson and Earle note.

The resulting safety net has been called by various anthropologists "social security," a "life insurance policy," and a "savings account." The diversity of labels illustrates that, even today, people argue over whether this function belongs in the public or private sector. The issue is partly a moral one, turning on judgments about whether the wealthy should help the poor. But to some extent the issue is technological. Modern information and transportation technology make it easier to do all kinds of exchanges without central coordination. Today a middle-class citizen of an industrialized nation can diffuse risk to the ends of the earth, buying mutual funds that invest east and west, north and south.

Two economists of differing ideologies could have a long argument about whether the Northwest Coast Indians, even with their primitive technology, couldn't have put some of the Big Man's functions in the private sector. But almost any economist would admit that, given the absence of money, these native Americans had a remarkable economy, with great specialization, large capital investment, and disaster insurance. All of this is a tribute to how steadfastly, even unconsciously, human nature pursues non-zero-sum gain, shaping social structure to that end.

In this case, the requisite social structure was elaborate: villages of up to 800 people, with dozens of families recognizing a single, central authority. To be sure, the Big Man's powers were hardly absolute, or even very formal; to keep the economy humming, he sometimes had to cajole or browbeat reluctant donors. Still, the Big Man system carries economic and political complexity to a level higher than any other society we've seen so far. The Northwest Coast Indians are testament to the arrow of cultural evolution, and to its guiding force.

FREAKS OF NATURE?

Yet they've often been depicted as the opposite. Anthropologists of a Boasian stripe have presented Northwest culture not as a natural progression toward modern social complexity, but as a freak of nature, proof that no universal evolutionary scheme can accommodate all cultures. The Northwesterners, wrote the archetypal Boasian Ruth Benedict, were a "vigorous and overbearing people" who "had a culture of no common order." Its "values were not those which are commonly recognized, and its drives not those frequently honored."

Actually, its values and drives are quite familiar—strikingly like those of the modern world. There was jockeying for status, an attendant accumulation of wealth, and thus an economy driven partly by demand for nonessential items. Indeed, there is every reason to believe that, if not corrupted by white men, Northwestern culture would have kept doing what it had done up until then—modernizing: producing wealth more and more efficiently.

The Northwesterners may have been on the verge of a de facto currency—strings of dentalia shells, used as a sign of prestige and occasionally as compensation for public service. Certainly the idea of abstractly embodied wealth was familiar. One Big Man, in exchange for goods, issued tokens that entitled the holders to blubber from the next whale that washed up on his people's beach. If he had been born a century later, he might have founded the Chicago Mercantile Exchange.

Granted, the Northwesterners did, as Benedict suggested, have their oddities. Tossing handwoven blankets onto a bonfire seems more wasteful even than modern means of conspicuous consumption. But not by much. Besides, the absurdity of the Potlatch has been exaggerated. It became flagrantly wasteful only after white traders filled Northwest culture with new luxury goods and in other ways jostled tradition. In pristine form, the Potlatch had mainly served non-zero-sum ends. It was a time to share useful information, and, since the "generosity" was ultimately reciprocal, it was a rudimentary (and perhaps vestigial) way of using surplus to dull future risk. Even Benedict's mentor, Boas himself, had said that one point of the Potlatch was to assemble an audience for large-scale altruism between villages, thus ensuring that the debt was recorded in the public mind.

Of course, Boas didn't mean to affirm an evolutionary schema. In observing that the Northwestern economy was "largely based on credit, just as much as that of civilized communities," he was trying to throw a wrench into the evolutionary works. After all, weren't hunter-gatherers, in such standard evolutionary theories as Lewis Henry Morgan's, supposed to have communal economies?

Yes, they were. But one thing cultural evolutionists later decided was that Morgan had erred in defining his evolutionary stages in such tightly technological terms. In his scheme, for a culture to graduate from lower barbarism to middle barbarism it had to domesticate plants or animals (and to move from savagery to lower barbarism in the first place it had to have pottery). But, as the Northwest Coast Indians show, an affluent hunter-gatherer society can be more complex than some societies with domestication.

In the early 1960s, when the mid-century revival of cultural evolutionism was in full flower, Elman Service, one of Leslie White's students, proposed a new taxonomy: band, tribe, chiefdom, state. The four grades were demarked not by technology but by political and economic organization. The Shoshone, hunter-gatherers, were a band. The Northwest Coast Indians, also hunter-gatherers, were a chiefdom—because they had extensive economic specialization coordinated by a central authority.

Defining evolutionary stages in terms of social structure was a great advance, but even so things remain fuzzy. No two societies in the same "stage" are exactly alike. Besides, cultural evolution can move so gradually as to resist division into "stages" in the first place.

To be sure, there are thresholds that get crossed somewhere between the earliest hunter-gatherers and ancient Mesopotamia: the suprafamilial threshold, in which multiple extended families come under unified village governance, and the supravillage threshold. But even these concepts lack Platonic clarity. Among the Northwest Coast Indians, the strength of supravillage leadership firmed up during wars and loosened in peacetime. So scholars who consider such leadership a hallmark of chiefdomhood have dissented from Service's assigning the "chiefdom" label to Northwest culture. Allen Johnson and Timothy Earle, whose 1987 book *The Evolution of Human Societies* proposed a new seven-fold taxonomy, classify it as a "Big Man collectivity," just shy of a chiefdom.

Whatever. The main point is that, since Boas's day, when the Northwest Indians were considered so aberrant as to almost single-handedly refute cultural evolutionism, they have begun to seem less peculiar. One reason is this realization that technology doesn't work as the basic gauge of evolution, even if it is a basic impetus of evolution. But there are other reasons, too.

For starters, the ruthless generosity on display during the Potlatch has been found in a number of societies at roughly the same level of social organization. A Big Man in New Guinea, having bestowed piles of food and wealth on another Big Man, was heard to proclaim, "I have won. I have knocked you down by giving so much."

The satisfaction of delivering that one line seems meager compensation for the months it took the Big Man and his followers to scrape together the largesse. But the line impresses people, raising his social status. And social status, however ephemeral it sounds, has long brought tangible rewards. For example: Big Men, if Big enough, can attract multiple wives.

For that matter, a successful Big Man's stature can rub off on his lieutenants. At one feast in the Solomon Islands, observed in 1939, a Big Man named Soni parted with mounds of sago-almond pudding and thirty-two pigs at a feast attended by 1,100 people. Soni's closest followers, who had toiled long and hard toward this day, watched proudly, but, like Soni, ate nothing. They consoled themselves with this refrain: "We shall eat Soni's renown."

Whether their share of Soni's renown—along with any other patronage Soni passed along—justified the work they put into his elevation is an open question. Though every coalition has a non-zero-sum premise—the prospect of a win-win-win outcome for its members—every coalition also has its natural zero-sum dimension:

tension over how to divide the costs and benefits of collective action. In the end, even if the coalition attains its collective reward, some members may get such slim pickings that they'd have been better off not joining in the first place.

In cases such as Soni's, my guess is that this sort of out-and-out parasitism is the exception, not the rule. I say that not because I assume Soni is a nice guy, but because I assume his followers, being human, are naturally good at guarding their interests. In any event, regardless of whether their gain indeed warranted their labors, Soni's job was to convince them that it did. And in thus mobilizing a productive coalition of scores, even hundreds, of followers for a project whose payoff is distant and vague, the Big Man carries politics to a level beyond that of the Shoshone rabbit boss.

Indeed, though he lacks formal powers of office, and must rely on persuasion, the generic Big Man foreshadows the modern politician. He "is usually a good speaker, convincing to his listeners," writes Allen Johnson of the Melanesian Big Man. He "has an excellent memory for kinship relations and for past transactions in societies where there is no writing." The Big Man "is a peacemaker whenever possible, arranging compensatory payments and fines in order to avoid direct violent retribution from groups who feel they have been injured." But, if peacemaking fails, "he leads his followers into battle."

Central among the Big Man's political challenges is to keep different families united in a single polity. It is a loose polity by modern standards, but it is firmer than, say, the essentially leaderless communities of the Tareumiut. Crossing this threshold, to the centralized suprafamilial political unit, was assuredly no easy task. Affection and trust by nature come more easily, and less conditionally, within families than between them.

It is striking how often, around the world, Big Man societies have used the same cement to keep friction between families low: rituals and language that harness the natural emotional valence of kinship. Thus the various families in a Northwestern clan would celebrate their common, distant ancestor (sanctified by their totem pole), though it's not clear that this ancestor had in all cases actually existed. And in a New Guinea Big Man society, a men's organization that lent cohesion to the suprafamilial fabric was called "Brothers Under the Skin."

THE ABERRANT !KUNG

But if the native American societies of the Northwest Coast were not a freak of nature, and in fact represent a natural phase in cultural evolution, then why do so few hunter-gatherer "cultural fossils" record that phase? Why *is* it that most other Big Man societies on the anthropological record, including those above, had domesticated crops? Why are a large majority of known hunter-gatherer societies labeled by anthropologists as "egalitarian" (at least, *relatively* egalitarian), such as the !Kung and the Shoshone?

Maybe because that's all that was left by the time the anthropologists showed up. One reason the !Kung and the Shoshone are so culturally simple is that they live in barren lands. This means, for example, that they must move often to keep fed, and so can't accumulate the weighty status symbols that the more sedentary Northwesterners spent their time crafting. (Never set out on a fifteen-mile trek with a thirty-foot totem pole balanced on your head.) What about hunter-gatherers who live on choice land? Well, by the time anthropologists happened on the scene, most of those societies were gone. Their cultures had either evolved to a higher level (perhaps the ones the anthropologists had come from) or been overwhelmed by a culture that had. Agricultural societies, after all, are known to lust after good land, and hunter-gatherer societies are known for succumbing to their advances, willingly or not.

So hunter-gatherers who may have inhabited the fertile shores and valleys of Europe and Asia are not available for inspection. Similarly, in Africa, the Bantu and other agriculturalists long ago swept across the best land, erasing past culture. Even in North America, much of the richest land—along the Mississippi River, around the lower Great Lakes—was being farmed by the time white men arrived. And of the remaining affluent hunter-gatherer cultures, many—in Florida and California, for example—were pushed aside before nineteenth-century anthropologists could marvel at their peculiarity (though in both of those areas, white intruders got a clear glimpse of impressive social complexity before annihilating it). The Northwest Coast Indians, far from the main avenues of western onslaught, are one of the few hunter-gatherer societies closely observed by anthropologists in such a rich natural habitat. For all we know, they are typical.

Indeed, archaeologists have lately amassed evidence to that effect. On various continents, between 6,000 and 15,000 years ago, hunter-gatherer remains show signs of growing complexity. There are capital projects, such as storehouses. And "prestige technologies" sit alongside the familiar stone blades and arrowheads. There are bracelets, necklaces, finely sculpted amber pendants, and headdresses. Jewelry is made of bone or shells or malachite or volcanic glass. Even practical things—bowls, knives—get more ornate. And within a given society, some people are buried amid more of this wealth than others. In short, the remains of these societies look rather like the remains of the Northwest Coast Indians.

That is why it is not mere conjecture to call the Northwest Coast Indians "cultural fossils," rough exemplars of a particular stage in cultural evolution. Boasian anthropologists can spend all the time they want denying evolutionary pattern, but archaeologists have found a trend too widespread to be meaningless: the more recent the hunter-gatherer remains in a given region, the more likely they are to speak of social and technological complexity. And in particularly lush spots—along rivers, lakes, and oceans—the complexity often approaches that of the Northwest Coast Indians. As the archaeologist Brian Fagan has written, artifacts unearthed in recent decades speak of a "global trend toward great complexity in hunter-gatherer societies in well-defined regions as widely separated as northern Europe, southern Africa, Japan, the [American] Midwest, and coastal Peru."

The nineteenth-century evolutionists lacked these data. They can be forgiven for overgeneralizing from a biased sample, and type-casting hunter-gatherers as poor and communal. And Boas and Benedict, having this type-casting in mind, can be forgiven for considering the Northwest Coast's advanced culture a problem for evolutionary theories. Still, the fact remains that they were wrong. A mature evolutionary theory is bolstered, not undermined, by the parallels between complex hunter-gatherer societies and modern economies.

For that matter, evolutionary theory is bolstered by simpler hunter-gatherers, including the Shoshone. They, too, embody directional cultural change. When the first Americans crossed Beringia, they possessed Upper Paleolithic technology, which had begun to flourish around 35,000 B.C. This was a big advance over Middle Paleolithic technology, but it was still pretty basic: long stone blades, fine bone points for spears or harpoons, spear-throwing contraptions. Then, around 12,000 years ago, after Upper Paleolithic people reached America and started carrying this technology all over the New World, the Old World was swept by the next big thing: Mesolithic technology. This included various fishhooks, hunting and fishing nets, complex traps and snares, racks for smoking meat to preserve it, storage baskets, and mortars and pestles for grinding wild seeds. And the bow and arrow, invented at the very end of the Upper Paleolithic, now spread widely.

AN EVOLUTIONARY MIRROR

How closely was this Old World trend mirrored in the New World, our hemispheric petri dish? Very. Every society in the Americas, by the time the Europeans arrived, had reached or surpassed the Mesolithic level (which for obscure reasons of academic history is called the "Archaic" in its American manifestations). Even the "lowly" Shoshone had their rabbit nets, and commensurate institutions. Of course, not every American society independently reinvented everything in the Mesolithic tool kit. The bow and arrow may have come all the way from Eurasia via Eskimos a few thousand years ago. But much of the New World technology is too local in its utility to have come across the Bering Sea. All of this helps explain why we can call the Shoshone "cultural fossils": because, like the Northwest Indians, they correspond roughly to what the archaeological record depicts as a natural phase in a worldwide evolution from simple to complex social structure.

If there were a single continent, or even a single large piece of turf, that didn't reflect this trend, then the skeptics of cultural evolution would have ground to stand on. But the once-standard example of cultural stagnation—Australia, land of the aborigine—has now been swept from under their feet. Archaeologists have found a trend in Australian hunter-gatherer culture toward more subtle subsistence, featuring, for example, fishhooks and a neat trick for harvesting eels— digging dead-end ditches. (And when was the last time *you* invented anything as clever as the boomerang?) Meanwhile, trade was growing, as aborigines tried to get their hands on such fancy items as green-stone axes. This and other evidence of a shift comparable to the Mesolithic has undermined the "traditional static model of Australian prehistory," observes the archaeologist Harry Lourandos.

Maybe there's a limit to how flattered the aborigines should be by this revisionism, for here I'm using it to support the notion of an upward arrow of human history, and on this arrow their culture ranks "lower" than most others. Still, embracing this notion isn't as insulting as denying it. For to deny any directionality in cultural evolution is to say that the aborigines, or the Shoshone, or the !Kung, left to their own devices, would show no natural tendency toward higher levels of technological sophistication and social complexity.

This is of course ridiculous. Every known hunter-gatherer culture embodies a technological evolution that speaks of stubborn ingenuity focused on the resources at hand. The !Kung's one-liter ostrich-egg canteens, their four-liter antelope-stomach sacs, their bone arrow-heads coated with beetle-pupae poison paste—all have their counterparts in comparably smart hunter-gatherer innovation around the world. The aborigines use kangaroo incisors as chisels. They attach wooden handles to their stone knife blades by heating gum from eucalyptus to make a glue as strong as epoxy. The Andaman Islanders fuse vegetable gum and beeswax for the same purpose. They use bivalve shells for woodworking—a trick also discovered by the Alakaluf of southern Chile. The Copper River Eskimo took copper from the river, heated it, and forged it into daggers and arrowheads with stone hammers. The Greenland Eskimo used sandstone to saw chunks of iron from local meteorites and then pounded it into tools and weapons. The Chukchi of Siberia made webbed snowshoes, and the Australians of the Great Sandy Desert made webbed sand-shoes. And again and again—as with the Northwest Coast Indians—we find technology that goes beyond the demands of subsistence. The Ainu hunter-gatherers of Japan made "mustache lifters" to keep their soup facial-hair-free.

All of these cultures show longstanding, incremental progress. That this trend would continue indefinitely, had it not been interrupted by outsiders, is a central tenet of cultural evolutionism—or, more precisely, of the kind of cultural evolutionism championed in this book: a hard-core kind that sees the coming of the modern world as having been all but inevitable. To deny that tenet is to deny, explicitly or implicitly, the unity of humankind, the fundamental equality of aptitude and aspiration among people of all races.

Not that all of the aspirations are noble. Among the drivers of cultural evolution are the quest for status, the pleasure in showing off, and the thirst for material goods, ranging from key survival tools to needless gadgetry. The archaeologist Brian Hayden, having lived with indigenous peoples in Australia, North America, the Near East, and the Far East, has this to report: "I can say categorically that the people of *all* the cultures I have come in contact with exhibit a strong desire to have the benefits of industrial goods that are available. I am convinced that the 'nonmaterialistic culture' is a myth. . . . We are all materialistic."

Of course, notwithstanding the psychic unity of humankind, different peoples have moved along the arrow of history at different speeds. And there must be a reason. But if there is one lesson to take away from native American cultures, it is that race doesn't seem to be that reason. The various peoples living in the New World when white men showed up were not genetically homogenous; both biological and linguistic evidence suggest that they came from Asia in three successive waves. Still, the distinctions were not large; to the extent that the concept of "race" has coherence, the native Americans were all in the same "race." Yet, within that race all basic levels of social evolution were represented: from the Shoshone—the "irreducible minimum"—through the complex hunter-gatherers in the Northwest, through the agricultural chiefdoms to the east and south, to the state-level societies in what we now call Latin America.

Indeed, if we take linguistic affinity as an index of genetic affinity—as it usually, though not always, is—we can see how little genetic differences mean. The Shoshone and Nahuatl languages are close relatives—both members not just of the Amerind subset of the New World languages, or of the Central Amerind branch of Amerind, but of the Uto-Aztecan sub-branch of Central Amerind. Yet the Shoshone were at the "bottom" of the ladder of cultural evolution, and the speakers of Nahuatl—the Aztecs, who built pyramids and had hieroglyphic writing—were at the top. Blood would seem to mean little compared to environment, historical contingency, and cultural legacy.

Or consider, again, the peoples comprising the Northwest Indians. All belonged to what anthropologists recognize as a basically unified, though hardly homogenous, culture. Yet recent studies show them to be as different genetically as any group of Indians living contiguously in a comparably small space. And they speak languages so different as to suggest that some are descended from the first great migratory wave from Asia and some from the second. Yet they had a common habitat and intertwined cultural histories, and these proved decisive.

Okay, so if genes aren't the answer, what is? What exactly was it about the Shoshone that so handicapped them? So far in this chapter we've more or less assumed that an environment of plenty, such as the Northwest Coast, conduces to advance, while barren land does not. But, as we'll see in the next chapter, that explanation is incomplete at best. To be sure, fertile environments often accelerate cultural evolution. But not for the reasons you might think—not just because they yield natural affluence or permit the accumulation of surplus. The real key to cultural advance is more subtle. And, as we'll see, it is a powerful key, capable of explaining broad patterns in the past. For example, not just why the culture of the Shoshone was "slower" than that on the Northwest Coast, but why the New World in general had "slower" culture than the Old World.

Chapter Four

THE INVISIBLE BRAIN

All thought draws life from contacts and exchanges.
—*Fernand Braudel*

Explaining the affluence of the Northwest Coast Indians seems simple at first. They lived amid natural abundance. And abundance, after all, *is* affluence. Presumably that's why so many of the affluent hunter-gatherer societies recently discerned by archaeologists lived near large bodies of water. If you want to be rich, settle on rich land.

But this explanation won't wash. For starters, population tends to rapidly reach the "carrying capacity" of the environment. Although it's true that an acre of Northwest territory offered more food than an acre of the Shoshone's desert, it's also true that an acre of Northwest territory had a lot more mouths to feed.

Granted, even on a per capita basis, a day's work brought in more food for the Northwest Coast Indians than for the Shoshone. Bushels of fish were hauled in and packed away for winter dining. But this efficiency was due to high technologies—the massive salmon traps, the smokehouses, the storage cellars—as well as the social structures that governed them. And the technology and social structure are part of the Northwesterners' advanced economic development—part of what we're trying to *explain* when we seek the causes of affluence. To make them the cause as well as the effect would be cheating.

Besides, the Northwesterners' affluence goes well beyond food, and thus can't be merely the direct outgrowth of a fertile homeland. The designer robes, the spacious homes—these things don't grow on trees.

Earlier this century, anthropologists thought it easy to explain such

arduously crafted wealth as the *indirect* outgrowth of a fertile homeland. The key was "surplus." (This scenario assumes that the "carrying
capacity" of the environment wouldn't quickly be reached—perhaps
because there's a limit to how many hunter-gatherers can peacefully
coexist in a small space.) With tons of salmon just begging to be
eaten, you could meet your daily food needs in an hour or two and
have plenty of time left over to weave robes and build homes. After
all, such industriousness comes naturally to people, no?

Apparently not. In 1960, the anthropologist Robert Carneiro published an influential paper about the Kuikuru, who inhabited the jungles of Amazonia and tended gardens of manioc, a staple food (once
you remove the poison, which they did) and the source of tapioca.
The Kuikuru could have doubled or tripled their manioc output,
Carneiro calculated, but they preferred leisure time. Since then anthropologists have found various hunter-gatherer societies that, similarly,
had time left over after their daily food gathering. And, as one scholar
tartly put it, they "rarely spend this time designing cathedrals or in
general 'improving their lot.' "

So much for the theory that potential surplus always equals economic development. Indeed, against the backdrop of the Kuikuru
and other apparently laid-back societies, it almost seems that the conventional wisdom about the Northwest Coast Indians might be right:
they weren't examples of a general trend in cultural evolution; they
were just weirdly ambitious.

If surplus isn't the ticket to wealth, what is? What *did* create all the
specialization and trade found in "complex" hunter-gatherer societies? What did the Northwest Coast Indians have that the Shoshone
didn't have? What was the key to prehistoric economic development?

WORKING OVERTIME

Maybe we should direct these questions to a noted authority on economic development (and on non-zero-sumness, though he long predates such terminology): Adam Smith.

Two factors, Smith noted in *The Wealth of Nations,* are especially
conducive to the growing division of labor that characterizes economic advance. One is cheap transportation. Spending your after-
noon making yarn for a Chilkat robe makes sense only if the finished
product can be transported at a cost acceptable to its buyer. The second factor is cheap communication. The costs of finding out what

buyers want—and the cost to buyers of finding out what's available, and at what price—have to be bearable for transaction to ensue.

Note that the costs of transport and communication apply not just to the final "purchase" of the robe—at the "retail" level—but to the links in its creation, such as getting cedar bark from the south and copper from the north. At all levels, the movement of Smith's "invisible hand" gets smoother as information and transportation costs drop. The lower these costs, the more highly non-zero-sum the relationship among the players—the more each can gain via interaction, the more productive, per capita, the web of exchange.

How to keep these costs low if your communications and transportation technologies are primitive? One way is to stay near your customers and suppliers. In other words: live in a society with high population density. This may be the key to the wealth of the American Northwest: not natural abundance per se—an abundance quickly diluted by thick population, anyway—but rather the thick population that does the diluting. Back before communications and transportation were sufficiently high tech to catalyze markets, the stimulus came instead from a habitat that would tolerate large, close populations. And, conveniently, such habitats were often near water, which could give both technologies an added boost. Goods and data sometimes travel better by boat than by foot.

Not only were patches of the Northwest thick with people, one of whom could make blankets while a next-door neighbor focused on woodcarving; larger stretches of the coast—hundreds of miles to the north and south—were thick with *peoples,* the various linguistic groups that constituted Northwestern culture. Their diversity of natural resources and of cultural heritage worked in synergy.[†] Trading for cedar bark and copper was feasible, and the robes made by drawing on local tradition struck nearby peoples as enticingly foreign.

Maybe, then, what the ostensibly carefree manioc growers of South America lacked was not driving ambition but population density. In the rain forests of the Amazon basin, settlements were small and sparse. Had there been more nearby peoples—especially peoples with handicrafts and natural resources different from those on the local menu—farmers would have felt inspired to grow more manioc for exchange.

In fact, this speculation has been confirmed. Though many scholars have cited Carneiro's 1960 paper in dismissing the surplus theory of economic development, few have noted how Carneiro clinched his argument that manioc production in the rain forests was far below capacity: When Europeans showed up with lots of neat gadgets to trade, manioc production skyrocketed!

So, all along, the difference between native Americans of Amazonia and of the Northwest Coast hadn't been their work ethic. The difference had been that the Amazonians weren't getting paid to work overtime. Neither were the Shoshone. The arid Great Basin was even less conducive to thick population than were South American jungles.

ADAM SMITH AMENDED

Maybe we should amend Adam Smith's trademark metaphor of the invisible hand. Smith's point, of course, was that a bunch of far-flung people pursuing individual gain can, without really trying, collectively orchestrate a large-scale social process. The ingredients of a beautiful robe just seem to magically congregate, assemble themselves, and then find a buyer, as if guided from above.

It's a nice image, and in some ways apt. After all, a "hand" can do more work if moving goods is easy—if transportation costs are low thanks to the proximity of the players. Still, this metaphor gives short shrift to the other kind of cost that Smith stressed: the cost of processing data and "deciding" where the various resources should go.

Hands aren't very cerebral, after all; guiding any invisible hand there must be an "invisible brain." Its neurons are people. The more neurons there are in regular and easy contact, the better the brain works—the more finely it can divide economic labor, the more diverse the resulting products. And, not incidentally, the more rapidly technological *innovations* take shape and spread. As economists who espouse "new growth theory" have stressed, it takes only one person to invent something that the whole group can then adopt (since information is a "non-rival" good). So the more possible inventors—that is, the larger the group—the higher its collective rate of innovation. All told, then, the Northwest Coast Indians outproduced and outinvented the Shoshone not because they had better brains (the sort of conclusion Franz Boas worried about) but because they *were* a better brain.[†]

The fact that population density and size lubricate economic and technological development has been largely ignored by archaeologists and cultural anthropologists. True, some of them, such as Marvin Harris, do stress population growth—indeed, some see it as the prime mover of cultural evolution. But they emphasize a different side of the growth: a downside, not an upside. "Irresistible reproductive pressures," writes Harris, have "led recurrently to the intensification of production," which in turn puts stress on the environment, leading to an ecological crisis that only new forms of technology and social organization can solve. In short: Innovate or die! Population density, in this view, drives technological and social development not by creating opportunities, but by creating problems.

This is not the place to ponder the relative importance of the "upside" and "downside" of population growth (though it is the place to suggest that the downside has been overemphasized). The key point is that these two scenarios are compatible. Even if new ideas flow mainly from the synergy of a large, dense invisible brain, environmental stress could also spur innovations—and, in any event, could make people more receptive to them. If, for example, one village, for whatever reason, had developed a serviceable salmon trap, and a means of managing it, chronic food shortage would make nearby peoples acutely receptive to the idea. Also, peoples who lacked the technology might perish, freeing up real estate for those who *did* adopt it. So whether or not stress often triggers the birth of technologies and social structures, it could certainly hasten their spread. To borrow some terminology from biological evolution: stress raises the rate at which cultural "mutations" proliferate; it raises the "selective pressure."

In both the upside and downside scenarios, non-zero-sumness looms large. Whether people are trying to add to their wealth or avert disaster, their rational pursuit of self-interest is leading to economic cooperation and social integration that make them better off than they otherwise would be. So either way, you expect population growth to foster upward cultural evolution. And since the human population has grown with few interruptions ever since it was human, the impetus behind cultural evolution would seem to be strong.

If this view is correct—that one way or another, more and denser population means more advanced technology and more complex social structure—then there should be a close connection between population size and density on the one hand and technological and social complexity on the other. And there is. Consider the indigenous

societies that once flourished on the variously sized islands of Polynesia. The larger and more dense the island's population, the greater its
division of economic labor, the more advanced its technology, and
the more complex its polity.

The story told by these and other such cultures on the ethnographic record—these "living fossils" of cultural evolution—is repeated by the archaeological record. During the Middle Paleolithic,
as the human population grew slowly, the rate of technological innovation (not just the total *number* of technologies) also rose slowly.
Then, around 40,000 years ago, both of these trends passed milestones. The human species was for the first time large enough to
encompass the Old World, occupying virtually all inhabitable parts of
Africa and Eurasia; population growth would hereafter raise population density. Meanwhile, cultural evolution reached a level that warranted a new archaeological label—the Upper Paleolithic. During the
Upper Paleolithic, the average rate of technological change would be
one innovation per 1,400 years, compared to one per 20,000 years
during the Middle Paleolithic. Then, after 10,000 B.C., during the
Mesolithic, with population growing faster than ever, the rate of
technological innovation reached one innovation per 200 years
(including such gifts to posterity as combs and icepicks). Hunter-
gatherer societies, as we've seen, reached a new level of social complexity.

Of course, correlations between population size and density, on the
one hand, and cultural innovation, on the other, don't prove that
population growth is the driving force. Maybe things work the other
way around. Maybe cultural advances are what allow population to
rise. In fact, that is undoubtedly part of the story. Take away the
Northwesterners' salmon traps (and the Big Man's guidance of their
construction and use), or the Shoshone's rabbit nets (and the rabbit
boss's leadership), and population would have to thin out.

But that, really, is the point: technological, economic, and political
development spur population even as population spurs them. In this
symbiotic growth lies the inexorable power of cultural complexification. Whether you stress the "negative" side of population growth or
the "positive" side—whether you stress problems or opportunities—
the link between that growth and cultural evolution is one of mutual
positive feedback. The more people, the more culture; the more culture, the more people.

The "negative" side of population growth—environmental stress that makes subsistence precarious—may or may not be a big part of the story, but it is certainly not the whole story. The gadgets that pile up at an ever faster rate as population grows are not just subsistence technologies. Even back during the Middle Paleolithic, more than 50,000 years ago, people were intrigued by ochres (for painting) and pyrite crystals. And, as we've seen, during the Mesolithic, such "prestige technologies" as jewelry became an appreciable chunk of gross domestic product.

Great effort went into getting these status symbols. They seem to have been traded over hundreds of miles, back in a time when hundreds of miles was nothing to sneeze at. Even by 30,000 B.C., long before the Mesolithic, beads made of pierced seashells were migrating 400 miles from their point of origin. Later, regular networks of exchange blossomed, linking local invisible brains to distant invisible brains. The faint outlines of giant regional brains began to form. And the driving force wasn't periodic environmental "stress" but a more constant force: human vanity, powered by the status competition that is part of all known societies and seems to be innate.

The fitful but relentless tendency of invisible social brains to hook up with each other, and eventually submerge themselves into a larger brain, is a central theme of history. The culmination of that process— the construction of a single, planetary brain—is what we are witnessing today, with all its disruptive yet ultimately integrative effects.

CONTINENTAL DIVIDES

Once you appreciate the importance of population size and density, the cultural gap between the Old and New Worlds starts to make sense. Twelve thousand years ago, as the occupation of the Americas was just picking up steam, population was larger and thicker back in the Eastern Hemisphere. Thereafter, so far as we can tell, the New World lagged behind the Old World in population by several millennia—which is roughly how far the New World lagged behind the Old World in reaching early technological thresholds, such as agriculture, and early political thresholds, such as the chiefdom. (Indeed, if anything, post-agricultural social evolution unfolded slightly *faster* in the New World.)

Michael Kremer, an economist who stresses the link between population size and technological evolution, and who has pointed to the

New World's lagging technical level as exhibit A, has also found exhibit B. The melting polar ice caps that severed the Old and New Worlds also severed Tasmania and Australia. Tasmania's small population promptly fell culturally behind the more numerous Australians. Modern explorers, on contacting the Tasmanians, found them to lack such Australian essentials as fire-making, bone needles, and boomerangs. Thus the four petri dishes created by the melted ice caps—Old World, New World, Australia, Tasmania—behaved in accordance with theory: larger and denser populations equal faster technological advance.

Certainly, as we'll see, social advance can be retarded by things other than scanty population. Still, population is vital. It helps explain why we can call a people "less advanced"—the Shoshone compared to the Northwest Coast Indians, native Americans in general compared to Europeans—without insulting them.

After all, an individual Shoshone's brain could house as much information as a European brain. A hunter-gatherer is a vast and general data bank, featuring arcane knowledge about local flora and fauna and a basic grasp of all known technology. The average European in Columbus's day knew much less about nature, and very little about most European technology. He or she used the cerebral space thus saved by specializing in one narrow economic task. It was the synergy of many such specialized European brains that created the technology with which Columbus and other such men intimidated Indians. Individual native Americans weren't stupider than individual fifteenth-century Europeans—they just had the disadvantage of being Renaissance Men.

If the mutually positive feedback between culture and population were the whole story of cultural evolution, then we could end our narrative here with the words "and so on." But it isn't. As we've seen, there are forbidding thresholds that got crossed somewhere between the Shoshone and modern America. Among them are the "suprafamily" threshold, crossed by the Northwestern native Americans, and the "supravillage" or "chiefdom" threshold, which the Northwesterners flirted with but may have never quite crossed.

The reason these thresholds are so dicey is that, typically, along with all the benefits of affiliation comes a loss of autonomy—subordination to a political leader. And clusters of families, like individual families and for that matter individual people, prefer autonomy to subordination, all other things being equal.

To be sure, human nature in some ways lubricates these transitions. Our species is "naturally" hierarchical in the sense that people tend to sort themselves out by social status. And the resulting facility with which leaders lead, and followers follow, has no doubt eased the evolution of social complexity. But according to modern Darwinian theory, social hierarchies didn't evolve for the "good of the group," so followers don't cheerfully submit to the leader for the sake of the public interest. People by nature like high status, and generally reconcile themselves to lower status only grudgingly (and temporarily, until a chance for advancement arises). Thus when any family, or any group, surrenders autonomy, submerging its identity in a larger body, subordinating itself to a central authority, some natural resistance has been overcome.

So far we've been talking as if economics alone can impel people across these barriers. The rational pursuit of wealth and economic security entails non-zero-sum interactions that—at least in societies which lack a currency—seem to call for political hierarchy. So political hierarchy materializes. But is the story really so simple? Can economics alone get people and groups to surrender their sovereignty? And if not, what else has helped to do that job? This is the grim subject of the next chapter.

Chapter Five

WAR: WHAT IS IT GOOD FOR?

> If we think how many things besides frontiers of states the wars
> of history have decided, we must feel some respectful awe, in
> spite of all the horrors. Our actual civilization, good and bad
> alike, has had past wars for its determining condition.
>
> —*William James*

Ah, Tahiti. The lush island whose carefree natives the painter Paul
Gauguin used as icons of primitive bliss. The serene culture which
Jean-Jacques Rousseau considered evidence that humans had been
"noble savages," peaceful and benign, before their corruption by civilization. Unfortunately, as the anthropologist Lawrence Keeley has
noted, Rousseau relied for this conclusion on reports of Tahiti that
omitted relevant parts of its history. For example: the custom in which
a victorious warrior would "pound his vanquished foe's corpse flat
with his heavy war club, cut a slit through the well-crushed victim,
and don him as a trophy poncho."

Time and again there have been reports of a truly peaceful primitive people. Almost always, the reports have not worn well. Remember the "gentle Tasaday," the isolated band of hunter-gatherers
discovered in the Philippines in the early 1970s—the people who had
no word for "war"? Their authenticity fell into doubt along with the
credibility of their discoverer, Manuel Elizalde. As the *New York Times*
would later note, "It did not help when members of a neighboring
tribe said Mr. Elizalde had paid them to take off their clothes and
pose as Tasadays for visiting journalists."

To be sure, there are hunter-gatherer societies that don't exhibit the
elaborately organized violence denoted by the term "war." But often

what turns out to be lacking is the organization, not the violence. The warless !Kung San were billed in the title of one book as *The Harmless People,* yet during the 1950s and 1960s, their homicide rate was between 20 and 80 times as high as that found in industrialized nations. Eskimos, to judge by popular accounts, are all cuddliness and generosity. Yet early this century, after westerners first made contact with a fifteen-family Eskimo village, they found that every adult male had been involved in a homicide.

One reason the !Kung and most Eskimo haven't waged war is their habitat. With population sparse, friction is low. But when densely settled along fertile ground, hunter-gatherers have warred lavishly. The Ainu of Japan built hilltop fortresses and, when raiding a neighboring village, wore leather armor and carried hardwood clubs. The main purpose of the raids—to kill men, steal women, and settle grievances, real or imagined—is a time-honored goal of primitive warfare. Even today it is part of life among the Yanomamo of South America.

The behavior of observed Stone Age peoples is hardly the only evidence that the Stone Age was a bloody time. In a cave in Germany, clusters of skulls more than 5,000 years old were found arrayed, as one observer put it, "like eggs in a basket." Most of the thirty-four victims had been knocked in the head with stone axes before decapitation.

Anyone hoping that cultural evolution always translates into moral improvement will be disappointed to hear that such evidence of violent death is especially common among remains of the more complex hunter-gatherer societies. And in the yet-more-complex agrarian societies on the ethnographic record, things are similarly grim. In south Asia, a young Naga warrior was not considered marriageable until he had brought home a scalp or a skull. In Borneo, a Dayak hero returning from war would be seated in a place of honor and surrounded by singing women, with the head of one his victims placed nearby on a decorative brass tray. The warriors of Fiji gave their favorite weapons terms of endearment; one war club was called "Damaging beyond hope," and a spear was dubbed "The priest is too late."

All of this forces us to confront the fact that, as Keeley has put it, "what transpired before the evolution of civilized states was often unpleasantly bellicose." Human violence has been around a long time, and often it has been not man against man, but group against group. Ever since the early stages of cultural evolution—the era of hunter-gatherer societies—that evolution has been shaped by armed conflict.

COMRADES IN ARMS

This would seem to throw a wrench into the analytical works. So far this book has mainly stressed the forces of human cooperation, the win-win situations. The thesis has been that the direction of history results largely from the playing of non-zero-sum games. But, presumably, once someone has decided that he wants to use your corpse as a poncho, the two of you are playing a zero-sum-game; his gain is your loss. So too with warring villages. When men from one village raid the other, kill the men and abduct the women, the air is rife with zero-sumness. And so on, up the ladder of cultural evolution: whether the contestants are villages, city-states, whatever—war is hardly non-zero-sumness incarnate.

Still, war isn't nonstop zero-sumness, either. One big reason is that, even as war is inserting zero-sum dynamics between two groups, *within* the groups things are quite different. If your village is beset by axe-wielding men bent on slaughter, your relations with fellow villagers can pivot quickly toward the non-zero-sum; acting in concert you may fend off the assault, but divided you will likely fall.

Much the same interdependence exists among the axe-wielding slaughterers; in unison lies their best hope for victory. So, whatever side you're on, you and your fellow villagers are to some extent in the same boat; your fate is partly shared. That, actually, is a good rough-and-ready index of non-zero-sumness: the extent to which fates are shared. War, by making fates more shared, by manufacturing non-zero-sumness, accelerates the evolution of culture toward deeper and vaster social complexity.[†]

This was a constant refrain of one early cultural evolutionist, the sociologist Herbert Spencer. He overdid it ("Only by imperative need for combination in war were primitive men led into cooperation"), but he was on to something.

Consider again the Northwest Coast Indians. We've already seen how their evolving technology of sustenance raised social complexity. Division of labor and capital investment grew, and leadership emerged in the form of the "Big Man," who handled the logistics and helped keep social life in harmony. But all of this heartwarming cooperation to harvest nature's bounty was not the only social cement, nor the only cause for the Big Man's authority. War among the various peoples

along the Northwest Coast was a chronic threat. Just look at their technology—not the fishhooks and salmon traps, but the daggers, battle-
axes, war clubs, and bone-head spears; the stockades, wooden helmets,
and coats of armor made of leather and wooden slats.

Aggression was a way to obtain land or slaves or women, or just to
do some quick plundering. (The Haida of the Queen Charlotte
Islands have been called the Vikings of the Northwest Coast.) But
whatever the cause of war, being on the losing end was bad news. So
being tightly knit was a good idea, as was having a leader. Someone
has to guide the peacetime amassing of military technology. And, as
Spencer observed, war, "requiring prompt combination in the actions
of parts, necessitates subordination. Societies in which there is little
subordination disappear." Again and again, societies have chosen subordination over disappearance. Faced with war, they fall in line. Walter Bagehot, one of Spencer's contemporaries (and an early editor of
The Economist), explained the consequent social harmony this way:
"the tamest are the strongest."

That zero-sumness promotes non-zero-sumness should come as no
surprise. The standard example of non-zero-sum dynamics, after all, is
the game-theory exercise called "the prisoner's dilemma." If two partners in crime cooperate—if each agrees to stay mum when interrogated by the prosecutor, rather than implicate the partner in exchange
for lenient treatment—both can benefit.[†] And the source of their
common interest is the conflict of interest between them and the
prosecutor.

In the realm of primitive warfare, this unifying effect can go
beyond the people of a single village. One great way for a village to
fend off assault, or to conduct assault, is to ally with another village, a
standard tactic among the Northwest Coast Indians. And, once this
alliance exists, any enemies have good cause to themselves find allies.
And so on: an "arms race" of organization that expands the social web
outward, weaving more and more villages together.

The speed with which hostility can thus move to higher levels of
social organization, leaving harmony in its wake, has been much
noted by anthropologists. The Nuer of Sudan, as studied by E. E.
Evans-Pritchard early this century, were an especially vivid case. A
Bor tribesman explained, "We fight against the Rengyan, but when
either of us is fighting a third party we combine." Evans-Pritchard
described the dynamic abstractly: "Each segment is itself segmented

and there is opposition between its parts. The members of any segment unite for war against adjacent segments of the same order and
unite with these adjacent segments against larger sections."

It sounds almost like a general law of history. And indeed, history
offers lots of examples: formerly contentious Greek city-states forming the Delian League to battle Persia; five previously warring tribes
forming the Iroquois League (under Hiawatha's deft diplomacy) in
the sixteenth century, after menacing white men arrived in America;
American white men, two centuries later, merging thirteen colonies
into a confederacy amid British hostility (and pithily capturing extreme non-zero-sum logic with the slogan "join or die").

In the short run, this impetus for aggregation may seem aimless.
Alliances shift, tensions come and go, and large social structures dissolve almost as often as they form. But in the long run, over millennia, the worldwide trend has been toward consolidation, toward
higher and higher levels of political organization. And one reason is
war—intense, essentially zero-sum games that generate non-zero-sum
games.

In past chapters, listing technologies that elevate non-zero-sumness,
I cited such contraptions as rabbit nets and huge salmon traps. The
fact of war expands this list to include technologies that aren't so
obviously conducive to cooperation. The spread of iron weapons near
the end of the second millennium B.C. may not sound like a recipe for
social cohesion. But it was. The Israelites encountered iron weapons
in the hands of the Philistines, who, according to 1 Samuel, brought
"a very great slaughter, for there fell of Israel thirty thousand foot soldiers." In response, the loosely confederated tribes transformed themselves into a unified monarchy. The Israelites were warned that this
would mean taxation and conscription, but they insisted: "We will
have a king over us" so that "our king may govern us and fight our
battles."

PUSHING AND PULLING

One could describe the congealing effect of war by saying it *pushes*
people together into organic solidarity; it poses an external threat that
impels them into closer cooperation. And one could describe the
causes of solidarity emphasized in previous chapters—the economic
causes—as *pulling* people together; opportunities for gain originate
within the society and draw people into closer cooperation. If you

think about this distinction long and hard, it is guaranteed to start seeming fuzzy.[†] Still, the words "push" and "pull" provide a handy, if rough, terminology for describing two basic kinds of non-zero-sum forces in cultural evolution.

The relative importance of "push" and "pull" has been the subject of much disagreement. The dispute tends to focus on the several major thresholds of cultural evolution. Consider the "chiefdom" threshold. It is one thing for neighboring villages to become trading partners or even to attain a measure of "supravillage" political organization via loose confederation. It is another thing for neighboring villages to grant real, ongoing power to a central authority—for one village's chief to become the "paramount chief." When this happens, a chiefdom has been formed.

The anthropological literature features many chiefdoms—from the indigenous peoples of Hawaii and Tahiti to the American Indian princess Pocahontas's people, the Powhatan. And the anthropological literature features differing interpretations of them. Some scholars say the villages in chiefdoms were initially pulled together by trade and other economic sinews. Others say they were pushed together, by war or the threat of war. And some "push" enthusiasts go further and say war was doubly important. Not only were villages often united to better fight wars; they were united in the first place by war, by raw conquest. Robert Carneiro has written, "Given the universal disinclination of human groups to relinquish their sovereignty, the surmounting of village autonomy could not have occurred peacefully or voluntarily. It could—and did—occur only by force of arms."

Carneiro—a fan of Herbert Spencer, and the editor of a volume of Spencer's writings—was a student of Leslie White, who did so much to revive cultural evolutionism at mid-century. As it happens, White's other star pupil, Elman Service, took a quite different, un-Spencerian view. Service (who died in 1996) envisioned chiefdoms often being formed when several nearby villages were bound by commerce; the village located at the nexus would naturally become the richest and would gradually grow dominant. It could all happen peacefully, Service believed. In his 1962 book *Primitive Social Organization,* he wrote, "It is, in fact, clear from the record in some cases and probable in many others that small neighboring societies, or parts of them, often join an adjacent chiefdom quite voluntarily because of the benefits of participation in the total network." Carneiro's reply, essentially, is:

Networks, schmetworks. "Force, and not enlightened self-interest, is the mechanism by which political evolution has led, step by step, from autonomous villages to the state," he wrote in 1970.

Carneiro and Service have come to serve as icons of the "push" and "pull" views of the world. A group of social psychologists, studying social evolution in the laboratory, even set up elaborate simulations of the "Service condition" (different groups crafted products and traded with other groups) and the "Carneiro condition" (groups could trade, but one of the groups was allowed to confiscate the products of other groups). For what it's worth, the experiment showed that leadership—an acknowledged paramount group—emerged in either event, and under the Service condition the standard of living was higher.

But in the real world, Service's theory has been harder to validate. When the political merger of villages has been observed, it has usually come by aggression or intimidation. And when consolidation has been voluntary, it has generally been to fend off external aggression.[†] On the other hand, Service might point out, economically driven integration is a slower process than conquest or military alliance; hoping to have seen it during the brief history of anthropology is like taking a few seconds to watch grass grow. Besides, since Service's death, some archaeologists have noted how little evidence of warfare they found amid the rubble that reflects the evolution of chiefdoms.

WAGING PEACE

In a way, the difference between Carneiro's "push" view of the world and Service's "pull" view isn't as stark as it sounds. Service did acknowledge, even stress, the frequency of war in prehistory. Indeed, in his view, conflict has been so persistent that to talk of the "causes of war"—as if it needed them—was almost to get the story backward. Better to think of war as a fairly natural condition and then examine the means by which people have avoided it; we should study not the "waging of war" so much as the "waging of peace."

Implicit in this view is the suggestion that the picture of war presented above—as a strictly zero-sum game—is too simple. And it is: war can be so mutually devastating that both sides clearly lose. In that case, it is a non-zero-sum game—specifically, you could call it a *negative*-sum game, in contrast to such positive-sum activities as economic exchange.[†] (In *The Art of War,* Sun Tzu, recognizing war's lose-

lose aspect, counsels commanders to leave their enemies a means of escape.) Hence the incentive to "wage peace."

Of course, societies have a stubborn tendency to *think* of war as a zero-sum game. They heedlessly launch it even though their own men will surely die.[†] What's more, once the war is launched, it is full of zero-sum dynamics; when opposing soldiers are in pitched battle, their fortunes are inversely correlated. For these and other reasons, I will continue to treat wars as mainly zero-sum games, and add nuance as necessary. Nonetheless, Service has a point. Even if individual wars are often essentially zero-sum, featuring a clear winner and a clear loser, war*fare*—endless intermittent back-and-forth battling—can be, in the long haul, very bad for both sides. And such persistent negative-sumness is indeed grounds for waging peace.

In Service's view, then, war is just another reason to value the harmony that comes from economic integration; it isn't to fight wars that society evolves, so much as to escape wars, to carve out broader and sturdier war-free zones. Perhaps, Service suggested, "not only the evolution of government, but the very evolution of society and culture itself, depends on the evolution of the means of 'waging' peace in ever-widening social spheres—by continually adding new political ingredients to the social organization."

In theory, the first step toward waging peace would be to recognize that ongoing warfare is indeed a lose-lose game. It doesn't take a game theorist to see this. In Papua New Guinea, one man observed: "War is bad and nobody likes it. Sweet potatoes disappear, pigs disappear, fields deteriorate, and many relatives and friends get killed. But one cannot help it."

The second step would be figuring out how to help it. And here some progress can be found in societies at all levels of organization. That includes the Northwest Coast Indians. When two peoples with a history of enmity could benefit through trade, a Big Man on one side might ritually bond with a Big Man on the other side. The two would exchange gifts and ceremonial names, becoming "brothers." Thereafter, wrote the anthropologist George Peter Murdock in 1934, "neither ever engages in war with the other, and the dwelling of either is a sanctuary where the other can always find refuge." Through this channel the Haida gave dried halibut, mats, furs, and canoes to the Tsimshian in exchange for grease, candlefish, copper, and blankets.

Among "Big Man" societies generally, perhaps the most common means of "waging peace" is the lavish feast. (One more reason not to consider the Potlatch irrational.) An anthropologist who studied two societies in Papua New Guinea noted how inter-village feasts were used to create "regions of peace and stability in an inherently danger-ous world."

Then again, to Carneiro and other "push" theorists, this sort of waging of peace looks suspiciously like a tool for waging war. Host-ing a feast can win a village a military ally. And even if the feast merely neutralizes a potential foe, that may provide enough leeway for conquest on another front, or defense on another front. The more you think about "waging war" and "waging peace," the more insepa-rable they seem: a single coin, with Carneiro's visage on one side and Service's on the other.

Consider the Yanomamo of South America, intensively studied by the anthropologist Napoleon Chagnon. In the middle of Yanomamo territory, where villages are crowded together, war is especially com-mon. Carneiro likes to stress that this core area shows signs of politi-cal evolution: larger villages and firmer political leadership, especially during war. On the other hand, Carneiro and Service might both cite—each with his own spin—the fact that this area also includes more inter-village bonding (designed, in Chagnon's words, "to reduce the possibility of warfare between the principals"). And both might note—again, with different spins—Chagnon's surmise that vil-lages sometimes cultivate a kind of phony trade in the name of peace. A village, Chagnon believes, will refrain from making a particular tool precisely to create interdependence with a neighbor that does make it.

What Carneiro and Service—and for that matter Chagnon—would agree on is that in one sense or another, an atmosphere of war can fos-ter the evolution of complexity. "Where warfare is intense and migra-tion out of the area not feasible," wrote Chagnon, "there is selection for larger local groups and more elaborate intergroup relations."[†]

As Service realized, the "waging of peace" long predates war per se. Feuds between two hunter-gatherer families don't qualify as war. Yet, like war, they impede the gains of concerted economic behavior. To enjoy the fruits of suprafamilial organization, feuds must be sup-pressed. Peace must be locally waged.

And, as it happens, this local broadening of amity is further encouraged by the existence of slightly less local enmity—contention between *clusters* of hunter-gatherer families. And so it goes, up the scale of cultural evolution: the crevices of social organization—the zones of zero-sum contention between families or villages or chiefdoms or states—keep getting filled in by the cement of non-zero-sumness; and the zero-sumness thus displaced keeps retreating to higher levels of organization. And from there it continues to have its paradoxically congealing effect at lower levels.

Still, as Service stressed, the congealing also has its own internal logic; whatever *caused* the expansion of peace—war, the threat of war, the farsighted avoidance of the threat of war—peace is ultimately its own reward. Just ask the Auyana of New Guinea. After they were pacified by Europeans, the men rejoiced in their newfound ability to go urinate in the morning without fear of ambush. The first step toward a productive day. Also the first step toward trading with the former ambushers, thus weaving a web of interdependence which can fortify the peace that's been waged.

Note how war—or at least the threat of it—narrows the range of choice. In the previous chapter I more or less assumed that people tend to harvest the fruits of non-zero-sumness; they just naturally like to raise their standard of living and to insulate themselves against risk. This fondness for bounty and security assures that they will try to realize positive sums, experimenting with new technologies and new forms of social organization; and this realization sustains the basic directional drift of history.

Whether I'm right in this claim about human nature certainly matters, but in a context of war—the context of human history—it matters less. For in that context people have little choice but to pursue economic and organizational advance. After all, unproductive societies tend to get squashed. An anthropologist described one group of Northwest Coast Indians, trapped between powerful neighbors, as being "ground to bits"—reduced to scurrying around nibbling on raw food lest campfires attract attention.

People who are conquered may live on after conquest, and they may be lucky enough not to be enslaved; but they will almost surely wind up adopting a new system, the system of the conquerer. And conquerers' systems tend to be productive, to involve, for example, a relatively advanced division of labor. One way or another, non-zero-sumness wins in the end.

All of this brings us back to Kant's emphasis on "unsocial sociability." The realm of "sociability"—the geographic scope within which peace reigns—has grown massively since our hunter-gatherer days. And commensurately massive quantities of unsociability have been overcome. Yet they are often overcome under the ironic stimulus of higher-order unsociability. To put this dynamic of cultural evolution in the Darwinian language of natural selection: what is "selected for" is larger and larger expanses of non-zero-sumness, but one of the main selectors is the zero-sum dimension of war. In this sense, waging war, in the end, *is* waging peace.

An authority on human behavior once remarked that if two people stare at each other for more than a few seconds, it means they are about to either make love or fight. Something similar might be said about human societies. If two nearby societies are in contact for any length of time, they will either trade or fight. The first *is* non-zero-sum social integration, and the second ultimately brings it.

Chapter Six

THE INEVITABILITY OF AGRICULTURE

The farmer takes a wife, the farmer takes a wife . . .
—*From the nursery song "The Farmer in the Dell"*

A favorite pastime of archaeologists is to invent competing explanations for the domestication of plants and animals, which first happened around 10,000 years ago. Perhaps, one theory has it, a hotter climate, by drying up once-fertile lands, made the hunter-gatherer lifestyle suddenly precarious, and people groped for a new livelihood. Or maybe the extinction of giant elk, woolly mammoth, and other big game had the same effect. Or, on the other hand, maybe the key was a more *benign* environment, a climate which happened to nourish certain plants that were good candidates for domestication.

And then there is a simpler theory: Farming was just a good idea. It was a good idea in the same sense that the various tools and techniques constituting the hunter-gatherer lifestyle had been good ideas, and thus had been added to the human repertoire.

This was the radical position taken in 1960 by the University of Chicago archaeologist Robert Braidwood. Reviewing his own fieldwork in the Middle East, where farming first appeared, he depicted agriculture's advent as merely "the culmination of an ever increasing cultural differentiation and specialization of human communities." So far as he could tell, "there is no reason to complicate the story with extraneous 'causes.'"

Braidwood is considered the founder of the modern study of agriculture's origins, but this particular opinion wasn't destined for veneration. Notwithstanding his injunction against complicating the story, archaeologists have continued to complicate the story. The above-cited "causes," and others, still jockey for preeminence. More than two decades after Braidwood insisted that agriculture needs no special explanation, an archaeologist, summarizing the consensus, declared that agriculture is "not yet satisfactorily explained." The search for causes continues. An air of mystery still surrounds the origins of agriculture.

Indeed, if anything, the air thickens. Some scholars now say that, paradoxically, early farmers would actually have had to work longer and harder to grow food than to just get it the old-fashioned way, by hunting and gathering. Thus the logic behind the origins of agriculture, we are told, is much less straightforward than it seemed back in Braidwood's day.

This view poses a problem for cultural evolutionists—or, at least, for hard-core cultural evolutionists, such as me. After all, if farming was such an unappetizing prospect, how could humanity have been virtually certain to take it up eventually? Shouldn't passage through this threshold be counted as a lucky break, a chance venture that could just as easily have been the road not taken? And, if so, doesn't that make all that followed farming—ancient civilizations, less ancient civilizations, and so on—look far from inevitable?

Plainly, before we can get on with the rest of this book we must dispel some of the mystery surrounding agriculture's origins and deflate the ongoing search for "causes." Conveniently, that will give us a chance to dispel some misconceptions that persist to varying degrees within the social sciences and in various ways sap enthusiasm for hard-core cultural evolutionism.

HAPPY HOUR

The case against agriculture's being a natural cultural advance began to gather momentum with the surprising discovery that hunting and gathering isn't such a bad way to make a living. The !Kung San, Richard Lee found in the 1960s, work just a few hours a day—hunting, digging roots, harvesting mongongo trees—and then it's Miller time. In 1972, the anthropologist Marshall Sahlins (a former cultural evolutionist turned skeptic of cultural evolutionism) dubbed hunter-gatherers "the original affluent society" on grounds that "all the people's material wants are easily satisfied."

And the problem isn't just that primitive agriculture may have been a regression in terms of sheer efficiency. The more populous villages that farming ushered in would presumably foment disease; and the low-protein, high-starch content of some staple crops might be unhealthy. Studying the bones of early farmers, some archaeologists have concluded that they had shorter lives, and more rotten teeth, than hunter-gatherers.

This brings us to *misconception number one:* that cultural evolutionists believe *change is guided by farsighted reason.* Actually, cultural evolution has involved little advanced planning. No prehistoric hunter-gatherers assembled a committee to decide whether a growing reliance on starchy foods would eventually promote tooth decay. Planted food slowly replaced wild food over many generations. And slowly the planted food became less like its wild ancestors; it got domesticated. The question isn't why hunter-gatherers "chose" farming, but why they chose the long series of tiny steps leading imperceptibly to it.

Part of the answer is that these hunter-gatherers were people. People are innately curious. They fiddle around with nature and try to bend it to their will.

Consider the Kumeyaay of southern California. Technically, they were a hunter-gatherer people. But, when encountered by the Spanish in the eighteenth century, they had transfigured the landscape. At high altitudes they planted groves of oaks and pines, whose nuts they harvested. Elsewhere they planted yucca and wild grapes. Near villages they planted cactus for liquid refreshment. The Kumeyaay burned off unwanted plants to pave the way for their favorites and razed dense shrubs to attract deer.

None of the plants they cultivated were domesticated. So this massive intervention didn't qualify as farming. Still, is it really likely that the Kumeyaay could have gone another 1,000 years without breeding juicier grapes?[†]

The Kumeyaay are far from the only "hunter-gatherers" who have given nature a helping hand. Australian aborigines replant the tops of the wild yams they eat. And remember the Shoshone of the Great Basin, often taken as paragons of the primitive? They burned off unwanted foliage, and some Shoshone planted wild food species. Some even used irrigation.

Hunter-gatherer societies that cultivate plants but haven't yet domesticated any are sometimes called "proto-agricultural." Dozens of such societies have been observed. You might think that anthropologists would look at all these societies and say, "The impulse to groom nature seems strong and widespread—maybe the coming of agriculture wasn't so unlikely after all." You would be wrong. Often, the reaction is the opposite. Proto-agriculture, we are told, just goes to show that many hunter-gatherers *knew* enough to become full-fledged farmers yet chose not to.

Often underlying such pronouncements is the unspoken premise that cultures are static; they have assumed final form; it wasn't just that the Kumeyaay *hadn't* taken up farming, but that they *didn't* take up farming—end of story. Thus an evolutionary view of culture is dismissed by assuming that cultures are not in the process of evolving.

THE MYTH OF EQUILIBRIUM

The assumption that primitive cultures are static is grounded in *misconception number two: the idea of intrinsic equilibrium*—the idea that cultures stay the same unless jostled by such outside forces as retreating glaciers or sudden drought. Happily, this notion has lost favor among many archaeologists and anthropologists. But it has more than its share of defenders—that is, more than zero—and has deeply influenced thinking not just about agriculture but about culture generally. A recent archaeology textbook asserts that cultures do not "change in any patterned fashion as long as they are successfully adapted to their environments and the environment does not change." It is the assumption of equilibrium that compels archaeologists to seek an external "cause" for any development as dramatic as agriculture.

Subscribers to the equilibrium fallacy underestimate the unsettling nature of human innovation—the extent to which new ideas and techniques spring from within societies and transform them. But downplaying our species' genius is not the only problem. As we've seen, the main impediment to farming isn't thought to be a lack of inventiveness, but rather a lack of necessity. As Marvin Harris has put it, "What keeps hunter-collectors from switching over to agriculture is not ideas but cost/benefits. The idea of agriculture is useless when you can get all the meat and vegetables you want from a few hours of hunting and collecting per week."

Here, aiding and abetting the "equilibrium" fallacy, is *misconception number three: that human societies are fundamentally unified,* devoted to meeting their *collective* needs. The mistake gets back to the romantic notion of hunter-gatherer societies as oases of communal bliss. All for one and one for all. And if all are getting enough food, then why should anyone bother trying something new?

The answer is that hunter-gatherers are in truth just like us. They're competitive, they're status-hungry, and, above all, they are individuals. In those hunter-gatherer societies that are proto-agricultural, the clusters of cultivated wild foods aren't typically community property; usually they are owned by a particular family or extended family that dispenses the harvest as it sees fit. Once you start thinking of hunter-gatherers as driven by the physical and psychic needs of themselves and their families, there is no shortage of reasons why they might cultivate plants in their spare time.

Consider (once again) the Northwest Coast Indians, whose lavish use of cultivated wild plants is now coming to light through the work of the geographer Douglas Deur. A Kwakiutl household might have its own salt-marsh garden for clover roots or silverweed roots (nutritional delicacies), and might tend plots of wild berries or edible ferns. In hard times—when, say, the salmon weren't running—the family might eat the entire harvest. But often the food would serve the family's interests more obliquely. Being a gastronomical delight, it could be swapped for candlefish oil, and sometimes crates of garden-grown food were paired with other foods and handicrafts to fetch a prized copper shield. Often such exchanges took place between villages, orchestrated by Big Men, but non-zero-sumness also welled up within villages. A household might "give" food to a needy neighbor, with a view to future reciprocation. In the meanwhile, the giver, in addition to having garnered an IOU, enjoyed some status elevation. And families chronically in a position to "give" enjoyed chronically high status, like philanthropists.

Even in modern suburbs and small towns, avid gardeners win local esteem by giving neighbors fresh tomatoes or flowers. This strikes most of us as normal behavior. But the possibility that people might behave the same way in a primitive economy—where both the gift and the ensuing IOU were of much greater value—seems rarely if ever to cross the minds of archaeologists as they ponder the mystery of agriculture.

The various benefits of gardening were an incentive to refine it. There's evidence that the Northwest Coast Indians were weeding out the less robust specimens, the first step toward domestication. And, to expand the level land in their uneven habitat, they built retaining walls, which had the added virtue of holding nutrient-rich soils. The Kwakiutl word for "garden" means "place of manufactured soil."

In addition to the Northwest Coast Indians and other proto-agricultural hunter-gatherer societies, there are "cultural fossils" further along in the evolution toward agriculture.[†] Various "horticultural" societies grow domesticated crops in gardens but still rely on some hunting and gathering. Most of these societies resemble the Northwest Coast Indians, with gardening a private enterprise that pays off at the family level.

Thus, a young Yanomamo man in the jungles of South America, having just gotten married, will clear a garden for plantains, maize, cotton, tobacco, and other crops. He is not doing this for the good of his whole village. Indeed, he may surround the coveted tobacco with a fence, even plant sharp bones as booby traps. When he shares his harvest, he will do so selectively, cementing friendships, incurring unwritten IOU's, repaying his own debts, amassing status.

THE FARMER TAKES A WIFE (OR TWO)

Could something as ephemeral as status really entice people into becoming agricultural innovators even when they face no regular shortage of food? The answer comes from looking at the top of the pecking order—at the Big Man or "Head Man," a version of which is found among the Yanomamo and horticultural societies generally.[†] Big Men tend to have not just big gardens, but big numbers of choice wives.

The idea here isn't that aspiring Big Men necessarily sketch out a systematic plan for acquiring multiple wives. During the biological evolution of our species, one of the benefits of male status was easier access to sex. (So too with our nearest relatives—chimpanzees, bonobos, gorillas.) Because of this correlation between status and fecundity, genes imbuing males with a thirst for status have fared well by natural selection.[†] The resulting drive to impress people needn't bring conscious awareness of its reason for being—any more than hunger entails a knowledge of nutrition. Status just feels gratifying; it seems to be its own reward, even if its ultimate evolutionary purpose was genetic proliferation.

On the other hand, conscious awareness of the sexual payoff for farming is, if not necessary, hardly out of the question. When Soni of the Solomon Islands (three chapters ago) was preparing those thirty-two succulent pigs that he wasn't going to get to eat, he no doubt knew that the more adroit Solomon Islands feast givers—that is, Big Men—got as many as five wives. Indeed, sometimes the link between amassing food and amassing wives is explicit. Among the Northwest Coast Indians and some other polygamous peoples, loads of garden-grown food could be part of the "brideprice" paid for a wife.

Archaeologists, faced with the observed correlation between a farmer's status and wealth on the one hand, and his number of wives and offspring on the other, have tended to get things backward. Big Men are said to seek multiple wives "since many wives produce more food than one wife" and to have many children "since many children produce more food than few children." To be sure, Big Men may value the labor provided by a large family. But, in terms of the ultimate logic of their quest—the Darwinian logic that selected the genes that fuel the quest—they are amassing food to amass wives, not the other way around. If the food pays off nutritionally, that's great, but even if it doesn't, it is valuable, because it raises their status relative to competing males. Among the Trobriand Islanders, one anthropologist reports, farmers aimed to "accumulate so many yams that they may rot in storehouses and stimulate the envy of rivals."

The problem with scholars mystified by agriculture's origins isn't that they are unaware of status hierarchies in horticultural and fully agrarian societies. The problem is that they tend to view the hierarchy as a *product* of domestication—in which case it couldn't be a cause. Hence, *misconception number four: the notion of the "egalitarian" hunter-gatherer band.*

We've already suggested that the venerable notion of the utterly communal hunter-gatherer band is suspiciously romantic; that the !Kung San, for example, are subtly permeated by selfishness. Are they also prone to social climbing? The answer isn't obvious, since wealth—even in the form of a little extra food—is hard to accumulate; they live in a desert and often relocate. But it's a good bet that if gardening were more practical, they would find that cultivating extra food was a good way to win wives and influence people. Of course, as the most industrious men exploited this fact most fully, accumulating wives and power, social inequality might grow. Still, social climbing would have been the cause of the farming, not just the result.

In a sense, this thought experiment has already been conducted—

in the form of the //Gana, nearby Bushmen who supplement their hunting and gathering with farming. Among //Gana men, the anthropologist Elizabeth Cashdan has noted, the allotment of sexual resources is quite unequal; one-fourth of the men have more than one wife. Writing in 1980, near the heyday of hunter-gatherer romanticism, Cashdan heretically argued that it would be wrong to see the //Gana's social inequality as having *emerged* with agriculture. After all, she noted, about 5 percent of !Kung men had more than one wife. The reason the struggle for status is so subtle among the !Kung, she contended, is their precariousness; shortfall could strike any given family, so it is in each family's interest to support an ethic of sharing, as insurance. The //Gana, Cashdan wrote, illustrate "the *lifting* of the constraints that produce strict egalitarianism among other Kalahari hunter-gatherers."

And full-fledged domestication is not the first step in that lifting. "Proto-agricultural" hunter-gatherer societies broadly are more likely than the average hunter-gatherer society to have conspicuous disparities in status. Apparently the Northwest Coast Indians aren't the only people who found that home-grown food is a social lever.

It would be an exaggeration to say that all archaeologists who ponder the origins of agriculture have ignored the quest for status. Brian Hayden has championed a maverick "competitive feasting" theory, inspired by the Potlatch and other, less famous, forms of inter-village feasting. The idea is that if in any society some aspiring Big Man—some Soni—can get fellow villagers to produce lots of food, he can use it to elevate his status in feasts with other villages. In the process he can acquire political influence within his own village.

So far so good. But Hayden describes the Big Man as a genetically distinct "personality type" present in all societies—the "aggrandizer." These aggrandizers are "empire builders; they seek to control human affairs for their own benefit and gratification." In short, they are bad guys, different from such good and innocent souls as you, me, and Hayden. This is in some ways a comforting worldview, but it is at odds with modern Darwinian theory, not to mention observed social reality. To be sure, some people, for whatever reason, are more ambitious than others. But there's a little Big Man in all of us. We are all social climbers by nature. Some just manage to climb higher than others.

What does it matter whether social ambition is a property of our whole species, rather than just of the Henry Fords and Margaret

Thatchers of the world? Well, the more widespread the urge to impress, the stronger the force that drives cultural evolution. If everyone is always striving for social status, then every increment in the evolution of agriculture, from the tiniest, scruffiest garden on up, is easy to explain; there's a kind of arms race with food as the weapon.

Actually, food is just one of the weapons. Political organization is another. From the early days of agricultural history, as Hayden's theory hints and the Sonis of the world show, coalition-building comes into play. Leaders who can harness non-zero-sum logic to draw people into cooperative effort prevail in competition for status and other social resources, inviting future leaders to do the same on a larger scale.

MILLER TIME RECONSIDERED

Once you realize that man does not live by bread alone—that status and sex are nice, too—the claim that hunting and gathering beats primitive farming as a subsistence technology begins to lose relevance. Of course, the logic behind agriculture would be even stronger if it turned out that this claim about hunter-gatherers was wrong, or at least exaggerated, in the first place. And it may have been. The seminal calculations of the !Kung workday—two or three hours, then party time—have been put to skeptical scrutiny and found wanting. The calculators forgot to include time spent processing the food, making spears, and so on. It now appears that *these* hunter-gatherers, at least, work roughly as hard as horticulturalists.

Further evidence that hunter-gatherer life is not a year-round vacation can be found in proto-agricultural societies. The Shoshone's planted wild foods, one anthropologist observed, were "insurance" crops, and "frequently served as crucial secondary staples." So too with primarily hunter-gatherer but incipiently horticultural societies, such as the Siriono of Bolivia. While trekking through the forest in search of game, writes another scholar, they would visit their scattered gardens, "depending on them as secure sources of food energy."

All of this suggests that the layperson's common-sense notions about life among prehistoric hunter-gatherers is on target: adversity was part of life, shortage loomed over the horizon, and fortune favored the prepared. Between the quest for status and the quest for sheer survival, we have a powerful impetus behind the evolution of agriculture.

The impetus gets even stronger when we add one more factor: our old friend from the previous chapter, war. How would war encourage agriculture? In primitive war, few things come in handier than sheer manpower. And agriculture supports much larger settlements than hunting and gathering does. One of the earliest known farm towns, the ancient, excavated village of Jericho, housed hundreds of people on around six acres. Not huge by modern urban standards, but compare it to what lies beneath: remnants of a hunter-gatherer camp one-fifth as large. Imagine a battle between these two villages, and you'll see that farming was a compelling lifestyle. Whether or not early farmers thought about the military edge their lifestyle offered, war would have helped the lifestyle spread.

Perhaps fittingly, Jericho is surrounded by a wall. At four meters high and three meters thick, with cylindrical watchtowers, this wall may have once been the largest capital project in the history of the world—a monument to the non-zero-sumness created by conflict between groups and thus intensified by farming.[†]

THREE STRUGGLES

In the end, then, the claim that agriculture is "not yet satisfactorily explained" is misleading at best. If anything, the coming of farming was "overdetermined"—there is a surplus, not a shortage, of plausible explanations: the struggle for status within societies, armed struggle between societies, and the struggle against scarcity. Of course, "excessive" explanatory power is no scientific vice when the three explanations are logically compatible.

The archaeological record bears the clear marks of the first two struggles—wars and status competition. During the Mesolithic, just before the emergence of farming, wood and bone armor appear, cemeteries contain lots of people who died violently, and artists start depicting archery battles. Meanwhile, *within* societies, status competition is getting more conspicuous, with more and more bracelets, beads, and amber pendants showing up in high-status graves. (Were people trading food for these, adding an incentive to expand food production? There's no archaeological way of knowing, but such exchanges have been seen among hunter-gatherers, as when the Pomo of northern California got acorns and fish for their beads. In any event, Jericho, the quintessential farm town, would eventually become a regional trade center.)

The third struggle—against scarcity—doesn't leave such clear records. But we can say this much: even assuming this struggle wasn't a central force behind the evolution of agriculture, it would have kicked in if given enough time. As the planet's population grew—and indeed it grew faster and faster as hunter-gatherer societies grew more and more complex—the day was bound to come when nature's cornucopia couldn't feed the teeming masses, however ingenious their hunting and gathering.

Whatever the relative importance of these three struggles in driving the evolution of food procurement technologies, the effect was evident before farming. These technologies evolved from the Upper Paleolithic, with its well-crafted stone blades, through the Mesolithic, with its sickles, bows and arrows, mortars and pestles, nets and fancy traps. During the Upper Paleolithic, the menu grew beyond traditional staples—nuts, roots, and big game—to include birds, dangerous animals (lions, boar), and smaller animals, such as rabbits. With the Mesolithic tool kit, the menu expanded further, encompassing snails, lizards, frogs, grass seeds, lots of fish and shellfish, and lots of plants, including poisonous ones that had to be detoxified. At one hunter-gatherer village near the Euphrates River around 10,000 B.C., people were processing 157 species of plants. (With this growing environmental mastery, there was less and less need to migrate, so gardening made more and more sense. Sedentism seems to have preceded full-fledged domestication in most, if not all, cases.)

This long, clear trend—the ever-more-intensive search for food—is rather at odds with the image of hunter-gatherers sitting around picking their teeth until some external change created a sudden need for agriculture. More specifically, it is at odds with the assumption that a hunter-gatherer band wouldn't embrace new food techniques unless they were clearly less arduous than the old ones. As the scholars T. Douglas Price and James A. Brown have noted, additions to the hunter-gatherer diet during the millennia preceding agriculture were often "more costly in terms of procurement and processing" than were existing foods.

All of this leaves agriculture looking less revolutionary than evolutionary. Hunter-gatherers had long been working hard to intensify their yield, getting more and more food from a given acre of land. Farming was "no great conceptual break with traditional subsistence patterns," in the words of Mark Nathan Cohen, one of the first anthropologists to voice doubts about the notion of a natural "equilibrium."

To be sure, agriculture would ultimately prove revolutionary, a technology that would restructure society. Indeed, the rate of social change after agriculture so surpassed the more sedate pre-agricultural rate that it is fair to speak of a kind of "equilibrium" being disrupted. But the point is that the disruptor wasn't some external and whimsical force, such as drought or retreating glaciers, but rather internal and inherent forces, such as social striving and population growth.

Moreover, however "sudden" the changes wrought by farming, the *nature* of the changes was nothing new. Agriculture's ultimate social implication—sharply elevated social complexity and non-zero-sumness—had long been manifesting itself more slowly. Toward the end of the hunter-gatherer era there were more storage huts and other capital projects requiring political leadership, more long-distance alliances, and more trade—not to mention more kinds of food and tools than ever before.

In short, Robert Braidwood was right to dismiss the "mystery" of agriculture back in 1960 by depicting farming as merely "the culmination of an ever increasing cultural differentiation and specialization." Standard attempts to explain domestication as a response to epic change—a suddenly more barren landscape, or a suddenly more fertile landscape—are indeed unnecessary. Certainly environmental changes can *add* to the logic of farming, and help explain why it arose in one area before another. But if the question is why farming evolved *at all,* we needn't delve into the details of climate, flora, and fauna. Given enough time, it was bound to happen.

This notion of a persistent and universal evolutionary logic behind farming helps explain an otherwise puzzling fact: farming kept getting invented, and once invented, it tended to spread. The consensus among archaeologists is that farming arose anew at least five times—three times in the New World, twice in the Old World—and possibly seven. Surely this is no coincidence.

Of course, different cultures reached this threshold at different speeds. We've already seen some reasons for lags in cultural evolution, and there are others. The biologist Jared Diamond, in his book *Guns, Germs, and Steel,* has explained many such disparities via geography. For example: some areas are more blessed with readily domesticable species than others. And species spread east–west more easily than north–south because the climate changes less, so Eurasia was a better place for crops to diffuse than were the Americas or Africa. Even so, areas that are in these or other ways handicapped often surmount their handicaps.

In fairness to archaeologists, it should be noted that few would deny directional pattern in the archaeological record, and some would even agree that the advent of agriculture was quite likely, given long enough. Still, once you start throwing around words like "inexorable," or "virtually inevitable," almost all archaeologists grow skeptical, if not disdainful.

There is an irony in the refusal of so many scholars to embrace hard-core evolutionism, and concede the stubborn force behind culture's ascent to higher levels of organization. As we've seen in this chapter, it is an overly "integrated" view of human society that blinds many of them to this integrative power. By thinking of hunter-gatherer societies as tightly organic, naturally and deeply cooperative, devoid of envy and one-upmanship, they overlook the subtle but strong, and ultimately productive, forces of competition within any human society—Kant's "unsocial sociability." The harmony they wrongly perceive leaves them deaf to ongoing harmonization.

At the outset of this chapter, we suggested that perhaps farming arose simply because it was a "good idea." But "good" in what sense? In the sense that it helped people avoid starvation? In the sense that it helped people win wars? In the sense that it helped people gain status? Yes—in the sense that it helped people do the things that people try to do. And, by virtue of thus satisfying people, the idea of farming was "good" in another, very fundamental, sense: it was good at getting itself spread; in cultural evolution's war of all against all, the concept of farming was a survivor.

Chapter Seven

THE AGE OF CHIEFDOMS

When the philosophers of the eighteenth century made religion out to be an enormous error conceived by priests, at least they were able to explain its persistence by the interest the sacerdotal caste had in deceiving the masses. But if the peoples themselves have been the artisans of these systems of erroneous ideas, at the same time that they were their dupes, how has this extraordinary hoax been able to perpetuate itself throughout the course of history?

—*Émile Durkheim*

Three centuries ago, when Europeans in North America encountered the chief of the Natchez Indians, they couldn't help but notice his high self-esteem. One Jesuit priest observed that he "knows nothing on earth more dignified than himself." And, since the chief knew nothing in the heavens greater than the sun, it seemed only natural to deem himself "brother of the Sun." This logic made sense to the sun-worshiping Natchez people, who vied for proximity to the chief's divine aura. Upon his death, those who had the honor of accompanying him into the afterlife would swallow enough tobacco to lose consciousness and then be ritually strangled.

From a modern vantage point, it is hard to relate either to the chief or to his followers. Few politicians today consider themselves gods or demigods—or, at any rate, few would admit it. And few citizens aspire to spend eternity in the company of politicians. It's tempting, indeed, to dismiss the Natchez people as a bizarre aberration. But they were actually pretty typical—typical of human beings living in a particular phase of cultural evolution: the chiefdom, in which numerous villages are subordinated to firm, centralized political leadership, and that leadership is distinctly institutionalized.

So far as we can tell from the archaeological record, all the ancient state-level societies were preceded in cultural evolution by chiefdoms. So far as we can tell from the ethnographic record, the leaders of chiefdoms have routinely claimed special access to divine force. And, remarkably, their people have typically considered this claim plausible.

How can we say with confidence that "chiefdom" is a standard phase of cultural evolution, a natural transition between the "Big Man" society and the states of the ancient world? Since the rubble of prehistory by definition holds no written records, what lets us discern the social structure of a long-lost people? Here the chief's characteristically large ego becomes a good source of illumination.

We know from chiefdoms observed over the past few centuries that chiefs go to great lengths to underscore their chiefliness. Some Polynesian chiefs turned their entire faces into ornate works of art, enduring a painful, tattoo-like engraving process that leaves the skin looking like the leather on a fancy cowboy boot. Other chiefs have force-fed their wives into obesity, creating vivid testament to their affluence. Unfortunately for archaeologists, fat cells and engraved skin don't fossilize well. But other common forms of chiefly self-advertisement are more enduring, such as monumental architecture, often built in tribute to (and as a reminder of) the chief's distinguished lineage.

Hence the huge mounds built in North America as tombs for past chiefs. Or the pyramid-like temples on Tahiti, or the earliest ziggurats in Mesopotamia. The giant stone heads on Easter Island, up to ten meters tall, also suggest social organization beyond the Big Man level. Using these and other hallmarks[†] of a chiefdom, archaeologists have found a clear pattern: After agriculture first spreads across a region, chiefdoms tend to follow.

This doesn't mean that farming is a prerequisite for a chiefdom. Natural abundance, and attendant population density, will occasionally do the trick. As we've seen, the Northwest Indians were on the verge of chiefdomhood. And the Calusa of Florida, also coastal hunter-gatherers, were a full-fledged chiefdom, whose leader dispatched an armada of eighty canoes (not enough) to battle Ponce de León.

Nor, on the other hand, are we saying that chiefdoms inevitably follow fast on the domestication of plants and animals. In the jungles of Amazonia or New Guinea, farming doesn't become very productive very fast. But given a friendly environment and a millennium or two, widespread agriculture does seem to propel social organization into the age of chiefdoms.

Thus, farming and cattle ranching come to England around 4000 B.C., and within a thousand years "megaliths"—orderly arrangements of boulders, as at Stonehenge—start appearing. The same pattern—first farming, then chiefdoms—is found earlier in continental Europe. (Julius Caesar would happen upon chiefdoms when he ventured into Germany and Gaul.) In Mesoamerica—Central America and the south of modern Mexico—farming villages were common by 2000 B.C., and within a thousand years, immense stone heads, in the Easter Island genre, had been carved. And so on. Chiefdoms, the scholar Randolph Widmer has written, "were at various times the most common form of society found throughout Europe, Africa, the Americas, Melanesia, Polynesia, the Near East, and Asia." Around the world, with the multiple invention and rapid spread of agriculture, cultural evolution marched on. Chiefdoms sustained the basic trend toward larger and more complex social organization.

They seem to have flourished in part by harnessing large quantities of non-zero-sumness. The chief, like the Northwest Indians' Big Man, orchestrated much of the necessary coordination. But the orchestra was larger—thousands, even tens of thousands of people, sometimes spread over diverse landscapes, with diverse resources. So economic integration could be deeper and broader, with more division of labor and larger swaths of regular economic intercourse. Capital projects could be more ambitious—irrigation systems, even the occasional dam.

Sounds wonderful. But it poses two puzzles.

First, how could the cold logic of non-zero-sumness thrive in a hotbed of ridiculous superstition? How, if at all, did things like sun worship and ritual strangulation translate into economic efficiency?

The second puzzle is how the stereotypical chief could be a faithful steward of the public good. Chiefs, after all, aren't known for their sensitivity to the welfare of others. Just ask four sixteenth-century Calusa village leaders who were subordinate to the paramount chief. Not subordinate enough, apparently. He cut off their heads and displayed them at a party. The followers of Chief Powhatan (father of the Indian princess Pocahontas) were described this way by the Englishman John Smith: "At his feet they present whatsoever hee commandeth and at the least frowne of his browe, their greatest spirits will tremble with feare." This attitude is not a good antidote to a politician's self-aggrandizing tendencies.

One standard response to this puzzle is simple: Chiefs actually *didn't* serve the public; they duped the public into serving them, and religion was part of the duping. As one archaeologist puts it, "Chiefs coopt the religious authority of the community for themselves." In this view, a chiefdom's division of labor and its public works did yield positive-sums—more output than the same people could have produced working alone—but the chiefs then appropriated the gains rather than returning them to the people whose synergy created them. Chiefs, in short, were parasites.

Here we revisit a venerable debate we've already touched on, a debate that applies to much of human history: the question of exploitation by ruling elites. At one extreme are Panglossian optimists, often of a rightward political bent, who can find the sunny side of the most gratuitous social inequality. At the other extreme are those—typically on the left, and sometimes Marxist—who see exploitation everywhere they look.

One place to seek evidence in this debate is Polynesia. This vast stretch of the south Pacific, dotted sporadically by islands, is a laboratory of chiefdomhood. Between 200 B.C. and A.D. 1000 settlers leapfrogged from island to island, starting new societies. In some cases, such as Hawaii (settled around A.D. 400), these social experiments were thereafter isolated from the others. Such seclusion notwithstanding, a general pattern emerged across Polynesia: the blossoming of chiefdoms that grew more complex over time. The question is: Was it "good" complexity, fairly equitable in its benefits, or was it "bad" complexity? And where did religion fit in?

THE POLYNESIAN CHIEFDOMS

The generic Polynesian chief had plenty of sacred clout. He was an earthly representative of the gods, the conduit through which divine power, or *mana,* flowed into society. Indeed, he possessed *tapu*—such sanctity that commoners were not to come in direct contact with him. (Hence the modern word *taboo.*) Some chiefs were carried around on litters and had trained spokesmen, "talking chiefs," who handled the dirty business of public communication. The Polynesian chief, observed one western scholar, "stands to the people as a god."

At first glance the chief's sacredness would appear to give him nearly infinite license to exploit. But to some extent, at least, he seems to have served the people, solving such classic non-zero-sum problems as risk diffusion and public works. His spiritual aura helped him compel contributions to the communal pool of starch paste that, on various islands, was then centrally stored in case of famine. Other donated foods—tax proceeds, that is—went to feed laborers who built irrigation systems. In Hawaii, chiefs laced the coast with more than four hundred saltwater "fish ponds," set off from the ocean by stone walls. Chiefs also handled the making of canoes and the training of navigators. And then there were great, taxpayer-financed feasts. Through this "redistributive" ritual the commoners ate delicacies that they themselves didn't grow, playing a non-zero-sum game that people in a market economy take for granted.[†]

Sound like paradise? Not so fast. Especially in the large, highly stratified chiefdoms, the division of labor the chief fostered as a proxy for Adam Smith's invisible hand sometimes did little good for the laborers. In Hawaii, the feathers, dogs, and bark cloth collected for "redistribution" seldom trickled down to their level of origin; they were largely a kind of patronage for the chief's key subordinates, many of whom, conveniently, were close relatives. And many of the fish in those arduously built ponds were destined for elite dinner tables. In Tonga, the official stonemasons, fed by the toil of commoners, spent their time making chiefly tombs.

Still, if power corrupts, and absolute power corrupts absolutely, we must wonder why a man who "stands to the people as a god" wasn't even more self-aggrandizing than Polynesian chiefs generally were. The question isn't so much why chiefs weren't wholly equitable as why they bothered with equity at all. What might keep a chief on moderately good behavior, notwithstanding his awesome stature and the greed inherent in human nature?

For starters: fear. Surprisingly, demigods can lead a precarious existence. As Elman Service put it, "the 'rise and fall' of chiefdoms has been such a frequent phenomenon that it seems to be a part of their nature."

There are two main sources of chiefly demise. One is losing wars. Here are some descriptions, by various anthropologists, of various Polynesian islands: "a chronic state of war" (Samoa); "in a state of more or less incessant warfare" (Niue Island); "constant fighting and warfare" (Tongareva). In this regard Polynesian chiefdoms parallel other chiefdoms, such as those in the Cauca Valley of Colombia, where war was "universal, acute, and unending."

War, as we saw in chapter 5, works against social structures that lack certain properties. One property is economic vigor—enough wealth to make weapons and canoes, enough food to let large numbers of men live close together, in formidable density. Thus does war encourage chiefs to do a good job of imitating the invisible hand. And one key to the invisible hand's success is rewarding people for their labor. That is: return non-zero-sum gains to the workers who produced them, as an incentive to produce more; resist the parasitic temptation.

But why worry about the commoners' incentives? Why not just command them to work harder—since you are, after all, a demigod? That brings us to the second source of chiefly demise: popular discontent. One of the great misunderstandings about evolved human nature is that people are sheep; that, because we evolved amid social hierarchy (true), we are designed to slavishly accept low status and blindly follow the leader (false). People by nature seek the highest status they can attain, under the circumstances, and they accept leadership only so long as it seems to serve their interests. When it doesn't, they start to grumble. The Tahitians had a phrase for chiefs who "eat the power of the government too much."

The upshot is that certain kinds of theoretically possible outcomes of non-zero-sum games are not much seen in the real world. Consider the following "payoff matrix" for a game we can call the utter-exploitation-of-the-hapless-commoners game. If the chief and five commoners don't play the game cooperatively (that is, don't constitute a smoothly functioning chiefdom), then the chief gets zero points and the commoners get zero points each. If chief and commoners do play the game cooperatively, then the chief gets five points and the commoners still get zero. Even though this game has no win-win outcome, it qualifies as non-zero-sum because the players' fates aren't inversely related; the *total* payout isn't fixed at zero, but rather can go up via cooperation. Still, I would argue, this is the sort of positive-sum outcome that history doesn't much feature, and one reason is that human nature won't permit this degree of exploitation—pure parasitism—under any real-world conditions short of literal slavery.

In fact, human nature doesn't often permit this sort of exploitation under fake-world conditions, either. In one classic game-theory experiment, a pair of subjects is offered a collective windfall—money for nothing. The first subject (the chief, in this analogy) decides how to divide up the money between the two, and then the second subject (the commoner) chooses between accepting his allotted share and vetoing the deal, vaporizing the windfall for both of them. Time and again, commoners veto deals that are radically unequal; an offer of $20 out of $100 gets vetoed about half the time, and any offer of less than that is probably doomed. Imagine: college students turning down real money for no work! Apparently there are some kinds of non-zero-sum games that people just won't play.[†] And this pride is found cross culturally; experiments in Japan, Slovenia, the United States, and Israel yield the same basic results.

To say that people naturally resist extremely raw deals isn't, of course, to say that raw deals don't happen. In 1994, the game theorist John Nash won the Nobel Prize for, among other things, rigorously exploring how various circumstances could weaken one's bargaining position, so that "logical" outcomes of non-zero-sum games may not be what most of us would call fair. Thus, when people of different income levels bargain over how to divide the benefits of their joint and equal labors, the richer person is in a stronger position; the player who needs the money less can more credibly threaten to drop out of the game altogether.[†]

In chiefdoms, the commoners' bargaining disadvantage went beyond their low incomes. At sub-chiefdom levels of social organization, people can often "vote with their feet." Northwest Coast Indians peeved at their Big Man could shift allegiance, signing on with another Big Man. But when you live in a chiefdom, there's often no easy way out; the village next-door has the same boss as your village. (Powhatan's chiefdom covered more than 100 villages.) Besides, the chief may not have an open emigration policy. This bargaining disadvantage—not being able to go find another game altogether—helps explain why social inequality became stark and sometimes rigid with the advent of chiefdoms. The Natchez, for example, divided their society into "Sun People," "Nobles," "Honored People," and "Stinkards."

Nash's work—and the Natchez class system—is a reminder that non-zero-sumness, though a mainly good thing, isn't goodness itself. That it tends to grow naturally during history doesn't mean that common conceptions of justice and social equality will magically prevail "in the end" without extra guidance. Still, human nature does tend to place *some* limit on injustice. For a tiny elite to monopolize the fruits of mass labor is not generally feasible. The commoners grow restless.

God knows the average chief would be willing to brutally suppress discontent. But his power is finite. He may have some reliable henchmen, but he doesn't have the sort of large police force or army found in state-level societies. (One standard distinction between a chiefdom and a state is that a chiefdom lacks a monopoly on the legitimate use of force. Victims of crimes, in concert with kin and friends, may take justice into their own hands, though they're well advised to do so with the chief's blessing.) Thus, tyrannical though chiefs may be, chiefdoms ultimately have a certain diffuseness of power. Indeed, the tyranny is in part to compensate for this fact. When South American chiefs decorated the fronts of their homes with the impaled skulls of past enemies, lest the awe of a passerby should flag, this was, in a sense, a sign of insecurity.

Even in the highly stratified chiefdoms of Hawaii, Tahiti, and Tonga, observes the archaeologist Patrick Kirch in *The Evolution of the Polynesian Chiefdoms,* "chiefs were still expected to work for the communal welfare, and an overly bloated chieftainship might raise the spectre of rebellion." Consider the Hawaiian storehouses full of tax revenues: craft items and food. They served some genuinely non-zero-sum functions—social insurance, capital to support public works, and so on. But from the chief's perspective, this economic function served a deeper political function. As one nineteenth-century Hawaiian chief explained, the storehouses were "designed . . . as a means of keeping the people contented, so they would not desert the king." (After all, "the rat will not desert the pantry.") This is why leaders serve the public interest: not because they are public-spirited, but because neglecting the public welfare can diminish their own welfare. That same nineteenth-century chief said that a number of Hawaiian rulers "have been put to death by the people because of their oppression of the *maka 'ainana'* [the commoners]."

A FEW KIND WORDS ABOUT CHIEFS

Scholars who stress only the exploitative side of chiefdoms underestimate the difficulty a chief faces in trying to get things done in a moneyless economy—the difficulty of being a one-man invisible hand. One archaeology textbook says that chiefdoms arose as chiefs achieved the "restriction of access" to critical resources and exploited the power thus gained. But of course, markets, too, restrict access to critical resources. You don't have money, you don't get access, and the way to get money is to work. A common tactic of chiefly restriction, the textbook says, is the extension of "control of water" into "control of people." But sometimes, at least, that's a questionable description. What can we say about the Hawaiian chief who rewarded those who helped build a dam with parcels of land watered by the resulting irrigation? Well, first of all, he got the dam built. Second, he overcame an impediment to non-zero-sum gain, the free rider problem; he made sure you couldn't benefit from the project unless you helped pay for it.[†]

Given the good done by chiefs, it's dubious to assume, as some archaeologists have, that the ornate Polynesian "grand houses, assembly places, and temple platforms" signify wealth being commandeered toward "the apex of the socio-political pyramid." Temples and public assemblies are an integral part of a religion that, while abetting some exploitation, also fosters some public welfare. Where in the accounting books they should go is thus arguable. As Marvin Harris has written: "Viewed within the living context of a redistributive system, tombs, megaliths, and temples appear as functional components whose costs are slight in comparison with the increased harvests which the ritualized intensification of agricultural production makes possible."

Enough kind words about chiefs. By and large, they seem to have been ruthlessly self-serving, power-hungry monsters. Then again, aren't we all? Or, at least, wouldn't we be if we found ourselves playing for sufficiently high stakes? In any event, politicians often have been—in chiefdoms, in authoritarian states, in democracies. In assessing how exploitative different governments are, the key question is: How much greed can leaders get away with before it comes back to haunt them?

The answer depends partly on the level of cultural—especially technological—evolution. The technology of money, for example, eventually came along and made it easier for ordinary people to enjoy non-zero-sum gain via markets, with less meddling from on high. But in the pre-monetary economy of the chiefdom, much non-zero-sumness flowed through central channels, inviting exploitation.

On balance, this book will argue, the technological evolution of the past 10,000 years has been bad news for centralized parasitism.

Indeed, the liberating upshot of some new technologies—information technologies, in particular—is one of the cheerier themes in the unfolding of cultural evolution. But for now the point is just that, whatever the prevailing technology, however dependent people are on centralized guidance, self-aggrandizement has its limits.

Indeed, in the case of Polynesia, the religion had a kind of built-in safeguard against disastrously autocratic government. *Mana* was not monopolized by the chief. Through him *mana* flowed to his subordinates, who possessed less of it than he, and then to their subordinates, who possessed less of it than they. (Thus *mana* was what rank is to the modern military: a degree of authority geared to your degree of responsibility—a useful property in a top-down, "command" economy.) What's more, the chief's king-sized share of *mana* was supposed to be manifest in good governance. So if a chief, say, suffered military defeats, then, as Kirch has written, "the stigma of low *mana* would attend him, and his authority and power might well be challenged." He might be usurped by a warrior whose triumphs had evinced high *mana*.

Thus *mana* was, among other things, a feedback mechanism, a way of ushering inept chiefs offstage. It wasn't as smooth a conduit for discontent as regular elections, but it wasn't the handy tool of subjugation that the chief might have liked. The standard cynical view of religion in chiefdoms—that, as one archaeologist put it, chiefs simply "invent supernatural sanctions . . . to strengthen their authority"—makes the theologizing, and the politicking, sound easier than they were.

A FEW KIND MEMES ABOUT MEMES

The academic fountainhead of cynical views of religion is Marxism. In tempering the cynicism, I don't mean to dismiss the whole Marxist view of the world. Marx gets nothing but applause from this corner for depicting religion as a mediator of deep economic imperatives. Kudos, too, for the more general Marxist claim that a society's "superstructure" (religion, ideology, morality) reflects an underlying "infrastructure" (technology, and the relations of economic power implied by the technology). Even a certain amount of Marx's specific cynicism about religion is hard to argue with. Religion *does* sometimes function as the opiate of the masses, and elites *do* try to use their power to shape ideas to their ends. Marx just went a bit too far. And this century more than a few archaeologists and cultural anthropologists joined him.

Toward the end of the century, a new source of excessive cynicism about religion sprang up—not Marxist in orientation but Darwinian. Arch-Darwinian Richard Dawkins titled the last chapter of his 1976 book *The Selfish Gene* "Memes: The New Replicators." Trying to accent the parallel between cultural and genetic evolution, Dawkins posited the existence of "memes"—units of cultural information that can spread through a culture rather as genes spread through a gene pool. It's worth taking a quick look at the Dawkinsian view of cultural evolution—in part to see where it goes awry in depicting religion, but also for a larger reason: this initially disorienting, topsy-turvy view of cultural change is ultimately fertile, and will surface repeatedly in this book.

A meme can be just about any form of non-genetic information transmitted from person to person: a word, a song, an attitude, a religious belief, a mealtime ritual, an engineering concept. Bodies of memes can be whole religions or ideologies or moral systems or technological systems. The positing of basic units of cultural information, analogous to genes, is far from new. ("Idene" was among the previously proposed labels.) But, more than most past thinkers, Dawkins was willing to see memes as active—even, in a certain sense, alive. He asked us to invert our usual worldview. Don't think of songs, movies, ideologies as passive bodies of information that you, the active agent, choose. Think of them as competing for access to your brain, which they use to propagate themselves. When you whistle a favorite song, that song—that meme—has successfully manipulated your brain to its ends.

You may object that memes surely aren't *conscious*—they don't actually calculate stratagems for penetrating your mind. True enough. Then again, genes aren't consciously calculating, either. In fact, in a certain sense, genes aren't even active; their chemical environment just reacts to them in predictably constructive ways. Yet biologists find it useful to view genes as active agents that "replicate themselves" and "compete" for precious space in the gene pool; biologists talk about effective "strategies" of replication from the "gene's perspective." The justification for this metaphorical shorthand is that natural selection preserves those genes that happen to act *as if* they were pursuing a strategy. And so it is with memes. Songs that affect your brain in a way that causes you to whistle them—songs that deftly "manipulate" your brain—are the kind of songs that evolve. You are their breeding ground, like it or not. Short of committing suicide or living in a cave, there is no way to avoid that role.

So far so good. Now here comes the problem. Dawkins compares memes not just to genes but to viruses. Memes hop from one person to another much as viruses do. Moreover, memes, like viruses, can be bad for the people who help spread them. The meme of injecting heroin is so pleasurable that a person may do it repeatedly, and eventually die from it. The meme then dies with him, but that's tolerable from its point of view so long as some of his friends have picked up the habit from him. The heroin-shooting meme, like the AIDS virus, can thrive even while killing its host, so long as it waits long enough for the execution and transmits copies of itself in the meantime.

The problem here isn't that there aren't prolific memes that are like viruses. (Heroin shooting is indeed one.) The problem is that there aren't many. Human brains, having spent the last couple of million years of their biological evolution in a cultural milieu, are pretty good at selectively retaining memes that are good for them, while aggressively repelling memes that are bad for them. This is one problem with the idea of ruling elites whimsically imposing whole ideologies on brain-dead common folk.

Notwithstanding the rarely viral nature of memes, "mind virus" has now become almost synonymous with "meme." And this parasitic view of culture gets applied with particular zeal to religion. The philosopher Daniel Dennett writes of "the religious memes themselves, in effect, parasitically exploiting proclivities they have 'discovered' in the human cognitive-immune system." Dawkins himself (whose hostility toward religion approaches religious intensity) has compared belief in God to a virus.

Well, it depends on the god. Members of the Heaven's Gate commune, who in 1997 got all dressed up and committed mass suicide, indeed seem to have indulged a virulent theology. But many, if not most, religious people are happy and productive, and enviably free of existential angst. Meanwhile, the theology of Heaven's Gate doesn't seem to be catching on.

NON-ZERO-SUMNESS WINS AGAIN

The casual ascription of "viral" or "parasitic" properties to religion often rests on the conflation of two separate issues: truth and value. Religious doctrines have indeed often entrenched themselves in people's brains notwithstanding the fact that they are probably false. (Heaven and hell, for example.) But being false is not the same as being bad for the believer. Though all religions can have unpleasant side effects (neurotic aversion to "sin," say), it is hardly clear that religious belief is on balance worse than the various alternatives (heroin addiction, say).

Maybe the biggest problem with the "viral" view of culture is the way it ignores or at least downplays the various levels of social organization at which memes do battle with one another. Cultural evolution isn't just memes leaping from person to person; often memes leap from group to group. Chiefdoms fight each other, and the culture most conducive to victory tends to prevail. Meanwhile, within a chiefdom, villages vie with other villages for status, clans vie with clans, families with families, and finally individuals with individuals. Since this competition is typically nonviolent, people don't die, but memes do, because successful individuals and families and clans and villages get imitated. Their memes displace other memes through cultural selection.

A premise of this book is that memes which manage to pass through this gauntlet of cultural selection, and come to characterize whole societies, often encourage non-zero-sum interaction. After all, a common reason that groups of people get emulated—families, clans, villages, baseball teams, corporations, sects, nations, whatever— is their productive and (relatively) harmonious interaction. So memes that bring productive harmony get admired and adopted.

Consider, again, the heaven and hell memes. Almost all religions have the functional equivalent: good or bad consequences that are said to result from good or bad behavior. And, almost invariably, the "bad" behavior includes cheating in one sense or another: stealing your neighbor's property, lying about your contributions to the communal effort. By discouraging such parasitism, these religious memes help realize non-zero-sumness.

Cultural evolution, as various scholars have noted, is quite different from genetic evolution—in particular, faster and messier. Cultural innovations—new memes—can be introduced purposefully, not just randomly, and can spread like wildfire. And defining particular memes is famously difficult, given the fluidity of cultural information. Still, memes can leave distinct footprints that help us track them. The footprints may be in the earth—styles of pottery that spread across early

Europe, say. Or the footprints may come in the form of words. When languages evolve over millennia from a single, common source, linguists can reconstruct the vocabulary of that mother tongue by comparing its living descendants. For example, we know from studying languages in Europe and India that the ancient speakers of proto-Indo-European had horses and harvested grain and mined metal.

When a language family spreads across a large area—as Indo-European did, as the Polynesian languages did—it is tempting to look at the reconstructed proto-language for keys to the culture's fertility. It is not surprising, for example, to find that proto-Polynesian contained words for sail, paddle, cargo, and voyage. These cultural elements, these memes, no doubt helped move the larger Polynesian culture across the south Pacific and keep it robust.

What else might have helped propel and sustain Polynesian culture? It turns out that proto-Polynesian also contained words for *mana* and *tapu*. The details of the concepts have no doubt changed, and indeed came to differ from island to island. Still, the basic ideas proved robust over two millennia of manifest cultural fertility; societies that survived the frequent warfare of Polynesian life were societies that took *mana* and *tapu* seriously. This robustness suggests that these concepts were doing more than burnish the vanity of chiefs, and were not simply "parasitic" on the societies that hosted them. (It suggests, you might say, that the memes and their host societies had a positive-sum relationship.) So does the fact that notions strikingly like *mana* and *tapu* have evolved separately in various cultures, ranging from the Tiv of Africa to the Iroquois of North America.

CHIEFS AS SOUL SAVERS

There is a sense in which many Polynesians—and many residents of other chiefdoms—owed their very lives to the prevailing system of governance, complete with its religious underpinnings. People, remember, were not designed to live in close proximity to many other people. *Homo sapiens* evolved in small groups on sparsely settled land. When a hunter-gatherer band exceeds critical mass, tensions typically force a "fission" into two separate residential groups.

The population density on Polynesian islands—and in mainland agrarian societies, too—was in this sense well beyond a "natural" level. Such density wouldn't have been generally possible but for the coming of farming, and even then it wouldn't have been feasible but for a form of governance that subdued social frictions. One of the Polynesian chief's vital functions was heading off disputes—between individuals, between clans, between villages. For example, amid droughts, when people get edgy, he had to allocate scarce water in a manner perceived as equitable. (The Hawaiian word for "law" means "pertaining to water.")

Had the chief not possessed religious authority, he might not have been able to solve this non-zero-sum problem—or to "wage peace" more generally; a mere Big Man—who lacks formal, divinely sanctioned powers of office and must cajole his followers into following—wouldn't have been up to the job. Thus the chiefdom form of government—whatever its brutalities, its inequalities—did have the saving grace of cramming more and more souls onto the planet.

In addition to being good news for the souls in question, this was good news for social vibrance. The residential density of the chiefdom meant a big drop in the costs of transmitting information. An "invisible brain" could now have more tightly packed neurons—and more total neurons—than ever before.

Cultural evolutionists customarily call farming an innovation in "energy technology"—in the way people obtain the fuel that keeps them alive. And it was. But, given these lowered costs of communication in agrarian chiefdoms, it may make sense to think of farming as an advance in information technology as well. Indeed, farming may have ushered in the first true revolution in information technology.

Among the functions of the chief was to harness this new technology. The economy he presided over was a more complex information-processing system than the Big Man economy; its signals—many of them sent by him—orchestrated a greater division of labor, the production of a greater array of goods, and the undertaking of more ambitious capital projects.

But perhaps more important than what a large and thick invisible brain did for the daily workings of the economy was what it could do for the long-run evolution of culture, spawning new technologies, even new ideas about how to run a society. In the New World, it took little more than a millennium—sixty, eighty generations—to get from the first chiefdoms to the first state-level society. In China and the Middle East, the pace was comparable. Once political organization reached the multivillage level—once the age of chiefdoms dawned—great things were possible.

Chapter Eight

THE SECOND
INFORMATION REVOLUTION

Their function as a stamp of ownership on this item or that was mundane, but the best of them carry images of astonishing vivacity and refinement.
—An art critic's view of ancient cylinder seals

The oldest surviving written reference to King Solomon's Temple is an inscription on a shard of clay from the seventh century B.C. What might the inscription be? Lines from a prayer? A paean to divinity? No. The inscription is a receipt. Someone donated three shekels of silver to the temple, and the gift was duly recorded.

Writing has been an instrument for some of the highest expressions of the human spirit: poetry, philosophy, science. But to understand it—why it came into being, how it changed the human experience—we have to first appreciate its crass practicality. It evolved mainly as an instrument of the mundane: the economic, the administrative, the political.

Confusion over this point is understandable. Some scholars have equated the origin of "civilization" with the origin of writing. Laypeople sometimes take this equation to mean that with writing humanity put aside its barbarous past and started behaving in gentlemanly fashion, sipping tea and remembering to say "please." And indeed, this may be only a mild caricature of what some nineteenth-century scholars actually meant by the equation: writing equals Greece equals Plato; illiteracy equals barbarism equals Attila the Hun.

But, in truth, if you add literacy to Attila the Hun, you don't get Plato. You get Genghis Khan. During the thirteenth century, he administered what even today is the largest continuous land empire in the history of the world. And he could do so only because he had the requisite means of control: a script that, when carried by his pony express, amounted to the fastest large-scale information-processing technology of his era. One consequence was to give pillaging a scope beyond Attila's wildest dreams. Information technology, like energy technology or any other technology, can be a tool for good or bad. By itself, it is no guarantor of moral progress or civility.

If there is even rough validity in equating writing with civilization—and there is—it lies along a different plane. "Civilization," in a more technical sense of the word, is sometimes used to denote societies that have reached the state level of organization. And while writing doesn't guarantee statehood, it is a helpful ingredient. It opens up whole new realms of non-zero-sumness, and greatly lubricates the transition from chiefdom to state. Around the world, the evolution of state-level societies was intertwined with new ways to record and transmit information.

The advance of cultures to the level of the state, to civilization in the technical sense of the word, did in some sense pave the way for civilization in the layperson's sense of the word, civilization as an arena for civilized behavior. With the state would come, for example, the rule of law, which mandated that citizens treat each other with some minimal respect and systematized the punishment for failing to do so.

Indeed, one might argue that, as a general but not rigid rule, writing has made life better in various ways—even, eventually, eroding the power of tyrants. And one might make similar arguments about other thresholds in data processing, such as the advent of the printing press and of the Internet. But first we must understand the evolution of writing, for the printing press and the Internet are in some ways extensions of this ancient revolution in data storage and transmission.

TELLING FORTUNES

Like most truly epic cultural innovations—farming, for example—writing appears to have arisen independently in several places. Scholars long insisted otherwise. Back in the nineteenth and early twentieth centuries, western historians, especially, took a "monogenetic" view of civilization's spread, insisting that Chinese and New World cultures were largely derivative. In the case of the New World, especially, this claim took some ingenuity, but Eurocentric scholars were up to the task. A nineteenth-century French anthropologist, pondering remnants of Mayan writing, theorized that residents of the mythic continent of Atlantis, having abandoned their homeland before it sank, sailed to the New World, bringing literacy.

Today, there is no good evidence even for more sober scenarios of east-west contact that could explain American writing—which existed before the time of Christ—as an import from the Old World. And within the Old World, the origin of writing in China after 2000 B.C. was probably independent of its origin in the Near East around 3000 B.C. The academic consensus is that writing arose at least three times independently. And, just as there are "proto-agricultural" and "horticultural" societies—illustrating agriculture in the process of evolving—there are examples of writing in mid-evolution. Easter Island featured a primitive and apparently indigenous script called *rongorongo*.

Some scholars talk as if writing arose in the Near East, the New World, and China for very different purposes. The simplest version of the stereotypes runs something like this: The Sumerian script of the Near East was heavily economic in function; the Maya were more inclined to history, politics, and religion (including an elaborate astronomy-cum-astrology); the Chinese used their script to tell fortunes. But such generalizations turn out to rest largely on the assumption that in each culture the earliest known example of writing represents the earliest instance of it.

Consider the claim, quite common until recently, that Chinese writing arose as a means of divination. It's true that the earliest Chinese examples of true writing are etchings on "oracle bones" made during the second millennium B.C. Questions were engraved in the shoulder blades of sheep, cattle, or pigs. The bones, after being heated, yielded supposedly prophetic cracks, whose interpretation might be recorded as well. ("It should be Fu Hao whom the king orders to attack Jen." Or this gem of reassurance: "In the next ten days there will be no disaster.") But, as scholars have begun to realize, shoulder blades probably weren't the medium of choice for more casual jottings. The Chinese presumably wrote on less durable things—bamboo or silk or wood—that have since decomposed.

Mesoamerica has the same problem. Surviving examples of early Mayan hieroglyphics come in durable media: monumental hunks of stone. But of course, things that societies write on government edifices are not generally representative of things that societies write, as a stop at the Lincoln Memorial, followed by a glance at the *New York Post,* will attest.

THE EVOLUTION OF WRITING

Only in the Tigris-Euphrates river valley, in the land now known as Iraq, is there much hard evidence about the earliest evolution of a script. There the medium for early writing was not silk or bamboo but soft clay that, once dried, preserves symbols long enough for archaeologists to find them. Clay tablets were so abundant that whole trash heaps of them have been found. So we can speculate with some confidence about the early evolution of what may have been the world's first true system of writing: Sumerian cuneiform, the sinew that bound what may have been the world's first civilization.

The most widely accepted theory about the birth of Sumerian writing was developed by Pierre Amiet and documented by Denise Schmandt-Besserat. It begins with little clay tokens that show up in the eighth millennium B.C., as agriculture is coming to the Fertile Crescent. The tokens stand for particular crops. A cone and a sphere, for example, represented grain in two standard quantities, the *ban* and the *bariga,* roughly a modern-day liter and bushel, respectively. The tokens seem to have been used for accounting, perhaps recording how much a particular family had given to the granary, or how much it owed.

As cities were forming, in the second half of the fourth millennium, there came more complex tokens, elaborately shaped and marked. Often they symbolized products of an urban economy: luxury goods such as perfume and metal, processed foods such as bread and beer.

The shift from these three-dimensional symbols to two-dimensional, written symbols illustrates just how plodding cultural evolution can seem when observed up close, on a time scale of decades rather than millennia. Sometimes records were kept by storing tokens in large clay envelopes about the size of a tennis ball. Five clay cones might be sealed inside an envelope to record a debt or payment of five *bans* of grain. As a convenience, the tokens were pressed against the soft surface of the envelope before being enclosed. That way a person could "read" the contents of an envelope without having to break it open. Two circles and a wedge would mean the envelope contained two spheres and a cone. Apparently it was some time before a key insight dawned: the two-dimensional imprint on the outside of the envelope had rendered the three-dimensional contents superfluous. The envelopes could now become tablets.

This was the beginning of Sumerian cuneiform. The system evolved for millennia, growing more abstract and powerful. Thus the tokens for a little grain and a lot of grain—the cone and sphere—became, in two-dimensional form, general numerical symbols: a wedge meant one, and a circle meant ten. These signs could now be placed next to the symbol for an object to indicate its quantity. Eventually, the symbols for objects—and for people, and actions, and so on—came to stand for sounds, steering western civilization toward the modern phonetic alphabet.

Some scholars find the link between tokens and cuneiform unconvincing. But even they agree that the earliest examples of Sumerian writing are economic: tabulations of livestock and food and goods. And most agree that such data helped orchestrate division of labor and public works. A farmer brings barley to the temple, the payment is duly recorded, and the barley goes to pay men who build a canal—a canal from which the farmer may in one sense or another benefit.

For all we know, Chinese and Mesoamerican script received similarly strong early assistance from non-zero-sum logic. Certainly, in both cases, there were symbols for numbers by the time durable written records were being left. Still, the point here is not that written numerals universally preceded written words, or that the earliest writing was everywhere economic in function. The point is just that everywhere writing seems to have been a practical technology, figuring in economics or politics or both.

Yes, there were historical narratives in China, the Near East, and Mesoamerica. But their function was largely to buttress the authority of the ruler—to establish his noble, perhaps divine lineage; to selectively preserve his feats of government and conquest; to convey the vastness of his empire, the deadness of his foes. Yes, the Maya recorded incredibly precise astronomical observations, but this was part of a religion that, like other state religions, helped keep the people obedient and harmoniously productive.

WRITING AND TRUST

To say that writing transformed the potential for non-zero-sum interaction comes close to redundancy. For the link between information and non-zero-sumness is so basic that it's hard to imagine deep changes in the former that wouldn't deeply change the latter. Indeed, it is hardly an exaggeration to say that non-zero-sum dynamics are the reason information gets transmitted in the first place. As the pioneering game theorist Thomas Schelling has observed, in a purely zero-sum relationship, there is no rational reason to communicate. Opposing coaches have no cause to speak before a game. If you see them in extended conference, they're probably talking about some non-zero-sum realm, where their interests partly overlap. Both want to avoid injured players, and may decide to reschedule a game in bad weather. But neither coach has an interest in honestly communicating anything about the game itself.

As we'll see later in this book, Schelling's point is applicable to the origins of communication in the broadest sense of the term. Primordial communication, back before evolution had produced animals that talked (or for that matter animals, period), happened because of what can aptly be called a non-zero-sum relationship between bits of genetic material—bits that, by virtue of sitting next to each other, were in the same boat, their fates tightly linked. But for now let's stick with information in the familiar sense—words, numbers. By following the logic of game theory a bit further, we can see how new ways of storing and sending these symbols expanded and enriched the social fabric.

Consider again the textbook non-zero-sum game, the prisoner's dilemma. Two partners in crime are being separately interrogated. Each will be better off if neither rats on the other than if both do. But, though cooperation is in their mutual interest, there are two great barriers to it. One is a lack of communication; you can't agree on a joint strategy if there's a wall between you and your accomplice. And if you overcome this barrier, you face a second one—lack of trust; if you think your accomplice is going to renege on the deal, and rat on you after all, then you're better off copping a plea and ratting on him. Somehow, this fear of being cheated must be overcome for things to work out well.

If indeed barriers to information are one of the two basic impediments to non-zero-sum gain, then obviously new information technologies might unlock some positive sums. Yet the more dramatic effect of writing may have been to overcome the second barrier, the trust barrier. In ancient Mesopotamia, a lender didn't have to fear that the borrower would deny all recollection of the loan, and the borrower didn't have to fear that the lender would exaggerate it. There was an attested record, such as one from Babylon noting that a man had borrowed "ten shekels of silver" from the "priestess Amat-Shamash"; the man "will pay the Sun-God's interest. At the time of the harvest he will pay back the sum and the interest upon it." If you doubt the value of such peace of mind, consider how hard people in nonliterate societies work to etch financial obligations in the public memory. The ostentatious Potlatch seems less absurd when viewed as a way to assemble a large audience to witness the incurring of a large debt. And, on a smaller scale, when one family of Northwest Coast Indians would give food to a needier family, public ritual was de rigueur.

Writing was hardly the only thing in the ancient states of Asia, the Middle East, and Mesoamerica that helped solve the trust problem. Another was the systematization of justice: the assurance that cheaters will be punished. But even here, writing helps; legal codes carry more precision and heft when etched in something solid. The code of the Mesopotamian city of Eshnunna—written a century before the more famous code of Hammurabi—left no doubt what would happen if you paid a man a shekel to harvest your field and he never got around to it: he would pay you ten shekels.

Ten shekels, by the way, was also the punishment for punching a man in the face. If you went further, and severed his nose by biting it, that would cost you one full mina (60 shekels) of silver; severed fingers were cheaper—two-thirds of a mina each. These laws, though not governing economic behavior, still served productivity; they made urban living, with all its potential efficiencies (including the data-processing efficiencies of dense population) tranquil enough to bear. The informal justice system of a chiefdom just wouldn't do now that daily life involved so many close encounters with people who were neither relatives nor acquaintances. So the government had to build a new anti-cheating technology, a new technology of trust—trust not just in economic justice, but in the larger social contract, the mutual nonaggression pact that, by relieving people of fear and suspicion, smooths all kinds of cooperative efforts.

This is one sense in which "civilization" in the technical sense—a state-level polity, typically featuring writing—often leads to civilization in the nontechnical sense: walking around without fear of getting your nose bitten off. Still, this comfortable civility has so often accompanied writing not because literacy soothes the savage soul, but for earthier reasons. Societies that fail to use writing to solve various dimensions of the "trust" problem, that fail to create space for non-zero-sumness, typically fall, often at the hands of societies that better harness writing's potential. In the long run, ancient states had no more "choice" about whether to adopt new information technologies than they had about whether to adopt chariots, bronze shields, or iron swords. In all such cases, you use it or lose.

BUREAUCRATIC BRAINS

The justice system was only one part of the bureaucracy that emerged in ancient states and that is one of the hallmarks of a state. Bureaucracy has since gotten such a bad reputation that one of its dictionary definitions is "administration characterized by excessive red tape and routine." But, technically, bureaucracy is just government by distinct functional units, each run by a specialist—division of labor in the processing of information. So it's no surprise that new information technologies usually played a role. In Mesopotamia, for example, bureaucrats had ornate cylindrical seals that, impressed on a clay tablet, served as a majestic signature and carried official weight, shoring up trust in loans and other transactions.

And then there were the tablets themselves. Today a big hunk of clay seems like a non-optimal way to store data. But in its day this information technology revolutionized the large-scale coordination of matter and energy. In a single year of the Ur III dynasty, around 2000 B.C., bureaucratic tablets recorded the processing of 350,000 sheep, brought in by taxation and other means and used for, among other things, paying the salaries of government workers (including bureaucrats, no doubt). Government laborers were also paid in bread, fish, oil—all such disbursements precisely registered, as were the hours spent earning them by, for example, digging canals. And when the work was done, the government, like governments today, took conspicuous credit. In the Babylon of Hammurabi's day, one canal was named "Hammurabi is the Prosperity of the People." Your tax dollars at work.

All of this bureaucratic accounting required standard units of measurement—a kind of information technology in their own right. One of the most widely found artifacts in the early Near East is the "bevel-rim bowl," which is thought to have been a measure for foods paid to workers. It shows up in Mesopotamia in the fourth millennium B.C., as do cylindrical seals, writing, and city-states.

Ancient city-states had common interests—goods they could trade, mutual enemies they could jointly annihilate. But to reap the fruits of this non-zero-sumness they needed first to breach the information barrier. And back when the fastest mode of transport was the donkey, frequent conferences among kings weren't feasible. Couriers, on the other hand, were a dime a dozen. And, with the advent of writing, they could carry long, precise messages.

What's more, these missives served as hard evidence of deals forged and promises made. They thus made a dent in the trust barrier. And in doing so they spared no rhetorical expense. An ancient peace treaty reads: "He who shall not observe all these words written upon this silver tablet of the land of the Hatti and of the land of Egypt, may the thousand gods of the land of the Hatti and the thousand gods of the land of Egypt destroy his house, his country, and his servants."

Of course, the trust barrier is a stubborn thing. A certain wariness inevitably remains between people who have had few face-to-face encounters. One time-honored wariness reducer, going back to the days of the chiefdom, was to use the bonds of kinship; potentates sealed alliances with intermarriage of daughters or sons. Yet another approach, going back to the days of the Big Man, was to use the *vocabulary* of kinship, along with lavish professions of devotion. During the third millennium B.C., the king of Ebla in the Middle East wrote to the king of Hamazi: "You are my brother and I am your brother, fellow man, whatever desire comes from your mouth I will grant, just as you will grant the desire that comes from my mouth."

Kingly relations were sometimes lubricated with "gifts" so expensive that they amounted to de facto trade—reciprocal exchange of the exotic goods in which each kingdom specialized. This economic ingredient was alluded to by a Babylonian king who wrote, "Between Kings there is brotherhood, friendship, alliance and good relations—if there is an abundance of silver and an abundance of gold." This sort of cynicism needn't cause despair. Another way of making much the same point is to say that where non-zero-sum logic leads, amity often follows. Given the power of non-zero-sum logic, this would seem to bode well for the long-run expansion of amity.

PILES OF CORPSES

A more legitimate cause for despair is that this amity is often enlisted in the cause of enmity. The point of alliance was often to get aid during war (or, at least, to ensure the neutrality of a potential meddler). The royal archives of Ebla, so replete with those professions of brotherhood, are also replete with a phrase recorded by the king again and again, apparently with pride: "Piles of corpses I raised."

If a city-state succeeded in dominating another—by conquest or threat of conquest—and thus formed an empire, then communications technology was of course vital. The Ur III dynasty built a "donkey express"—roads with stations housing messengers. Other civilizations had their rough equivalents, such as Aztec relay runners with messages tucked into the forked ends of batons.

Analogies between societies and organisms go back to the beginnings of cultural evolutionism. A state bureaucracy is a bit like a brain, and Aztec runners, sending commands to military outposts or distant farmers, are a bit like nerve impulses. That these analogies are easy doesn't mean they're worthless. Just as you can't imagine an organism as complex as a human being lacking vast data storage and fast data processing and transmission, it's hard to imagine a state-level society without a significant information technology. Scholars sometimes speak of "nonliterate" civilizations—the Inca in South America or such West African states as the Ashanti of the early seventeenth century or the Dahomey of the eighteenth and nineteenth centuries. But on close inspection, such societies seem always to have some good information technology. They may not be able to record their poems, but they can handle more vital data, such as numbers, just fine.

The Dahomey, for example, took a census to aid taxation and military mobilization. Their database consisted of a room full of boxes containing pebbles that signified the number of men and women, boys and girls, in each village. Updating was continuous, via the registering of every birth and death (including the cause of death) throughout the land. For similar purposes the Inca used the *quipu*, the variously knotted and colored strings that only specialists understood. (Today they are understood by no one, but the best guess is that they recorded not just numbers but historical events.) Much of the Inca nervous system consisted of "roads," or at least footpaths, that snaked through and around the Andes, sometimes across suspension bridges; the roads were long enough, all told, to just about encompass the world. Runners were stationed one to five miles apart (the flatter the stretch, the longer the gap), and might hand off either *quipus* or oral messages, ritually repeated during the handoff to suppress error. Data could travel 150 miles a day.

But that's nothing compared to the Ashanti, who sent data hundreds of miles in a few minutes with a network of "talking drums" that could summon political leaders, warn of danger, mobilize the military, announce deaths, or (on a less urgent note) broadcast proverbs. Differences in tone had meaning, as in the Ashanti language itself.

This book has made little use of such familiar phrases as the "Stone Age" and the "Bronze Age." The reason, clearly, is not an aversion to "technological determinism," but rather a belief that metallurgy makes for a bad version of it. The Maya, the Aztecs, and the Inca were basically Stone Age people—what metal they had was used mainly for jewelry and the like, not swords or shields. Yet they had much more in common with Egypt or China in the Bronze Age than with, say, the Stone Age Shoshone of their own hemisphere.

A more useful technological dividing line between the Shoshone on the one hand and state-level societies on the other comes from energy and information technologies. All state-level societies farmed, and all had (relatively) sophisticated means of handling data.

Even with energy and information technologies, the word "determinism" is a bit much. A lot more goes into a state-level society than farming and processing data. For that matter, the term "state-level society" is itself misleading in its seeming firmness. It's true that the core criteria of statehood—including centralized, somewhat bureaucratic government, the power to draft an army, a monopoly on the legitimate use of force, regular taxation (often in the form of labor or goods)—tend to be found in clusters. But they don't show up simultaneously, or always in the same order. Hence ongoing arguments about borderline cases. (Did Hawaii, after European contact, cross the threshold between chiefdom and state?)

Still, such fuzziness notwithstanding, some things are clear. Social complexity tends to grow beyond the level of the chiefdom, toward the level of the state. And intimately involved in the growth is information technology: advances in the storage and transmission of data and in the processing of data, including the invention of bureaucracy.

By happenstance, we have a fairly clear view of this process in only one part of the world: Mesopotamia. And there growth in the complexity of society precisely mirrors growth in the complexity of the symbol system: as the social structure becomes more elaborate, so do the tokens, because the symbols and the structure feed off one another.

ONE MAN'S NZS IS ANOTHER MAN'S ZS

To say that writing eroded the two big barriers to non-zero-sumness isn't to say that writing turned society into a cornucopia of mutual benefit. Indeed, according to one common academic stereotype, ancient states carried repression and exploitation to new heights, making chiefdoms look populist by comparison; workers toiled away so that elites—having precisely recorded the toil—could precisely underpay them. How can we reconcile this scenario with my paean to the benefits of writing? Where's the win-win dynamic?

To begin with, the dynamic exists among elites. When Mesopotamian big shots traded wool cloth to big shots in other lands for the wood and stone needed for Mesopotamian construction, big shots on both ends of the deal benefited. To be sure, the Mesopotamian women who spun the wool into yarn—their output and wages tallied on clay tablets—may have benefited as well. After all, the ability to market yarn internationally raised its value, some of which could in theory trickle down to the spinners. And presumably the spinners benefited from various forms of bureaucratically realized non-zero-sumness—the diversity of available foods and crafts, the irrigation canals and other capital projects. Still, it doesn't take extreme cynicism to imagine that in general elites skimmed more than a bit off the top. (Tablets from a state brewery in Mesopotamia show one administrator walking off with thirty-five jars of beer. I'm suspicious.)

That the benefits of ancient writing may have clustered near the top of the social pyramid shouldn't surprise us. Power from new technologies tends to accrue to the people who wield them. Ancient wielders of the written word became gatekeepers of the wealth it unlocked.

The fewer the gatekeepers, the more power they had. Ancient Mesopotamia had an estimated literacy rate of less than 1 percent. It's hard to say whether this reflected an attempt by elites to monopolize the technology, but in any event scribes were a small and esteemed class, complete with an official deity (aptly, the goddess of fertility).

Entry to the class—via lengthy instruction at the "tablet house"—was granted mainly to the privileged. A Sumerian text describes a rich man giving his son's writing teacher food, a robe, and a ring to ensure a passing grade in spite of his son's indiscipline.

Many scribes were mere transcribers, and didn't themselves call the shots. Still, they seem to have reveled in the power emanating from their art. Some Egyptian scribes opined that the lower classes, lacking in brains, had to be driven like cattle. Actually, what the lower classes lacked was their own personal scribe.[†]

It isn't just economic power that information technology confers. Concerted political organization—to resist oppression, to lobby for lower taxes, whatever—is a form of non-zero-sum interaction among people who share an interest. As such, it calls for communication. African slaves in America would later demonstrate this fact by organizing slave revolts via talking drum. And writing, of course, would come to play a role in revolt as well. (During the U.S. Civil War, most southern states made it illegal to teach slaves to read and write.) But in the ancient states, with literacy rare, we find only glimmers of its future subversive use. A piece of graffiti from Egypt in the third millennium B.C. reads: "You arrested me and beat my father. . . . Who are you now to steal from me?"

Some scholars, comparing ancient states to chiefdoms, have argued that writing led to concentrations of wealth and power. But, strictly speaking, what they mean is that the *concentration* of writing abilities led to a concentration of power. The question of how far economic and political power would eventually spread beyond the upper classes was partly a question about the future of literacy. How strong were the forces favoring its expansion? This is a question to which we'll return. (Here's a clue: strong.) Meanwhile, even as elites held their monopoly on information technology, there was still some limit on feasible exploitation, including the constraints we saw at work in chiefdoms: in addition to facing revolt, leaders face other polities, and both forms of pressure punish highly parasitic regimes.

Writing isn't the only elemental information technology that evolved in ancient states. Money—a standardized currency—is an information technology. It is a kind of record of your past labors, of their value as judged by society. And when you spend the money, it becomes a kind of signal, confirming your wants and conveying them, however obliquely, to the various people involved in satisfying them;

passing from hand to hand to hand, money flows through the nervous system of the larger invisible hand, informing suppliers of demand.

In modern times, much kvetching has been done about money. Some consider it a tool for oppressing the downtrodden. But, in historical perspective, money looks more like a solvent of oppression. By invigorating market economies, it offered an alternative to a command economy dominated by the literate few. If an economic information technology is going to be wielded on your behalf, it's usually best to do the wielding yourself.

Money in truly convenient form—coins, portable and widely respected—didn't show up until the seventh century B.C., courtesy of the Lydians. If you don't think coins were a major advance, consider Homer's description, centuries earlier, of the value of Glaucus's armor: it was worth a hundred oxen, compared to nine oxen for Diomedes' shoddy stuff. (Imagine the armor store during the holiday shopping season.)

Actually, even before coins, ancient states were moving toward de facto, if still inconvenient, currencies. Barley or weighed-out silver was used in Mesopotamia, wheat in Egypt, cocoa beans by the Aztecs. And, when these were lacking, simple barter was always an option.

Indeed, the old idea that ancient states were "pre-market" is now dead. They had mixed economies, if often with a strong bias toward the government part of the mix. In Mesopotamia, private merchants conducted distant trade, a fact that probably worked to the ultimate advantage of commoners. Political power, too, now seems to have been dispersed more widely in Mesopotamia than the stereotype of the totalitarian ancient state would have it. Still, by today's standards, there was room for progress.

And progress there would be. Slavery, human sacrifice, and milder forms of exploitation would diminish over time. Today civilization is more "civilized"—in the everyday, nontechnical sense of the word— than the ancient civilizations. And a primary reason is the way money and writing would over time evolve and, as we'll see, interact.

Chapter Nine

―――――――――――――

CIVILIZATION AND SO ON

Whenever rulers and military classes tolerated merchants and refrained from taxing them so heavily or robbing them so often as to inhibit trade and commerce, new potentialities of economic production arising from regional specialization and economies of scale in manufacture could begin to show their capacity to increase human wealth.

—*William McNeill*

There is an old joke about the standard instructions on American shampoo containers, "Lather, rinse, repeat." A man takes the directions literally and spends the rest of his life in the shower—lathering, rinsing, lathering, rinsing, lathering, rinsing.

Sometimes it seems as if ancient civilization followed similar instructions. Rise, fall, repeat. The rulers and dynasties and peoples may change, but all seem locked into the same endless cycle of conquest and expansion, fragmentation and collapse.

Ancient history thus seems like little more than a parade of strange-sounding names. There's Uruk—not to be confused with Ur (or Ur II, or Ur III). There are Akkadians, not to mention Achaemenids. Eventually the Minoans and Mycenaeans arrive (or is it the other way around?), and then, finally, come the really familiar names: Greece and Rome.

Meanwhile, in China, there is conflict among the Ch'i, the Ch'in, the Chin, and the Ch'u (this during the late Chou). Finally the Ch'in win, and consolidate China, then quickly fall apart.

Over in the New World, civilization begins to stir long before the famous classic Mayan period. There are Olmec and Zapotec, and, by the time the Inca and Aztecs occupy center stage, we've also seen Huastec, Mixtec, and Toltec, not to mention Chimú and Chincha and Chichimec.

It all seems a blur. But really, the problem is that it is not blurry enough. The reason that ancient history seems chaotic is that we are using a zoom lens, focusing on small regions and small time frames. If we relax our vision, and let these details go fuzzy, then a larger picture comes into focus: As the centuries fly by, civilizations may come and go, but civilization flourishes, growing in scope and complexity.

The key is to take the "history" out of ancient history. Historians tend to dwell on differences. How was ancient China different from Sumer? Why was it different? Good questions, interesting questions, questions we'll get back to. But first let's ask: How were early states, in the various regions where they evolved, alike? This is one way to simplify ancient history—by realizing that, fundamentally, the same thing was happening everywhere you look.

THREE PETRI DISHES

Archaeologists speak of six "pristine" civilizations—states that arose indigenously, and weren't merely copied from a nearby civilization, or imposed on the populace by conquest. The standard six are: Mesopotamia, Egypt, Mesoamerica, South America, China, and the civilization of the Indus River valley (about which relatively little is known) in south Asia. Some scholars throw in West Africa as well.

Calling West African civilization pristine is something of an exaggeration, given earlier contact with states to the north. Then again, calling some of the standard six "pristine" states pristine is a bit of a stretch. Indus script (still undeciphered) may have been inspired by Mesopotamia, which was exchanging memes with Egypt as well. And some diffusion, however thin, probably linked South America (the Inca and their cultural ancestors) and Mesoamerica (Aztecs, Maya, and others).

Still, even after granting these early and occasionally momentous contacts, we are left with three large realms of ancient civilization, quite removed from each other: China, the Near East, and the New World. The scholarly consensus is that each developed its energy and information technologies—farming and writing—indigenously. And each then underwent its early civilizational history in essential isolation from the others.

Yet in all three cases, the same thing happened. Namely: more of the same. The trend that had gotten humanity to the verge of civilization—bands getting big enough to qualify as villages, which then got bigger and more complex and combined to form chiefdoms—continued. The chiefdoms' villages evolved into something more like towns, which themselves then got bigger and more complex. In all three regions, loosely defined city-states—urban cores surrounded by farmlands and villages and towns—seem to have evolved (though in some places, such as Egypt and the Andes, the "city" part of the state may have been so small as to stretch the definition of the term). And these city-states merged, forming multicity states, and these multicity states grew into empires. Sure, there were setbacks aplenty—droughts, barbarian hordes, and other catalysts of epic collapse—but in the long run the setbacks proved temporary. (Indeed, the setbacks attest to ongoing progress; their increasing vastness charts the growing magnitude of the systems that are being set back.) So there you have it—ancient history in a nutshell: onward and upward, to higher levels of social complexity.

Paragraphs such as the previous one, asserting simple patterns in history, and suggesting that they reflect general laws, are a fat target for criticism. They seem, as one philosopher of history has put it, to derive a "linear law" from "the alleged trend of the historical process as a whole." And, "since the process is unique, this looks, at best, like generalizing from a single case."

But in *ancient* history, at least, the process is not unique. There are at least three separate cases to study—and four if you assume, as many do, minimal early contact between Mesoamerica and the Andes. Granted, if we're going to call the patterns that these cases separately evince "laws," we should explore the mechanics of the "laws" and show why they are powerful. We've already done some of that, and we'll do more. But in the meantime, let's at least establish that in the several cases available for study, the basic pattern—deeper and vaster social complexity, more and more non-zero-sumness—indeed holds.

ONE CRADLE OF CIVILIZATION

In high school textbooks, Mesopotamia—the land of the Tigris-Euphrates river valley in modern Iraq—is called "the cradle of civilization." Ancient rulers from China and the Americas could be excused for rolling over in their well-appointed graves on hearing such Eurocentric propaganda. Their civilizations had their own cradles, domestically manufactured, thank you. Still, we have to start somewhere, and Near Eastern civilization did precede the other two civilizations.

In the Mesopotamian vicinity, the story of civilization begins, as elsewhere, with farming and attendant social complexity. By 4000 B.C. there are the familiar hallmarks of chiefdoms—temples, other capital projects (irrigation systems and what appears to be a granary), and, of course, special burials for big shots, complete with precious copper and ceramic knick-knacks. The chiefdoms' villages get bigger and bigger and at some point cross that blurry line between villages and towns.

Around 3500 B.C., though true writing had yet to appear, the stirrings of the first information revolution were evident: the cylinder seal, complex tokens, the bevel-rimmed bowl. As writing evolved, growth toward civilization was brisk. In southern Mesopotamia between 3500 B.C. and 2900 B.C., the number of villages grew from 17 to 124, the number of towns from 3 to 20. The number of "urban centers"—125 acres (50 hectares) or larger—grew from one to 20. By 2800 B.C., the city of Uruk covered 617 acres (250 hectares), and its temples, mounted on massive ziggurats, were visible from miles away. Surrounded by, and interdependent with, farming villages and towns, Uruk came to anchor an amorphous city-state. Comparable clusters evolved elsewhere in Mesopotamia.

Relations among city-states featured that double-barreled source of non-zero-sumness, Kant's "unsocial sociability." Polities traded and fought, traded and fought, and the result was, as usual, a strong argument for political unification. The logic isn't merely that a two-city state is stronger than its one-city enemy by virtue of size. The strength derives also from the fact that trade between its two cities can now proceed without the disruption of periodic warfare and untrammeled marauding, and without the burden of mistrust. The basic idea—creating large zones for the free play of non-zero-sumness—is akin to Elman Service's notion of "waging peace."

But peace was often not waged peacefully. Though municipal rulers might agree on the virtues of a multicity mega-state, they rarely agreed on who the mega-ruler should be. Like chiefdoms, multicity mega-states tended to get formed with the help of aggression, or at least the threat of it.

The first large multicity state in Mesopotamia was the Akkadian empire, formed around 2350 B.C., when Sargon of Akkade conquered Sumerian cities in southern Mesopotamia. Sargon's conquests came with a divine seal of approval; having toppled a city, he asked local priests to declare his victory the will of the Mesopotamian god Enlil. Perhaps to facilitate clear thinking on their part, he exhibited the vanquished local king in neck-stock. As a further aid to theological interpretation, Sargon installed his daughter as high priestess of the goddess Nanna at Ur, the religious capital of southern Mesopotamia. Having subdued much of the population in and around the Tigris-Euphrates valley, Sargon declared himself, a bit provincially, "King of the Four Quarters of the World."

Meanwhile Egypt, though slower than Mesopotamia to develop cities (and never as vastly urban as Mesopotamia), moved more quickly, and more enduringly, toward regional statehood. The threshold was crossed not long after the information revolution arrived. By 3100 B.C. Egyptian hieroglyphics were in use, recording ancestral lineages and property ownership. By 3000 B.C., give or take a century, Egypt was politically unified and was evincing the generic human combination of lucrative trade, warm alliance, and devastating enmity.

Egyptian pharaohs had high self-regard even by the lofty standards of ancient rulers. On an unusually consistent basis, they convinced the populace not just that they represented the will of the gods, but that they *were* gods—direct offspring of the Sun God Ra. The attendant subservience may partly explain the stunning embodiments of state power: the pyramids, which soaked up unimaginable labor so that pharaohs could have nice roomy tombs and spend the afterlife surrounded by shiny artifacts in garage-sale quantities. Divinity aside, the pharaohs relied heavily on a vast bureaucracy, whose detailed workings are obscure, but which certainly didn't suffer from a shortage of titles: Overseer of Granaries, Overseer of Works, Overseer of the Treasuries, Overseer of the Scribes of the Great Records, Overseer of the Great Mansions (the courts), and so on.

ANOTHER CRADLE OF CIVILIZATION

In east Asia, farming seems to have evolved a millennium or so later than in the Middle East, but its consequences followed just as surely: bigger villages, more artifacts, more trade, vaster conflict, bigger buildings, bigger realms of political control, starker status hierarchies (jade and bronze being favored upper-class grave accoutrements). An age of chiefdoms seems to have been reached by the late fourth millennium B.C., and in the second millennium B.C. came testaments to state-level organization: writing, cities, a king who could lead 13,000 men into battle and oversee epic engineering. The fortifications surrounding one urban compound of temples and palaces would have taken 10,000 workers eighteen years to build—assuming a one-day weekend and no paid vacations. Royal graves were roomy—thirty, forty feet deep, with terracing and easy-access ramps. Kings were buried with slaves and human sacrifices (sometimes neatly decapitated) and lots of wealth. Even one king's *consort,* though consigned to a modest grave, was surrounded with 468 bronze items, 775 jades, and 6,880 lovely cowrie shells.

All of this belongs to what is known as "the Shang civilization," but the suggestion of homogeneity may be misleading. Some scholars now dissent from the long-accepted Chinese view of a unified national past, and envision the Shang as much like early Mesopotamia: individual, perhaps amorphous, city-states that trade and battle, ally and fall out. There is even some question as to whether the Shang had quite reached the state level of social organization, or was more like a precocious chiefdom. Who knows? The main point is that the story in China moves in the same direction as the stories elsewhere. The Shang's successor—the Chou, who dominated the first millennium B.C.—forged a vast state with many cities.

But control was diffuse, and Chou principalities—Ch'i, Ch'in, Chin, Ch'u, and others—finally fell into open warfare. The Ch'in eventually prevailed, carrying Asian political unity to unprecedented scope. Hence the name China.

One key to the Ch'in triumph had been non-zero-sum reforms. The Ch'in made the law firmer and fairer, less partial to the powerful. They standardized weights and measures and the writing system. Having conquered their rivals, they extended these principles across China. The nation was further bound by a single currency, lots of canals, and 4,000 miles worth of new roads (traveled by carts whose wheels were of standard, government-mandated gauge—so the ruts worn into dirt roads would be one-size-fits-all).

Shih Huang-ti, who oversaw this unification of China and is thus known as the nation's first emperor, was by many accounts a nasty and parasitic man. He feared dissent, and reputedly burned texts that ventured beyond such kosher subjects as agricultural technique and divination. His idea of a valuable use of government workers was the crafting of 7,500 life-sized terra-cotta warriors, each unique, to accompany him to the grave.

His caprice and oppression no doubt encouraged the rebellion that followed his death. Still, his infrastructure for non-zero-sumness—all the roads and canals, and the various standards that smoothed the movement of people and goods and data—endured. This legacy would ease the still-onerous task of his successors, the Han, as they struggled to keep China politically whole.

Meanwhile, back in the Near East, more names had come and gone, and the regions they represented had continued to get bigger and bigger, if fitfully: the Assyrian empire dwarfed the Akkadian (the one that had covered the "four quarters of the world") and was in turn dwarfed by the Persian empire (and its "king of this great earth far and wide"), which was then overcome by Alexander the Great (the "son of God" and "general governor and reconciler of the world"), whose Macedonian empire would soon be overshadowed by the Roman Empire (its emperor being "the savior of all mankind").

AMERICAN CIVILIZATION

If in 200 B.C. the Han, or the Romans, had magically gotten a peek at life in the undiscovered New World, they would have been unimpressed. A casual glance across the Americas would have suggested a hemisphere full of savages and barbarians; almost everywhere, social structure fell somewhere on the spectrum from simple band to chiefdom. But here and there, visible on close inspection, were cradles of civilization, small pockets where culture was crossing the hazy line between chiefdom and state.

As the archaeologists C. C. Lamberg-Karlovsky and Jeremy Sabloff have noted, the first known city in Mesoamerica, Monte Alban (in southernmost Mexico, not far from Guatemala), is reminiscent of the first big city in Mesopotamia, Uruk. In both cases, the city-to-be was at first a mere town, outshining its neighbors in size and architecture, and dominating them politically, in the classic fashion of a chiefdom's hub. In both cases war and trade helped drive complexity upward, and in both cases information technology and urbanization proceeded hand in hand. In Monte Alban by 300 B.C. there were calendrical notations, and glyphs used to label sculptures of dead enemies. By 200

B.C. the population had grown to 5,000, and it would eventually surpass 30,000. But Monte Alban was destined to be outclassed by Teotihuacán, a trading partner to the north that by A.D. 550, with 125,000 residents, would be one of the six largest cities in the world, unbeknownst to the other five.

Teotihuacán is not to be confused with the nearby city of Tenochtitlán, the Aztec capital that, when seen by Cortez in 1519, housed around 200,000 people (more than any European city) and anchored a state twice the size of Portugal. Cortez called Tenochtitlán "the most beautiful city in the world," and compared it to Venice. Built on islands in a saltwater lake, it was laced with canals and bridges and adorned with floating gardens, a zoo, and an aviary. The city's waterborne commerce involved tens of thousands of canoes, and its central marketplace, according to Cortez, could accommodate 60,000 buyers and sellers.

Of course, the Aztecs had their unpleasant side. Just ask any of the captives—hundreds each month—who, shortly before being rolled down the temple steps, would have their hearts torn out so the Sun would not lack nourishment. At the main temple in Tenochtitlán, one of Cortez's men counted—or, at least, estimated—136,000 skulls.

Then again, human sacrifice was not uncommon in ancient civilizations, New World or Old. (Even classical Greece, that acme of early enlightenment, seems to have indulged.) And Aztec government was in other ways progressive. The commoners were well off by ancient standards, able to swap homemade wares for exotic imports. In simple adobe homes in the provinces, archaeologists have found obsidian knives, jade jewelry, bronze bells.

One reason for this affluence is that the government did a good job of breaching the trust barrier that can impede exchange. Inspectors prowled the urban markets in search of unscrupulous commerce, ranging from false measurement to passing counterfeits (wax or dough) of the cocoa-bean quasi-currency. Aztec law, more than most ancient legal codes, seems to have treated rich and poor alike. Judges were sometimes punished—hanged, in one case—for favoring nobles at the expense of commoners. Torture wasn't used to induce confession—a fact that, one scholar has opined, makes "the Indians appear in a better light than their European conquerers."

"Aztec" is what most people think of if they think of pre-Columbian Mesoamerica at all—a shining gem in the desert, a miraculous exception to the primitive norm of American Indian life. But Aztec civilization wasn't really so special—just the next step in a millennia-old regional ascent whose other rungs included Teotihuacán, the Zapotec of Monte Alban, and many others, such as Toltec, Mixtec, and Huastec. And these, in turn, all had their antecedents. The Maya, though not at first densely urban, had reached statehood around the third century B.C., a bit before the Zapotec did the same. And earlier, there were the Olmec, with their mammoth sculpted Easter-Islandesque heads and a society complex enough that their academic champions have occasionally suggested a promotion from chiefdom to state.

I could go on, naming more and more obscure Mesoamerican cultures, in a procession stretching all the way from the Aztecs back to the origins of Mesoamerican agriculture. I could also try to make a neat picture of this process, drawing textbook lineages of cultural descent: the Aztecs were heirs of the Toltec, who were heirs of Teotihuacán, and so on. But such charts mislead. From the days of the Olmec and early Maya, back in 1200 B.C., cultural influence was subtle and profuse, and with time it got only more so, as Mesoamerica's population grew denser and cultural contact, via trade and war, expanded. The whole region came more and more to resemble a single social brain, testing memes and spreading the useful ones.

True, distinct polities and peoples rose and fell ad nauseam, but these seemingly pointless cycles of growth and decay added up to a larger arrow of cultural evolution. The arts of writing and agriculture and handicraft and construction and government advanced. The Aztecs, like the Romans, were administrative and engineering whizzes. They had their well-oiled bureaucracy, their bridges and their aqueducts. With the sluice gates on their ten-mile dam they controlled the level of the lake surrounding Tenochtitlán.

But the Aztecs weren't exceptional *people,* and neither were the Romans. Both were just like the people who had come before them—human beings muddling through, incrementally adding to their cultural inheritance.

So too with the Inca, down in South America. In the popular mind, they get credit for the vast road network that smoothly moved goods and data, binding a sixteenth-century empire of 12 million people. But many of the roads were built by their predecessors. Construction had started by 500 B.C. under the Chavin, and the infrastructure was expanded by such societies as the coastal Moche, who reached statehood around A.D. 100. Moche roads, like Inca roads, were traveled by relay runners (who, some scholars believe, conveyed data not just orally but by symbols etched on lima beans). And, like Inca roads—and Roman roads and Chinese roads and other ancient roads—the Moche roads were used to coordinate both military and economic activities. Hence a twofer: There were enough prisoners of war to keep Moche warrior-priests busy with ritual throat-slittings, and then, when it came time to drink the blood, there were finely wrought metal goblets ordered just for the occasion.

Under various cultures—the states of Chimú and Huari, for example—the web of South American roads continued to grow, as did irrigation works. After this infrastructure had been laboriously expanded by millions of laborers over a couple of millennia, the Inca came along and said, "Why, thank you!" Conquering chiefdoms here, states there, they carried South American political unity to unprecedented extent, and by deft bureaucratic governance they held it more or less together. As proud as Sargon, and no more worldly, they called their empire Tahuantinsuyu, or "Four Quarters of the World." The new scope of political organization, by subduing the frictions of war, brought new productivity, rather as the Pax Romana had in the Old World.

Both Mesoamerica and the Andes illustrate how much you can do with limited materials. Bronze metallurgy was nascent, scarcely applied to weapons and tools. There were no chariots or wagons—indeed, there were no wheels and no horses; nature seems to have blessed the New World with few readily domesticable animals. But, regardless of natural endowment, there are always means of storing and transmitting data, and thus the means to run a bureaucracy and control a big army. Such is the power of data processing that advances on this front can almost single-handedly carry cultures over the threshold of statehood, notwithstanding stagnation in other realms.

And so great is the power of cultural evolution that such stagnation doesn't last forever. In Mesoamerica, by Aztec times, the principle of the wheel was understood—but, in the absence of draft animals, was applied only to toys, such as red clay rolling animals. Meanwhile, down in South America, the llama had been domesticated and was used as a pack animal. If Europeans hadn't intervened, cultural diffusion almost surely would have brought the wheel south or the llama north, and people would have put two and two together. Century by century, America's two biggest social brains were getting bigger and heading toward merger. Indeed, there is grim evidence that filaments had begun to link north and south: The smallpox the Europeans brought to Mesoamerica reached the Andes by land—apparently killing one Inca king shortly before Pizarro arrived by sea to kill another.

With the arrival of Pizarro and Cortez and other conquistadors, the long American experiment in autonomous social evolution was over. In the Old World, by contrast, the natural expansion of early civilizations, and their ultimate interconnection, had not been short-circuited by murderous aliens. By the first century A.D., the process had reached a culmination of sorts. The tendencies that had carried China and Rome to their glory—growth in the degree and scope of social complexity—had been at work in the lands between them. Just to the east of the Roman Empire was the Parthian empire, around modern-day Iran and Iraq. East of Parthia was the Kushan empire, from modern-day Afghanistan through northern India. And to its east was the western extremity of China under the Han. Eurasia was now wall-to-wall empires. In the terminology of the historian William McNeill, the "Eurasian Ecumene" had been closed. One could travel from the Atlantic to the Pacific, one-third of the way around the world, while passing through only four polities. And commerce did so, along the Silk Road.

Commerce was of growing importance not just among states but within them. Rulers increasingly found that trying to minutely control the creation of wealth was not the way to maximize it. By the first century A.D., McNeill observes, military and political power had come to depend heavily "on materials and services supplied to the rulers by merchants who responded to pecuniary and market motives more readily and more efficiently than to bureaucratic command." Slowly and fitfully, the basic chiefdom model of top-down, state-controlled economics, which seems to have lingered into the early phase of ancient civilization, was ceding ground to the logic of the market.

What caused the shift? One good candidate is the growing practicality of decentralized data processing. A phonetic alphabet, much more user-friendly than the old ideographic scripts, evolved in the Near East during the second millennium B.C., and was transmitted widely, in part by the traders who used it. Then, during the first millennium B.C., coined money emerged and spread via the same conduit. These developments jibe nicely with McNeill's observation that the tendency of markets to outproduce command economies "was beginning to be discovered in the second millennium B.C. and became normal and expected in the course of the next millennium."

The belt of commerce across Eurasia didn't create deep interdependence. The Silk Road, as the name suggests, was mainly for luxury goods. But within empires, an earthier division of labor now existed. Romans got wheat from Egypt, figs and salted meat from Spain, salted fish from the Black Sea. Even if imported fish weren't exactly daily fare for peasants, the benefits of non-zero-sumness were slowly beginning to reach below the ruling class.

NAGGING QUESTIONS

So there you have it—ancient history in a nutshell: onward and upward.[†] This sort of simple summary tends to inspire objections. Such as:

Complaint #1: What about the quirks? This view of history, intent on generalizing, ignores the fascinating and consequential differences between civilizations.

For people who lodge this complaint, there is a simple reply: You're in luck! There are thousands of books you'll love reading. Narrative histories tend to focus on the differences, the particular, often to the exclusion of the commonalities, the general.

To combat this bias, to spend a book dwelling on the common, is not to deny the differences, which are indeed large and interesting. Ancient Egypt, as we've seen, seems to have fused religion and government more than many other early civilizations (no mean feat). And China managed to keep an unusually large piece of turf politically unified under unusually long-lived dynasties.

It is always tempting (if you're me, at least) to try to explain such differences technologically. Is China's vast unity partly due to the use of a script that—because it was largely ideographic—allowed speakers of different dialects to comprehend a single written "Chinese" language? And might the flip side of this script—the difficulty of learning it—have kept power from diffusing rapidly beyond the ruling class?

And as for the intensely divine status of Egypt's pharaohs—well, who knows? Not even an ardent technological determinist tries to explain *everything* in terms of technics. The main point is that acknowledging these differences doesn't detract from the commonalities, and in fact, in a sense, underscores them. Early China was just an unusually good example of the general rule that all early civilizations draw *some* unity from their information technology.

Similarly, the pharaohs' divinity points to a general trend: church and state have grown more distinct over time, worldwide. There are today no states, not even so-called theocracies, run by people who declare themselves gods. And there are no economically advanced states in which leaders even call themselves divinely ordained. Whatever the causes of Egypt's pure theocracy, it was a relic in the making.

Other contrasts among ancient civilizations also hint at larger patterns. Markets played a larger role among the Aztecs than among the Inca, and in Mesopotamia than in Egypt. But all of these civilizations had an economy that harnessed non-zero-sumness through capital investment and division of labor; a command economy and a market economy are two routes to this universal imperative, even if one of them had a brighter future (especially in light of coming trends in information technology).

Complaint #2: What about the Greeks? This chapter, supposedly about the birth of civilization, hasn't even mentioned classical Greece, which in many minds is *synonymous* with the birth of civilization. Shouldn't we have paused for a paean to Socrates and Sophocles, Pythagoras and Archimedes?

Fine men, all of them. Smart, too. Still, from the perspective of *world* history, they don't deserve to hog the spotlight. Great literature and philosophy are not western monopolies. The ethics of ancient sages in India (e.g., the Buddha) and China (Confucius) hold their own in comparison with Greek moral philosophy, and were massively consequential. And as for Pythagoras and Archimedes: far be it from me to minimize mathematics—or science or technology. But we should certainly minimize the importance of any one person in these fields, because all three are on autopilot. The bent for innovation is so deeply human that progress doesn't depend on anyone in particular.

Pi was calculated with precision by Archimedes, but also, independently, by the Chinese. And the Pythagorean theorem, it now seems, had been grasped in ancient Mesopotamia. The concept of zero was invented not in Greece but in India—and also, independently, in Mesoamerica, by the Maya. The histories of math, science, and technology are chock-full of such independent inventions. If Pythagoras— and Archimedes and Aristotle—had died in the crib, the long-run picture in math, science, and technology would not have changed appreciably. Therefore, neither would the long-run course of cultural evolution.

None of this is to say that Greece shouldn't hold a special place in our hearts. For one thing, the Greeks helped test a thesis that the previous chapter hinted at: that, as a society's information technologies become broadly accessible, the result can be not just economic vibrance but political freedom. The Greeks added vowels to the phonetic alphabet, carrying it to its height of accessibility. They grasped the virtues of coins and started minting their own. And to this mix of information technologies the Greeks astutely added the ingredient of trust, making it easy for private parties to strike legally binding contracts. On balance, the results of the test were encouraging. Classical Athens, in its better moments, was economically vibrant, broadly literate (by ancient standards), and democratic (ditto).

Actually, the general notion that economic decentralization disperses political power had gotten some support from earlier phases in cultural evolution, as well. The (relatively) market-oriented Aztecs had their unusually egalitarian legal code. And in (relatively) market-oriented Mesopotamia justice was sometimes administered by citizens' assemblies.

Indeed, the written remnants of Mesopotamian civilization provide a virtual play-by-play account of how information, economics, and politics might benignly co-evolve. In the early third millennium, with writing a new and elite craft, still the province of scribes, records reflect mainly state-controlled transactions. But a millennium later, in northern Mesopotamia, a profusion of clay contracts speaks of a robust private sector, with, for example, traders sending tin and textiles to Anatolia in exchange for gold and silver. How did private citizens reach such heights? One clue may lie in the simplified, less esoteric cuneiform script used in these contracts; whereas professional scribes generally monopolized the ancient writing business, some archaeologists think these traders had broken that tradition, becoming literate themselves.

It is in this period, when the diffusion of information technology seems to have helped carry economic power well beyond the control of kings and priests, that we find evidence of something like democracy. The documents from community assemblies now show them not merely meting out justice, but assuming a deliberative, quasi-legislative function. There are even references to a "city hall."

Of course, meanwhile, elsewhere in ancient "civilization," there was tyranny aplenty, and there were ham-handed government attempts to control the economy. Still, this example from northern Mesopotamia was auspicious. If economic freedom harnesses non-zero-sumness, bringing the wealth that makes states powerful, and if economic freedom tends to entail political freedom, then history might turn out to be on the side of political freedom. After all, powerful states have a tendency to prevail over weaker ones. What's more, maybe later information technologies would strengthen this theoretical logic behind freedom. Still, at this stage in history, we find only a glimmer of evidence for such hopes.

Complaint #3: Where's the chaos? This picture of civilization's ascent has been a bit selective. It's all very well to talk about the Silk Road, or the seaborne commerce along the southern coast of Eurasia, as if such non-zero-sum sinews grow longer and stronger under divine providence. But what about all the disruptions? What about the pirates? What about the wild-eyed marauders from the primitive north who swooped down on Silk Road caravans?

First of all, travel during the first century A.D. was sufficiently civilized so that the rewards of trade warranted the risks, in the judgment of ancient traders. So non-zero-sumness did survive the parasitic assault of zero-sum ambitions. But it did more than survive; it prevailed. Disruptions of trade spurred the evolution of governance.

After all, pirates and marauders are just another form of the "trust" barrier that can block mutually fruitful exchange. They erode your faith that when you send goods eastward, silk will show up in return. And cultural evolution, given long enough, reliably devises means to breach the trust barrier, notably expanded political control. One of the early goals of Roman expansion, in the third century B.C., had been to squelch pirates, and thus defend Italian commerce, by drawing the Dalmatian coast of the Adriatic Sea into a Roman sphere of influence. And one of the ongoing rewards of subsequent expansion was to dampen such disruptions of commerce (and that other classic disruptor, war). That is one reason the Pax Romana brought new affluence. It created not just a war-free zone but a (relatively) brigand-free zone.

This pattern is hardly peculiar to Rome. In ancient times, commerce persistently ventured beyond political bounds. (When an official from Han China ventured past the western border into terra incognita, he was shocked to find, on reaching Afghanistan, Chinese goods for sale.) And political control often caught up with commerce, strengthening its logic. By making passage easier and safer and extending the reach of a uniform legal code, governance lowered both the communications and trust barriers. Indeed, it is largely the surmounting of these two barriers that separated the dominant civilizations from the rest of the pack. Ask a historian to name two things that made Rome great, that served as paragons for posterity to emulate, and there's a fair chance you'll hear: "Roman roads and Roman law."

Imperial expansion isn't the only way to fight pirates. There are also international accords. And in the late Middle Ages, as we'll see, there were private-sector solutions. But all of these fixes, as we'll also see, amount to a kind of expanded governance. Pirates, however you handle them, are just one example of how turbulence and chaos often turn out to be harbingers of new forms of order.

And so it is today. New information technologies, bringing new kinds of international commerce, bring new kinds of disruption. A thief can sit in one nation and steal money from banks in another. The solutions to such supranational problems inherently amount to small steps in the direction of supranational governance. How far we'll walk down that path is quite arguable, but the path's basic direction is less so. Whenever technology has expanded the envelope of non-zero-sumness, new zero-sum threats have materialized, only to be combatted by larger governance in one sense or another.

Complaint #4: You've missed the point of complaint #3. Pirates and brigands are hardly the only form of chaos. What about large-scale chaos? What about decade-long droughts, and deadly plagues? And what about barbarians—not bands of highway robbers, but whole hordes of raping and pillaging brutes? After all, the Roman Empire did eventually fall before their onslaught, right? Indeed, didn't the whole

"Eurasian ecumene," the vast belt of civilization that had evolved by the first century A.D., begin to fall apart? Didn't the Silk Road spend much of its life in tatters? In that light, what does it matter if civilizations tend to get bigger and more complex? The bigger they are, the harder they fall. The barbarians who sacked Rome don't seem to have had as their motto "onward and upward."

Actually, that assertion is debatable, as we'll see in the next chapter.

Chapter Ten

OUR FRIENDS THE BARBARIANS

We have to remember that the annals of this warfare between "civilization" and "barbarism" have been written almost exclusively by the scribes of the "civilized" camp.

—*Arnold Toynbee*

In A.D. 410, the Visigoths sacked Rome. Saint Jerome, who had studied in Rome and had translated the Bible into its language, was in Bethlehem when he heard the news. He wrote to a friend, "What is safe if Rome perishes?" He answered his own question: "The whole world perished in one city. . . ."

This was a somewhat insular view. As Romans watched the Goths wreak havoc, the Maya, for example, went about their business as usual. Still, there is a sense in which the stakes of the barbarian assault were indeed larger than the city of Rome, and larger than the whole empire. The belt of civilization that had spanned the Eurasian continent by the first century A.D. thereafter began to unravel in various places, thanks largely to "barbarian tribes"—Huns, Goths, Vandals, and others. China battled marauding nomads often, and sometimes lost. The Gupta empire of northern India fell under assault from Huns and finally crumbled. Sassanid Persia barely kept the Huns at bay, sometimes becoming, in effect, their vassal state. In the New World, budding civilizations faced the same problem; citadels of urbanity were besieged by rapacious bumpkins, and some fell.

All told, barbarians had enough success to raise the question: What if they had prevailed? What if their devastation had been more thorough and widespread? Can we really be so sure that the basic thrust of cultural evolution would have resumed any time soon, or indeed ever? Did barbarians stand a real chance of ending the world's basic movement toward vaster and deeper social complexity?

No. Indeed, the existence of barbarians, far from impeding cultural advance, may have, on balance, promoted it. This fact is illustrated even by the most famously devastating barbarian triumph: the fall of the Roman Empire.

COMMON MISCONCEPTIONS ABOUT BARBARIANS

What is a barbarian? To the cultural evolutionists of the nineteenth century, as we've seen, "barbarian" denoted a stage between "savage" (a simple hunter-gatherer band) and "civilized" (a state). This is indeed the level that most barbarian tribes of ancient times had reached; today we would call them "chiefdoms," though some were unusually mobile chiefdoms.

Historians use "barbarian" more loosely: Barbarians are peoples with a culture less advanced than their neighbors', and perhaps with a tendency to violently exploit their neighbors' advancement. Sometimes the exploitation—the pillaging—was done by swooping down on horseback, though this luxury was not available to New World barbarians.

To the Romans, "barbarian" was a less technical term. Its origins sound innocent—it came from the Greek word for "foreigner"—but its connotations were decidedly disparaging. Some Romans referred to the land within the empire's bounds as *oikoumene*—"inhabited land." The Roman view of barbarians—as uncouth, perhaps depraved, even subhuman—lingers on, making barbarians one of the most misunderstood and unjustly maligned of groups. Several misconceptions, in particular, need dispelling.

Misconception #1: Barbarians are less "civilized" than their neighbors in a moral sense—less decent, less humane. Behaving less humanely than the Romans would be hard. It was Roman cavalrymen who informed their nemesis Hannibal of the outcome of a recent battle by tossing his brother's head into his camp. It was Romans who avenged an early defeat at the hands of the Goths by taking Goth children—seized as hostages years earlier—and marching them into public squares in various towns, then slaughtering them. The emperor Nero bound Christians, smeared pitch on them, and ignited them, purportedly to light his gardens at night. One of his successors, Titus, celebrated his brother's birthday by publicly killing 2,500 Jews—pitting some against each other in combat, pitting others against wild animals, and burning the rest. On a smaller scale, of course, this sort of spectacle was a regular form of Roman entertainment.

Sacking cities was standard Roman procedure, and, indeed, common in ancient wars generally. Saint Augustine, reflecting on the Goths' looting of Rome, wrote, "All the destruction, slaughter, plundering, burning, and distress visited upon Rome in its latest calamity were but the normal aftermath of war." What was unusual, he observed, was that "fierce barbarians, by an unprecedented turn of events, showed such clemency that vast basilicas were designated as places where refugees might assemble with assurance of humanity." The Goths had burned only a few buildings, over a few days, before moving on.

Misconception #2: Barbarians lack culture. If by culture you mean fine sculpture, Greek tragedies, or eating salad with your salad fork, this charge has merit. But if by culture you mean what a cultural evolutionist means—*all* products of the human mind, especially practical ones—then barbarians needn't hang their heads in shame. Given their dearth of formal education, they've contributed a fair amount to humankind's great, upward-flowing stream of memes.

It was a barbarian plow that opened the heavy soil of northern and western Europe to farming during the Middle Ages. It was Asian barbarians who gave the stirrup to the Chinese and ultimately to the West. And back during the Chou era, barbarians assaulting the state of Ch'in had displayed a new style of warfare, based on the horse archer. The Ch'in used the technique not only to fend off later barbarian waves, but to conquer rival states and unify China.

Romans used to complain about the smell of barbarians, but that just goes to show there's no accounting for taste. It was the Romans who didn't use soap, and the barbarians who invented the stuff by boiling fat in alkali.

Misconception #3: Barbarians are beyond true edification. Granted that they've thought up a few neat ideas (often related to riding horses and killing people), when it comes to imparting culture to barbarians, you might as well be talking to a stone. Actually, barbarians, being human, are receptive to the same kinds of memes that people in general are receptive to. They like functional things, novel things, glittery things. Roman emperors used to dissuade Attila from pillaging by sending him gold. He would then do what any normal human would do after getting a big paycheck: go shopping. So the Huns wound up with silk, pearls, gold platters, silver goblets, gem-studded bridles, comfy sofas, linen bedsheets, and of course sturdy iron swords, with which to extract more gold from Romans.

Even bookishness was not beyond the barbarian mind. By the end of the fourth century A.D. a bishop had converted some of the Goths to Christianity and translated the Bible into their language. This was the beginning of Germanic literacy.

The dispelling of misconceptions 2 and 3 suggests a larger truth. If barbarians are reasonably good at generating memes, and at absorbing memes generated by others, and are prone to travel, then you would expect them to be valuable meme spreaders and synthesizers. Indeed, they are veritable Mixmasters of culture.

Consider the Celts, the chiefdom-level people who touched various parts of Europe in the centuries before and after Christ. According to one archaeologist, the Celts were "nomadic, boastful, quarrelsome, sumptuous, wild, and warlike, and they were headhunters." But whatever you do, don't call them uncultured. They sold salt and metals to the Greeks and used the proceeds to buy wine, pottery, and metalworks. They transmitted Greek artistic motifs northward and conveyed ironworking technologies across broad swaths of Europe. Eventually Celts would popularize horseshoes, iron locks, and barrels. The Romans learned the virtues of the "short sword" the hard way—while the Celts were sacking Rome in 390 B.C. By Caesar's day Celts were coining money in the fashion of Romans. Some Celts mastered the Greek alphabet.

Thus above and beyond the Celts' erratic bursts of marauding and trading was a larger role: data processing and transmission. Amid the hubbub, memes—conglomerations of cultural information—got selectively preserved and replicated. They included one of the most important material technologies ever—ironworking—and two of the most important information technologies ever—writing and money. Thank you, head-hunting Celts.

The moral of the story is simple: When thinking about cultural evolution, don't get wrapped up in the particular people and peoples. Instead, *keep your eye on the memes.* People and peoples come and go, live and die. But their memes, like their genes, persist. When all the trading and plundering and warring is done, bodies may be lying everywhere, and social structure may seem in disarray. Yet in the process, culture, the aggregate menu of memes on which society can draw, may well have evolved. Eventually, social structure will follow, coalescing around the newly available technological base. It may take awhile for the social structure to catch up with the technology (see next chapter). But given enough time, it will.

Misconception #4: Barbarians are by nature transient and chaotic. It is true that the barbarians of Europe and Asia were sometimes seized by wanderlust. And understandably so. If parasitizing painstakingly constructed civilizations is your line of work, you have to travel. But parasitism was in fact not the vocation of most "barbarians" most of the time. When they found nice fertile land, or a nice nexus among trading peoples, they often settled down to earn a living—an honest living, even.

You wouldn't know this to read the dramatic Roman accounts of barbarians, but then they were based on dramatic encounters between Romans and barbarians, not on a random sampling of barbarian life. Archaeologists have since found that the Germanic barbarians north of the empire lived in "stable and enduring communities," their economy "probably essentially similar to peasant agriculture within the Western Roman provinces."

What's more, barbarian societies, whether nomadic or sedentary, tend to evolve, just like other societies, toward higher levels of organization. One reason the Romans felt growing torment during the fourth century was that their tormentors possessed increasingly deft administration (some of it copied from the Romans); the barbarians were growing "more civilized," in the words of one historian, and thus more terrifying. Another historian writes of the barbarians menacing north China: "They became really dangerous to the extent that they became civilized, and versed in the arts of organization, production and war." One such barbarian memorized Confucian scriptures and was fond of saying, "To be ignorant of even one thing is a cause of shame to the gentleman." His well-schooled son sacked the capital of the Chin in 311, an event roughly comparable to the sacking of Rome.

The Huns in particular, though nomadic, were organized on a vast scale, sometimes described as an "empire." Like ancient empires generally, the Huns violently subjugated peoples and exacted tribute from them. (Who were the Romans to complain about that?) Meanwhile, over in east Asia, a barbarian confederacy called the Toba was assembling its own empire during the fourth and fifth centuries. The Toba eventually found themselves ruling most of north China—and having to defend their turf against fresh waves of irksome barbarians.

That barbarians can be just as happy upholding a civilization as tearing one down is nowhere better illustrated than in the Roman Empire. For centuries, Romans had used Germanic tribes as mercenaries, and by the time Rome was sacked, some of the empire's finest generals were of barbarian extraction. By and large, the Romans discovered, you could do business with these people. With the successful invasions of the fifth century, the barbarian tribes made this clearer. Before them lay vast expanses of Roman farmland, tilled by peasants who paid stiff taxes to a government that was increasingly unable to defend them. How to exploit the situation?

One approach might have been to sweep across the agrarian countryside, battling Roman soldiers, slaughtering peasants, then taking their land. Another approach was to leave the peasants more or less alone and simply cut a deal with Roman officials under which you begin to replace them as tax collectors. The barbarians of legend would have taken the first path. The real-life barbarians took the second, thus realizing every person's dream: a high ratio of income to work. Over the fifth and sixth centuries, the Roman tax apparatus came to be, in the words of one historian, "under new management."

There is a famous putdown applied to public servants who begin their careers with high ideals and wind up corrupted: they "came to do good and stayed to do well." Of the barbarians who fought Rome we might say, "They came to do bad, and stayed to do well." They may have begun their invasions in a mood to pillage, but they eventually found a more sedate livelihood. This flexibility is one reason that by A.D. 500 western Europe had evolved fairly smoothly from a single empire to several large barbarian kingdoms, such as the Visigothic in Spain and the Ostrogothic in Italy.

And how did Greco-Roman culture fare at the hands of barbarians? The Goths weren't the sort to ponder Plato's dialogues, but they praised the texts of Euclid; eschewing squishy subjects, they stressed nuts-and-bolts disciplines: architecture (they restored some of the empire's monumental buildings); surveying (helpful in tax collection); mathematics (especially as applied to coinage and measurement); medicine (among the booty the Visigoths carried south after sacking Rome was the physician Dionysius). The Goths disdained the study of "rhetoric," which had loomed large in Roman law schools, but law itself was another matter. Employing Roman jurists, they adapted Roman law to the governance of their Roman subjects, and also formalized their own legal traditions, which had previously been oral. One barbarian-published tome was called "Roman Law of the Visigoths."

None of this is to suggest an easy continuity between the final century of Roman rule in western Europe and the centuries to come. The Goths and Franks and other "barbarians," however eager to be Rome's heirs, were ill-equipped to assimilate a whole body of advanced culture. Besides, counterattacks from the still-formidable eastern sector of the Roman Empire, based in Constantinople, were disruptive. (The emperor Justinian's "reclamation" of Rome did more damage than the original sacking by the Visigoths.) In the end, the great barbarian "kingdoms" did not endure intact. Still, the barbarians didn't send Roman culture through a paper shredder, either.

Misconception #5: Barbarians were a peculiar affliction that for some reason materialized in the age of Rome (and then recurred occasionally, as with the Mongol maraudings of the late Middle Ages). The very fact that cultural evolution proceeds unevenly from place to place means that, for millennia before Rome, civilizations had been surrounded by less advanced cultures. These have-nots, being human, had the capacity for predation, and sometimes exercised it. Middle Eastern civilizations seem to have been beset by at least two waves of "barbarian" devastation, near the beginning and end of the second millennium B.C. So why do we hear relatively little about these earlier barbarians? Several reasons.

First, the further back in time you go, the less recording of history there is. Archaeologists find the ruins of former civilizations, and signs of violent clash, but they seldom find clear testament to what presumably was the perspective of the afflicted civilization: that hostile, uncouth, inferior aliens were on a mindless rampage. Indeed, when the barbarian assault is successful, its history, if ever recorded at all, will probably be written by descendants of the barbarians. And you can't expect *them* to trumpet their ancestors' barbarism.

Consider the Aztecs. They started out as semi-nomadic brutes lingering on the periphery of more advanced cultures, raiding them here, serving as their mercenaries there. (Sound familiar?) When they had finally learned enough to found their own great city and conquer literate peoples, they destroyed texts that described their own past as primitive. They wrote new histories, depicting themselves as the sole legitimate heir to the late, great Toltec civilization. For that matter, the Toltec themselves had started out as semi-nomadic barbarians, soaking up the culture of the peoples they would then push off the local pedestal.

Indeed, if you explore the murky recesses of just about any famously civilized people, you'll find this dark secret: they started out as barbarians. The Romans weren't exactly hailed by the Greeks as cultural equals when they happened on the scene. In fact, even after generations of Hellenic edification, the boringly practical Romans didn't exude quite the cerebral air of classical Greece. Yet they were massively infiltrated by classical Greek memes, which they then spread across the wider world. In Horace's phrase, "The Greeks, captive, took the victors captive."

And, anyway, who were the Greeks to look down on intrusive barbarians? They had their own checkered past. Their lowly ancestors took a big step in the long trek toward snobbery by invading Europe's first bureaucratic monarchy, Minoan civilization on Crete, in the fifteenth century B.C. The early Greeks had a title of honor, *ptoliporthos,* that meant "sacker of cities." And Dorian Greeks may have been among the troublemakers who wreaked havoc near the end of the second millennium B.C.

You can play this game all day, going back and showing the ignoble social origins of what would later become dominant civilizations. But whether these "barbarians" sack cities, or hover on the periphery and trade with them, or ally with them in war or ally against them, one outcome is nearly certain: win, lose, or draw, the "barbarians" become vehicles for advanced memes. As William McNeill wrote in *The Rise of the West,* "The history of civilization is a history of the expansion of particularly attractive cultural and social patterns through conversion of barbarians to modes of life they found superior to their own." This century, "English" has been virtually synonymous with civilization and refinement, yet the word "England" means land of the Angles—a tribe that, back in the days of Rome, was just another bothersome bunch of barbarians.

Misconception #6: Barbarian eruptions, in their chaos and destruction, are ironic punctuation to the supposedly progressive flow of cultural evolution. We've seen that barbarians, in the long run, fall in line, assisting the upward flow of memes. But there's a second sense as well in which barbarians, however defiantly they may seem to swim against the stream of cultural progress, are in fact going with the flow: The reason they're so well equipped for the brief display of defiance in the first place is because the flow of memes is so inexorable.

When a civilization such as Rome dominates its neighbors, it typically possesses some sort of cultural edge: better weapons, say, or, better economic organization. Yet this dominance is hard to maintain precisely because these valuable memes tend naturally to spread beyond its borders, empowering its rivals. In the case of Rome, the barbarian-empowering memes included military strategy. But the exact memes will differ from case to case. As the historian Mark Elvin has observed, the diffusion of Chinese ironmaking technology to the Mongols during the thirteenth century would come back to haunt China. Elvin was among the first to clearly see that this is a general dynamic in history: the very advancement of advanced societies can bring the seeds of their destruction. As Elman Service put the matter: "The precocious developing society broadcasts its seeds, so to speak, outside its own area, and some of them root and grow vigorously in new soil, sometimes becoming stronger than the parent stock, finally to dominate both their environments."

The point can scarcely be overemphasized: the turbulence that characterizes world history is not only consistent with a "progressivist" view of history; it is integral to it. The turbulence itself—including the sometimes devastating empowerment of barbarians—is a result of the fact that technology evolves, with the fittest technologies spreading rapidly. Hegemony can bring stasis, such as the Pax Romana, but in the long run such imbalances of power naturally undermine themselves, and stasis ends. The ensuing turbulence may look for all the world like regression, but it is ultimately progressive; it reflects—and, as we've seen, often furthers—the globalization of new and improved memes, on which the next stasis will rest.

Misconception #7: Barbarians prey on innocent victims. The phrase "barbarians at the gate" conjurs up a Manichaean image. Inside Rome's walls, librarians are shelving painstakingly translated editions of Euripides when smoke starts seeping through the stacks. What had the Romans done to deserve this?

Plenty. For starters, the economy of imperial Rome, to an extent notable even by ancient standards, had been built on slaves. This may sound like a moral critique—and it is—but it is also something more. It is an evaluation of Rome by the basic gauge of cultural evolution: How thoroughly did Rome realize potential synergy among its people?

Not very. When a society keeps people in chains, and confiscates the fruits of their labor, it is trying to play a non-zero-sum game in utterly parasitic fashion—a strategy that, I've argued, has its pitfalls. First, oppression takes time and energy; Rome more than once had to put down slave revolts, and vigilance was constant. "No one can feel safe, even if he is a lenient and kind master," lamented Pliny the Younger (who inferred from this fact that "slaves are ruined by their own evil natures"). Second, slavery rather weakens a worker's incentive to work, thus making close oversight a prerequisite for efficient labor—and oversight is costly, so efficiency suffers in any event. Third, slavery keeps workers from becoming robust consumers. Fourth, by keeping labor artificially cheap, slavery dampens the society's incentive to develop more productive technologies. Rome's ruling class was famously indifferent to labor-saving innovation, and technological progress was unspectacular.

There is evidence that slavery waned toward the end of the empire. But at the same time, workaday peasants were becoming *less* free, more like medieval serfs—tied to land they didn't own. And the government started trying to stop craftsmen and shopkeepers from changing vocations; it even insisted that their children follow in their footsteps. One classical historian believes that "the Mediterranean world came closer to a caste system than at any other time in its history." Again, leaving aside the moral critique, this is just bad social engineering; it stifles the gains that can arise spontaneously from freedom of choice in a market economy.

Even Rome's great contribution to commerce—the large, low-friction zone created by the Pax Romana—was hardly an unmixed blessing. It had its element of simple exploitation, especially visible to those who had to be subdued for the sake of the Pax. "To robbery, slaughter, plunder, they give the lying name of empire," opined one Briton. "They make a desert and call it peace." After subjugation was complete, there were often unwarranted taxes and greedy administrators on the take.

By most accounts, this sort of parasitism grew as the political culture became more corrupt, oppressive, and dictatorial. In the late imperial period, emperors were claiming divinity and acting like pharaohs. They stayed secluded, cultivating a mystique, and Romans who were granted an audience had to start by kissing the hem of the emperor's robe. The Senate was by now impotent, with emperors chosen by the military, sometimes through a kind of civil war bake-off.

None of this is to deny Rome's celebrated legacy. Roman principles of law and administration were lasting paragons, even if in practice they were progressively adulterated. Still, once these principles were on paper, and Roman engineering had left its mark, the Romans had little else to give posterity. Whether you are a champion of moral improvement or just of cultural evolution, you might defensibly conclude that, by the time the barbarians descended on the western Roman Empire en masse, it deserved to die.

Obviously, any conclusion this neatly gratifying should arouse suspicion. What actually caused the demise of the western Roman Empire is still debated, and some conjectured causes—disease, depleted soil, chance geographic exposure to unusually large barbarian hordes—don't reflect on the quality of Rome's government. Whether the empire's decline really is the morality play that many historians make of it—an interpretation I've happily adopted—is an open question. Whole civilizations can rise and fall on the basis of chance events; it is only in the long run, and in the broad sweep of events, that basic dynamics of cultural evolution sustain history's direction.

Still, it is notable how many of the commonly posited causes of Rome's decline are blights on non-zero-sumness. An artificially frozen labor market; an increasingly unfair legal system; corruption in the delivery of public goods by officials; excessive taxes and tariffs, dictated by the costs of supporting an empire that is parasitic on its provinces—all of these weaken the fabric of mutual benefit that holds well-run societies together.

THE VERDICT OF HISTORY

So thank heavens for barbarians! If dominant civilizations are stagnant and decaying, contributing little if anything to the march of non-zero-sumness, it is just as well (from cultural evolution's standpoint) to have troublemakers nearby. Better to tear the system down and start over. And because barbarians turn out to be so partial to civilized memes, you don't have to start from scratch!

The barbarian role of cultural demolition crew is especially important when you consider how often cultural reconstruction is needed. Many of Rome's glaring defects—exploitation, authoritarianism, corrupt self-aggrandizement—flow from deeply human tendencies. Time and again they've transformed promising civilizations into decaying, oppressive monstrosities. Time and again, history seems to cry out: Bring on the demolition crew! And time and again barbarians cheerfully respond to the call. Their previous massive wreaking of destruction, near the end of the second millennium B.C., had come after civilization went through centuries of apparent ossification.

In a way, barbarians are just a special case of that general and potent zero-sum dynamic in cultural evolution: brutal competition among neighboring societies. This rivalry renders ossified cultures vulnerable to a makeover, minor or major. They may be taken over by a vast neighboring civilization, which will revamp them in its image. Or they may be infiltrated and perhaps even disassembled by barbarians, paving the way for future reassembly. Or they may revive and prevail—an example of the "challenge and response" dynamic stressed by Arnold Toynbee. In any event, the point remains the same: however deeply human the tendencies of exploitation, authoritarianism, and self-aggrandizement, cultures that surrender to them may not be long for this world.

But wait. When did exploitation and authoritarianism suddenly become political liabilities? Didn't most of the early states, as well as their precursors, the chiefdoms, employ terror when useful, take slaves when possible, and claim the mantle of divinity, or at least of divine blessing, to nudge the masses into compliance with central dictate? However morally reprehensible these tactics, why should they have become ineffective by Rome's day?

Part of the answer, as we've seen, may be that technology changes the rules of governance. With the coming of standard, universally accepted coins and a fully phonetic alphabet, complete with vowels, the potential existed for a more decentralized economy than ever before. So, for example, slavery—the ultimate in exploitation—now carried a higher cost in forgone productivity; the better lubricated the market, the more it can benefit from untrammeled participation. A mind, as they say, is a terrible thing to waste—even if, in an ancient economy, it is focused mainly on toiling the fields for the highest wage offered and then turning around and spending the money.

Though Rome had a largely market economy, there was from the beginning a certain obliviousness to its potential. When the Romans started minting coins in the fourth century B.C., the stated purpose was not to smooth commerce, but just to create a medium by which the *government* could buy things. (Of course, coins did flood the private sector, willy-nilly.)

The eastern half of the empire, which survived the collapse of the west, was less guilty of some of these sins. The east seems all along to have had fewer slaves than the west. In the east the economy was less afflicted by such stultifying policies as the virtual ban on changing vocations. And, for historical reasons, the east had a more integrated economy, which ably moved goods from one region to another. There were large old market towns, whereas many western towns were more like shells—administrative centers lacking an organic core.

Of course, in the battle for survival, the east had one other big asset: a shorter barbarian frontier. And certainly, in any event, the east, which shared many of the west's ossifying tendencies, was no paragon for a high-tech future. The entire empire was vulnerable, some parts just more so than others.

The historian Chester Starr once wrote, "Every so often civilization seems to work itself into a corner from which further progress is virtually impossible along the lines then apparent; yet if new ideas are to have a chance the old systems must be so severely shaken that they lose their dominance." This may strike some as teleological, even mystical—as if the god of progress looks down and weeds out civilizations that aren't prepared for coming ideas. But Starr's point sounds more reasonable once you view technological evolution as an active force in history. It is metaphorically true that cutting-edge technologies—economic technologies no less than military technologies—punish societies that don't embrace them and use them well, leaving those societies at risk of being "severely shaken." It is also metaphorically true that those technologies reward societies that employ them more profitably.

Of course, technology isn't some extraneous force, visited on the planet from outer space. It is selected by human minds through cultural evolution; people are the arbiters of technology. But technologies—in a broad sense, at least—are in turn the arbiters of social structure. The question for western Europe as of the fifth century A.D. was: to what kind of social structures would technology next give its blessing?

Chapter Eleven

DARK AGES

This is the old story: whenever one sets out to discuss collapse, one ends up by talking about continuity.

—*G. W. Bowersock*

In the 1969 book *Civilisation,* companion to the BBC television series of the same name, Kenneth Clark had a chapter called "By the Skin of Our Teeth." Its premise was that western civilization was lucky to be alive. The "Dark Ages," as some have called the early Middle Ages, truly had been dark; just barely had the smoldering embers of the west's classical heritage survived to illuminate the world another day. But for the labors of a few monastic scribes, carefully copying the great works, who knows what sort of cultural backwater Europe would be now?

This theme is a hardy perennial, more recently on display in Thomas Cahill's best-selling book *How the Irish Saved Civilization,* in which Irish scribes were singled out for special praise. Clark and Cahill have in essence looked back to Saint Jerome's despairing question—"What is safe if Rome perishes?"—and deemed it a very good one.

In the previous chapter we found reason to consider Jerome's question alarmist. Namely: barbarians are people, too. Unlike Conan the Barbarian—whose professed aim in life was "to defeat the enemy, see him run before you, and hear the lamentations of the women"—most real-life barbarians are eager to settle down and savor the fruits of civilization: to defeat the enemy, tax him, visit his doctors, marry his daughters.

Another reason to deem Jerome a bit melodramatic is that, as historians have come to realize, the term "Dark Ages" is misleading. Even in the early Middle Ages, and especially in the later, there was creativity and vibrance.

Still, it is true that the early Middle Ages must have seemed fairly dim at the time. There was indeed a collapse in the Roman vicinity. Roads that once carried goods safely to market fell into disrepair and were beset by outlaws. Towns contracted, and farmlands were reclaimed by the wild. Mines were abandoned, and metal production dropped. Reliable coinage, especially gold, grew scarce, and people found themselves in a nearly "moneyless" economy, bartering goods or paying with their labors. Population continued to drop. The scope of government vacillated; barbarian kingdoms grew and shrank, and sometimes their "kings" had little real power anyway.

So even if barbarians aren't deeply barbaric, and even if the "Dark Ages" weren't pitch black, things did look dicey in Europe for a time. Any basic tendency of cultural evolution to carry society to higher and higher levels of complexity was not vividly apparent. So those of us who believe in such a tendency must say a few more words about the "Dark Ages." What firm dynamics of history made their end just a matter of time? Why does reconstruction reliably follow collapse and disarray?

The question is not confined to the most famous dark ages, the European ones. Chinese cohesion suffered a big setback in the fourth and fifth centuries A.D., as barbarians poured in from the north. And the barbarian onslaughts at the end of the second millennium B.C. had wreaked havoc from the western Mediterranean to the Middle East. In the New World, there's the famous Mayan collapse, among others. And so on—a lengthy menu of regression to choose from.

In asking why collapses in general tend not to prove fatal, we'll focus mainly on Europe's Middle Ages. They are the best-documented example of major collapse and recovery and the most famous. They're also the most widely cited challenge to directional views of history. After all, if civilization really did survive by the skin of its teeth, then it must be pretty fragile.

There is one other reason to focus on Europe. In the "skin of our teeth" view of history, it isn't just civilization in the generic sense that was lucky to survive. Kenneth Clark might concede that, even without the monks, Europe would eventually have made leaps economically, technologically, politically. Clark's concern is more with *western* civilization. Would its distinctive emphases—on political liberty, for example—have survived if the barbarians had permanently broken the link with classical Greece, the cradle of western freedom and democracy?

This is a subtler question, with a more tentative answer. But as we'll see, there is reason to believe that many historians overplay the role of the peculiar "heritage" of the west in shaping its modern form, and underplay the role of universal forces of history, as played out amid the quirks of medieval history.

KEEP KEEPING YOUR EYE ON THE MEMES

The first step toward appreciating the inexorability of the west's resurgence in the later Middle Ages is to drain the early Middle Ages of the melodrama given them by the "skin of our teeth" view of history. And the first step in this melodrama reduction is to repeat the mantra from the last chapter: *Keep your eye on the memes.* In deciding whether a culture has collapsed in the first place, ask not what has happened to a particular people or a particular land; cultures can hop from person to person and place to place, leaving ruin behind yet staying healthy themselves. Well before the sacking of Rome, the Roman Empire's headquarters had been officially moved to Constantinople. There, in the eastern empire, in Byzantium, much of classical culture remained alive and well—in books, in minds, in practice—until Europe's "Dark Ages" had passed.

This is a common story with "collapses." Empty Mayan ruins are standard imagery in accounts of "lost civilizations." Seldom noted is that the Mayan collapse afflicted only part of Mayan civilization. To the north, whole cities survived, and kept Mayan culture, including its precious script, intact.

Similarly, the famous collapse of Indus valley civilization in south Asia during the second millennium B.C. is often depicted as an overnight vanishing. But people lived on and regrouped and retained key technologies—their advanced system of standardized weights, and possibly their script; one of the area's modern-day scripts may well be a descendant of the original, the family resemblance blurred by millennia of evolution.

The second great antidote to melodrama is to remember that political maps aren't always a good guide to social complexity. True, the Roman Empire had been a state-level society, a "civilization"; and, true, much of western Europe, including modern-day France and England, had technically been part of the empire, paying taxes and possessing elements of civilization—some roads, shiny coins, a few bureaucrats. But roads and coins don't turn a simple society into a complex one overnight. Marvin Harris has written that "transalpine Europe did not lapse back into the 'Dark Ages,' never having gotten out of them in the first place." This may be an exaggeration, but not a huge one.

Of course, the rustics of transalpine Europe, if not themselves fully "civilized," had grown dependent on Roman governance, so pulling this rug out from under them was an interesting experiment in cultural evolution. What would they do? Go into free fall and finally land somewhere near the Shoshone, scurrying around in small bands and digging up roots?

No. They not only held on to their agricultural technology; as if landing in a safety net, they fell back onto the remnants of their indigenous political system, the chiefdom—or, at least, a reasonable facsimile of it. As various cultural evolutionists have noted, European feudalism, though distinctive, had much in common with chiefdoms as observed from Polynesia to the Americas to pre-Roman Europe.

Under feudalism, peasants—serfs—got land to farm, but they also had to do some farming for the local chief—or, as he was called in Europe, their "lord." In return, the chief provided them not just with their land, but with public services; he offered military defense (no small thing during the Middle Ages) and administered justice (such as it was). Serfs are often depicted as virtual slaves, and they certainly weren't free in the modern sense. But, more than the slaves of the Roman Empire, they could expect protection and even some minimal respect in exchange for their subservience. "Serfs, obey your temporal lords with fear and trembling," one French cleric advised. "Lords, treat your serfs according to justice and equity."

The principle of mutual obligation—and of payment in services rendered, not in cash—extended upward. The local chief was himself subordinate to a higher chief; though lord over his serfs, he was a vassal to a higher lord. As a vassal, he had to serve as his lord's soldier, often on horseback in shining armor. This obligation was his payment for the land on which his serfs worked, and for the defense of that land by the larger army to which he belonged.

Though the details of feudalism differed from time to time and place to place (the above is a schematic simplification),[†] its upshot was indeed broadly reminiscent of the classic chiefdom. There was a cluster of local units—manors, sometimes distinct villages—each of which had its own boss. And one of these bosses ranked above the other bosses (was lord to these vassals), so the units constituted a larger, supravillage polity with centralized leadership. A key purpose of this hierarchy was the swift mobilization of men for war, but the lords also, like chiefs, orchestrated the local production of tools and other crafts.

With a classic, simple chiefdom, the story would end there: a bunch of villages with local chiefs, one of whom was paramount. With feudalism, the story often went much further. The paramount chief—the lord among lords—might himself be vassal to another lord, who in turn might be vassal to a still higher lord—all the way up to the ultimate lord, the king. It was as if several chiefdoms in a region formed a superchiefdom, and several of these superchiefdoms in turn constituted a super-superchiefdom. Sometimes the hierarchy went ten levels deep.

THE FRACTAL BEAUTY OF FEUDALISM

"Feudal" is today a pejorative term, but feudalism was in some ways well suited to a time of instability. Like the classic chiefdom, it kept food on the table without relying on a sound currency or on trade with distant peoples. It also kept warriors at the ready. This was especially impressive because, with the ascendancy of armored cavalry (due largely to the coming of the stirrup), equipping warriors got expensive. It was almost as if, today, no soldier could safely set foot on the battlefield without his own personal tank. The solution to this financial challenge—giving knights lordship over chunks of property they could subdivide for the use of peasants—worked well enough. And making each lord a governor of his immediate subordinates (not just peasants, but any other vassals) made for a decentralized government—a handy thing in a time of poor roads, low literacy rates, and other barriers to distant administration.

Perhaps most important, feudalism's nested structure, its long chain of mutual obligation, gave the system a kind of resilience. Each link in the chain was a simple and direct non-zero-sum relationship; a lord and his vassal both benefited from the deal, and had consecrated this interdependence with ornate oaths of devotion. So if for any reason the bonds at the highest level broke, the lower levels of the hierarchy tended to stay intact out of mutual self-interest. When kingdoms collapsed, they broke up into regional or local polities, not into anarchy. Moreover, because larger units were structurally identical to the smaller units constituting them—mathematicians call this a "fractal," or "self-similar," structure—subsequent reassembly could proceed readily.

Consider the Franks, in modern-day France and Germany. Of all the barbarian "kingdoms" that emerged with the fall of the western Roman Empire, the Frankish kingdom had proved the most vibrant. In A.D. 800, after aggressive expansion, its leader, Charlemagne, was declared by the pope to be emperor of a newborn Roman Empire. But the empire's coherence depended on Charlemagne's savvy and charisma (which in turn seems to have rested partly on his much-remarked-upon height, impressive enough to compensate for his also-remarked-upon squeaky voice). So his death in 814 boded ill for the continued unity of the empire. All the more so because his demise roughly coincided with the Viking onslaught from Scandinavia.

On paper, it might have looked as if the Vikings would establish an immediate rapport with residents of northwestern Europe. The Vikings were a Germanic people stigmatized as barbarians, and the residents of northwestern Europe were a largely Germanic people whose not-so-distant ancestors had been stigmatized as barbarians. But of course, the Vikings were not the sort to discuss their self-esteem problems, and, anyway, the northwestern Europeans had now reached a higher social station; to them the Vikings, in their wild-eyed savagery, seemed like members of a lower species, notwithstanding their ethnic affinity. (Just goes to show: it's all in the memes.)

In any event, the Franks now faced double trouble—hordes of pillaging Vikings on the one hand, and, on the other, the crumbling of Carolingian leadership as Charlemagne's heirs evinced their various shortcomings. Yet feudalism's links basically held. In northern France, counts—the second-to-highest ranking lords in the land—ceased to feel obligation to the highest ranking lord, the king. But they mostly retained the loyalty of their own vassals and set about expanding their domains by allying with or conquering other counts (thus fusing "counties"). So no sooner had the kingdom fragmented into building blocks than the blocks began to reassemble themselves. In other regions, disintegration proceeded further, below the level of counts, so the polities were smaller. Still, feudalism's fractal structure meant that polities of *any* size had the same basic structure, the same ingredients for internal cohesion, and the same potential for subsequent reassembly.

As a result, whether the Vikings confronted large military organizations or small ones, they always confronted organizations. And in the end the organizations—or at least the *organization*—won. Even where the Vikings were triumphant, they tended to melt into the fabric of feudalism rather than tear it apart. Like so many barbarians, they discovered that using existing social structures could be more profitable than destroying them, and less hazardous to your health.

Though European feudalism was peculiarly resilient, human society in general is good at regrouping under duress. When centralized authority has collapsed, true anarchy has seldom ensued. Political and economic reconstitution at *some* level is typically immediate. Sometimes the result is sufficiently reminiscent of the Middle Ages that scholars note its "feudal" elements. But regardless of how "feudal" the recoveries, they typically rest on the same basic cement that kept Europe orderly in the early Middle Ages: the human instinct for nonzero-sum relationship. People are good at finding zones of mutual self-interest and striking deals of mutual obligation. Greeks, during their own "dark ages," were a good example. After the collapse of the Mycenaean state at the end of the second millennium B.C., they regrouped into what appear to have been chiefdoms, which then evolved into the city-states that would make Greece famous.

Spontaneous renaissance was also visible in northern China in the fourth century A.D., when government dissolved under barbarian onslaught. Facing chaos, families clustered together in large camps, built fortresses, and agreed to submit to a common leader. Leaders of the camps conferred and agreed to do the same. Presto! Instant political structure, complete with improvised legal codes, economic self-sufficiency, and military might. The Chinese consecrated these bonds of mutual obligation by drawing on their spiritual heritage, Confucianism, whereas the Europeans sealed feudal bonds with Christian ceremony. But the same thing was happening in both places: human nature was ensuring that when structure collapsed, a safety net would materialize. And religion was adapting itself to this mandate, the mandate of non-zero-sumness.

THE WORLD MAKES BACKUP COPIES

By the beginning of the eleventh century, the Viking threat had subsided. Europe had weathered the storm. But it had done much more than that. It had gradually accumulated cultural capital and was now poised for a great leap forward.

This cultural capital, this precious stock of memes, had little to do with Europe's "classical heritage." In *How the Irish Saved Civilization,* Cahill gasps at what might have been lost in the barbarian invasions. "Had the destruction been complete—had every library been disassembled and every book burned—we might have lost Homer and Virgil and all of classical poetry, Herodotus and Tacitus and all of classical history, Demosthenes and Cicero and all of classical oratory, Plato and Aristotle and all of Greek philosophy, and Plotinus and Porphyry and all the subsequent commentary."

Well, them's the breaks. But what people of the early Middle Ages most needed wasn't a good stiff dose of Demosthenes. They needed mundane things, such as a harness that wouldn't press on a horse's windpipe. This new device, in use by A.D. 800, tripled the weight a horse could pull, and thus relieved European farmers from dependence on slow and lazy oxen, easing both transport and agriculture. Combined with other key advances—the heavy plow and later the nailed iron horseshoe—the harness drove an expansion of cultivated land.

These sorts of memes—nuts and bolts, practical technologies—are more durable than those generated by, say, Sophocles, most of whose plays were lost forever. There are several reasons. One is gut-level utility; literature is nice, but putting food on the table is nicer. A related reason is the ease with which practical technologies cross cultural and linguistic borders. Medieval Europeans didn't speak Greek, much less read it, so copies of *Antigone* would not have been in great demand even among unusually literary peasants. An iron horseshoe, on the other hand, speaks the universal language of utility.

The final reason that practical memes are so durable is that if they die they can be reincarnated. No one will ever write one of Sophocles' lost plays. But if the conceiver of the horseshoe had perished right after his or her epiphany, someone else would have stumbled onto the idea eventually.

The point isn't that any one useful idea is, strictly speaking, *certain* to spread, or *certain* to be reborn if extinguished. The point is that, the more useful the idea, the more likely both spreading and rebirth are. And as the spread of useful ideas raises the world's population, and raises intellectual synergy via improved communication and transport, these likelihoods grow all the more, until finally they do approach certainty. Increasingly, societies resemble large, thick brains, their neurons spreading incremental innovation rapidly and reliably, spurring further innovation.

Today this vast interconnectedness, on a global scale, is obvious. But even in the early Middle Ages, all of Eurasia and northern Africa had begun to constitute a single data-processing system. A slow system, yes, especially when trade would fall off after political dislocation—but a big system. The iron horseshoe and the windpipe-friendly harness seem to have been invented in Asia and then to have leapt from person to person to person—maybe hitching a ride with nomads for a time—all the way to the Atlantic Ocean.

One key to the resilience of this giant multicultural brain is its multiculturalness. No one culture is in charge, so no one culture controls the memes (though some try in vain). This decentralization makes epic social setbacks of reliably limited duration; the system is "fault-tolerant," as computer engineers say. While Europe fell into its slough of despond, Byzantium and southern China stayed standing, India had ups and downs, and the newborn Islamic civilization flourished. These cultures performed two key services: inventing neat new things that would eventually spread into Europe (the spinning wheel probably arose somewhere in the Orient); and conserving useful old things that were now scarce in Europe (the astrolabe, a Greek invention, came to Europe via Islam, as did Ptolemy's astronomy—which, though ultimately wrong, worked for navigational purposes). To an observer in Italy or France in A.D. 650, it might have seemed as if there was what we would now call a "total system failure"—as if the whole world's hard drive had crashed. But from a global perspective there was no cause for alarm, because the world makes backup copies. Useful memes replicate themselves en masse, insuring the planet against regional crashes.

Given this indomitability of technological evolution, it follows that cultural evolution more broadly—growth in the degree and scope of social complexity, and of non-zero-sumness—is similarly hard to stop.

If, that is, this social evolution depends fundamentally on technological evolution, and not on the chance preservation of particular works of literature or poetry or philosophy. In the European Middle Ages, as we'll now see, that seems to have been the case. Even stereotypically "western" features, such as the blossoming of personal liberty after centuries of serfdom, are in essence byproducts of technology.

AN ENERGY REVOLUTION

The revolution in agricultural technology—plow, harness, horseshoe—slowly but surely raised Europe's population, making its social brain larger. The result was more and more indigenous innovation—or, often, indigenous refinement of foreign innovations. As usual in cultural evolution (and for that matter in biological evolution), the most important innovations were of three kinds: energy technologies, information technologies, and materials technologies.

Horses aside, medieval Europe's key energy technology was the waterwheel. It had existed since Roman times, but the Romans used it rarely, and only to mill grain. Europeans improved the wheel and expanded its use: preparing malt for beer, crushing ore, pumping a blast furnace's bellows, forging iron, driving saws. Perhaps most important, by the eleventh century, mills were used to full cloth. This textile technology spread across Europe over the next two centuries, and finally provoked French fullers to violently protest the waterwheel's job-killing potential. This, a pre-Luddite Luddite protest, was an early harbinger of the industrial revolution, a sign that energy technologies were raising productivity. In the twelfth century Europeans invented the vertical windmill—perhaps having heard of the horizontal variety already used in the east—and it won favor on the plains of northern Europe, where waterwheels had a habit of freezing up in winter.

The textile business also profited from a new loom, perhaps descended from the Chinese silk loom, but now powered by pedals that freed the weaver's hands for subtler work. By the eleventh century, Flanders had started specializing in wool fabrics, and by the twelfth there were silk and cotton centers in Italy. Cotton was ginned with a device picked up from Arabs, who had picked it up from Indians.

Increasingly, making goods wasn't just wintertime work for farm families, but a calling in itself. In an Arthurian romance composed in the twelfth century, a character looks upon a town and sees: "This man is making helmets, this one mailed coats; another makes saddles, and another shields. One man manufactures bridles, another spurs. Some polish sword blades, others full cloth, and some are dyers. Some prick the fabrics and others clip them, and these here are melting gold and silver. They make rich and lovely pieces: cups, drinking vessels, and bowls, and jewels worked in with enamels; also rings, belts, and pins."

Pins indeed. The description hints at Adam Smith's famous analysis of the division of labor in a pin factory. Compare this town with the essentially agrarian Europe of a half-millennium earlier, when the division of labor, as summarized by one economic historian, was as follows: There were those who prayed, those who fought, and those who labored in the fields.

CAPITALISM MAKES THE WORLD SAFE FOR ITSELF

The difference between these two Europes lies not just in manufacturing technologies. One could in principle have specialized in polishing sword blades, or dying cloth, two millennia earlier. Why did almost no one do so, and why even during the heyday of imperial Rome does division of labor seem to have been modest compared to the towns of the thirteenth and fourteenth centuries? And why was late medieval *regional* specialization also sharp?

The answer has to do with the fact that, as we've seen, Adam Smith's "invisible hand" depends on an invisible brain. And invisible brains depend on information technology. Not just conspicuous information technologies, like the abacus (though its "rediscovery" around A.D. 1,000, after centuries of European neglect, did help) or writing (though the growing literacy of the Middle Ages helped, too) or money (though the revival of currency was vital). At least as crucial was information technology in a subtler sense: what you might call information metatechnologies—social algorithms guiding the use of such information technologies as money. In particular, what made the later Middle Ages a bridge between ancient times and the industrial revolution was the rudimentary metatechnology of capitalism.

Today we take for granted that the people who start a business and the people who provide the money for it are often not the same people. We have stock markets, limited partnerships, bank loans, and various other ways of turning John's profits into Mary's seed money. But this basic idea—efficiently converting savings into investment—had to be invented, and at the time the western Roman Empire fell apart, Europe's machinery for capital formation had been crude. The Middle Ages changed that.

In southern Europe the *contratto di commenda,* in use by the tenth century, allowed investors large and small to underwrite a ship's trading expedition. They were guaranteed in writing a share of "the capital and the profit which God shall have granted." The *commenda* was just one of many medieval tools for linking savings to investment. (By the fourteenth century, Venetian bankers had realized that they could lend out a fraction of their deposits, since depositers were unlikely to all withdraw their cash at once. The rest is European banking history.) All such instruments had this in common: they were distinct steps toward modern capitalism. Europe had long had markets for goods and services. Now it had markets for capital. It also had new ancillary metatechnologies of capitalism, such as insurance and double-entry bookkeeping.

All to the good. But does this really answer the question of why the economic complexity of the late Middle Ages had been so slow in coming? If the missing ingredients were the various metatechnologies of capitalism, then the question becomes: Why had *they* taken so long to evolve? After all, once you've got coins and writing, all that stands between you and capitalism is a little imagination. In principle, there could have been stock exchanges—which finally showed up in the seventeenth century A.D.—in the seventh century B.C., when the Lydians started coining money. Why did nearly two millennia pass before European capitalism got off the ground?

Maybe part of the answer lies in where exactly it did finally get off the ground: not in Byzantium, where ancient government and social structure endured, but in the west, where central rule collapsed, and tumult came and went and came, and authority was often dispersed. The average ancient ruling class, after all, would have blanched at full-fledged capitalism. For ordinary people to have potent ways of turning savings into investment meant a loss of leverage for aristocrats. What ships and grand buildings to build, and what to do with them— more and more of these decisions would now be made diffusely, by diverse investors and entrepreneurs, not centrally by the government and the entrenched elite. What's more, the fruits of investment would now be hard to monopolize.

This points to a general problem faced by ruling classes. To stay strong, a society must adopt new technologies. In particular, it must reap the non-zero-sum fruits they offer. Yet new technologies often redistribute power within societies. (They often do this precisely *because* they raise non-zero-sumness—because they expand the number of people who profit from the system and so wield power within it.) And if there is one opinion common to ruling classes everywhere, it is that power is not in urgent need of redistributing. Hence the Hobson's choice for the governing elite: accept valuable technologies that may erode your power, or resist them so well that you may find yourself with nothing to govern. Maybe it was western Europe's freedom from ancient elites that helped hasten the coming of capitalism.

To be sure, western Europe in the Middle Ages, like ancient Rome, had elites who disliked big power shifts. The instinct of feudal lords was to exploit the emerging class of merchants with tolls at bridges and feudal bounds. But it didn't take merchants long to sense their common interest. They united into guilds and demanded the freedoms necessary for commerce: not just freedom from outrageous tolls, but freedom to buy or sell property, freedom to enter into contracts—and freedom to decide what other freedoms they needed. Increasingly, in the eleventh and twelfth centuries, towns won the right of self-government, complete with their own courts and tax collection. What's more, feudal lords soon realized that local prosperity was good for them, and that prosperity required a bit of this freedom. Some started not just granting charters of self-governance to towns, but founding towns in order to grant the charters.

Why were these ruling elites more open to change than Rome's ruling elites? One reason, some historians say, was the decentralized nature of feudalism. Feudal lords often had the leeway to rewrite the rules in their territory, and they also had the incentive—competition with neighboring lords. As savvy lords tried to foster more prosperity than their neighbors, the many fractal units of feudalism became, in effect, laboratories for non-zero-sumness, competing with each other to raise productivity. This creative tension is what forced rural aristocrats to strike their dangerous bargain with capitalism: fostering the power of the merchant class as an asset in external competition, thus empowering upstart urbanites who might then turn around and challenge aristocratic dominance.

Municipal governance wasn't democratic in a modern sense (and it later became even less so, albeit temporarily). It was government "of the merchants, for the merchants, and by the merchants," as one historian put it. Still, compared to the stultifying class structure beyond the town walls—and compared to aristocratic Roman cities, where merchants might be tolerated but were hardly revered—towns were radically egalitarian. One backward twelfth-century nobleman deplored life in Italian towns, where society granted "honorable positions to young people of inferior station, and even to workers of the vile mechanical arts, whom other peoples bar like the plague from the more respectable and honorable circles." Within the towns, a different view prevailed. One urbanite wrote that "the countryside produces good animals and bad men."

The tension between the urban, more liberal future and the rural, oppressive past would take centuries to work itself out. Freedoms would wax and wane within city walls, and political struggles between the commercial and landed classes would go back and forth. But the good guys won in the end, and in the meantime the displays of inchoate capitalism's might were impressive.

Merchants in various German cities formed the Hanseatic League to subdue pirates, build lighthouses, and otherwise lubricate their livelihood. The league wound up defeating the king of Denmark in war and controlling maritime trade routes. In Italy, cities that had fast become city-states—complete with fighting among themselves—felt their freedoms threatened by the Holy Roman Emperor Frederick I. (One early clue: Frederick took an advocate of urban independence, burned him to death, and scattered his ashes in the Tiber River.) The cities put aside their differences, formed the Lombard League, and fought Frederick until he gave in: they would pay lip service to his supremacy but be free to govern themselves.

Commerce changed rural living, too. Some serfs migrated to towns. ("Town air makes one free" was a German saying.) Others became less serflike, as a money economy came to the countryside; rather than working for their land, they paid rent for it, and some turned a profit by filling urban stomachs. Slavery had faded during the Middle Ages, and now serfdom, the next worst thing to slavery, was fading, too.

The winning of freedom by medieval towns, the quelling of pirates by the Hanseatic League, and the humiliation of Frederick I by the Lombard League (albeit with papal assistance) were all early examples of a process that would continue for centuries and continues today:

capitalism making the world safe for itself. The power of this information metatechnology would time and again prove irresistible.

This pattern in turn fits into a broader and older pattern: non-zero-sum technologies making the world safe for themselves. The clay tokens that evolved into writing in Mesopotamia flourished because they lubricated economic exchange. Writing then persisted largely because it did the same, if at first in a roundabout, bureaucratic way. Both memes—tokens and then writing—empowered their host societies, fueling their own proliferation. The same is true of currencies—first de facto currencies, like Aztec cocoa beans, and then de jure currencies, like coins. They brought social synergy, and the resulting momentum carried them far and wide.

The metatechnology of capitalism then combined currency and writing to unleash unprecedented social power. Nothing so demonstrates this fact as the sudden shifts of political power from country to city during the High Middle Ages. Basically, a low-tech means of realizing positive sums—feudalism, geared to an age of little money, sparse literacy, and broken-down roads—gave way to a high-tech means, radically changing the power structure.

The medieval historian Joseph Strayer once noted "an interesting problem in the history of civilization. If there is steady progress anywhere, it is in the field of technology, and yet this kind of progress seems to have little connection with the stability of society. . . ." But, when you think about it, there is no reason to expect steady technological evolution to translate into the smooth evolution of social structures. Technology, time and again, has changed the balance of power within society. And people tend not to surrender power gracefully.

Marx saw this—that politics has an economic and ultimately a material basis, and that the evolution of technology therefore brings unsettling change. He just had some ill-fated ideas about the details of the process, and its direction.[†]

FREEDOM AND OTHER EFFICIENT TECHNOLOGIES

When medieval burghers carved out some breathing room for themselves, winning the right of self-governance, they were not spurred by the writings of Demosthenes, nor trying to revive their classical western heritage. They were just indulging their instincts for self-interest and collaboration, and embracing a productive information metatechnology: freedom. Freedom to buy and sell, to make contracts, to use one's savings as one sees fit—and the freedom of towns, more broadly, to define and fine-tune these freedoms—all these were fruitful algorithms of governance; they were the political technology that best energized the ascendant economic technology, capitalism.

If the monks who copied Greek classics deserve any credit at all for this expansion of freedom, it is not for copying the classics but rather for nurturing literacy generally. Contracts and records are a capitalist tool, so literacy was as well; notary became a thriving vocation during the Middle Ages. But literacy itself, unlike the Greek classics, was never in danger of disappearing. It was alive and well in the nearby Byzantine and Islamic worlds (and the faraway Indian and Chinese and American worlds), and would surely have engulfed Europe sooner or later, with or without Irish monks.

Nor did Demosthenes' paeans to democracy have much to do with the momentous evolution of national representative democracy. Here, too, the basic story is capitalism making the world safe for itself.

As some historians have noted, the Magna Carta, though rightly regarded as a milestone in democracy's history, looms too large in the popular mind. Issued by King John in 1215 under pressure from his barons, it guaranteed them the right to be consulted about taxation. But this was not so new. The English monarchy had consulted councils of nobles on major decisions for centuries. (And if that tradition was grounded in England's "heritage," the heritage was more likely from quasi-democratic Germanic tribes than from Greek orators or Roman law.) The bigger development was the expanding realm of representation over the following century, as the king's council, the parliament, came to include burgesses, representatives of the towns. As markets distributed economic power more broadly across British society, political power followed. That's how the world works.

Today the legacy of this early concession to capitalism's emerging power is Britain's House of Commons, created in the fourteenth century for burgesses (and also for the knights who administered counties). And the legacy of capitalism's growing power since then is the fact that the other House, the House of Lords, is today nearly powerless.

The Greek and Roman writings that championed freedom and

democracy are wonderful things. The freedom and democracy that dawned on western Europe are also wonderful things. But in the end there is no good reason to attribute the latter to the former.[†]

Indeed, we might just as well attribute the blossoming of freedom and democracy in Europe to the corruption and tyranny of the later Roman Empire. This desertion of Rome's earlier ideals may well have hastened the empire's collapse, bringing the fluidity that allowed the competitive experimentation which fostered capitalism. (Perhaps analogously, it is only after the famous Mayan collapse that archaeologists find evidence in Mayan culture of a "mercantile pragmatism," featuring the mass production of pottery, rising living standards for commoners, and the apparent demise of a theocratic elite in favor of a merchant class.)

The story of the Middle Ages is the story of new technologies of non-zero-sumness restructuring society in their image. Their upshot ran counter to, and ultimately prevailed over, the generic ambitions of ancient imperial regimes. Once the synergistic power of these technologies crystallized in the form of capitalism, they would allow the entire populace—including descendants of slaves and serfs—to play complex non-zero-sum games with people they would never meet. The bounty would not fall evenly across the populace, but it would fall more evenly than bounty in that part of the world was accustomed to falling. Political power would be more widely diffused than any ancient imperial government had been prepared to contemplate.

This basic drama—the aggrandizing instincts of powerful people versus the decentralizing tendencies of technology, especially information technology—would play out again and again. By that I don't just mean that free markets would clash recurrently with old regimes, and would win and finally permeate the world (though that seems to be one story line of the past half-millennium). I mean that new information technologies in general—not just money and writing—very often decentralize power, and this fact is not graciously conceded by the powers that be. Hence a certain amount of history's turbulence, including some in the current era.

Chapter Twelve

THE INSCRUTABLE ORIENT

"A Chinese novel," I said. "That must be rather curious."
"Not as curious as one might be tempted to think," replied
Goethe. "These people think and feel much as we do, and one
soon realizes that one is like them."
—*Goethe's conversations with Eckermann, January 31, 1827*

Western scholars have spent a lot of time puzzling over the East. Why, they ask, was it Europe, not Asia, that launched the industrial revolution? How could China and India and the Near East—all homelands of great ancient empires—be so outclassed, technologically and economically, during the past half-millennium?

Sometimes the answers have focused on spiritual matters. Max Weber touted Europe's "Protestant work ethic" and said India and China had been stymied by "magical traditionalism." Sometimes the answers have focused on politics. According to the theory of "Oriental Despotism," Asian civilizations, from Mesopotamia to China, were often built around large irrigation systems, which invited centralized bureaucratic control, leading to top-heavy governance that continued to stifle initiative for millennia.

But, whatever the various explanations for the pace of Asian development, their upshot tends to be that Asian cultures are strange things. The eminent economic historian David Landes, pondering China's erratic, seemingly futile pattern of technological ups and downs ("almost as though the society were held down by a silk ceiling"), declared it simply "weird."

Of course, weirdness is a relative thing. From the standpoint of nineteenth-century Asia, the industrial revolution, as heralded by menacing European ships, may have seemed weird. And since there were more Asians than Europeans, maybe this is the perspective that should prevail: maybe what needs explaining is not the apparent stagnation of Asia, but rather the oddly explosive advance of Europe.

This, at least, is a frequently drawn conclusion. Etienne Balazs, after pondering China's sluggishness, suggested that the series of events which led to capitalism in Europe—and thus "set in motion the industrialization of the entire world"—may have been "a freak of fortune, one of history's privileged occasions, in this case granted solely to that tiny promontory of Asia, Europe." In much the same spirit, E. L. Jones titled his influential study of economic history *The European Miracle.* As little Europe steamed along the highway of industrial progress, wrote Jones, the bulk of the Eurasian landmass was heading into "a demographic cul-de-sac"; had modernity not been imposed on China, India, and the Ottoman Empire, they would have faced "stagnation at the best, or Malthusian crisis at the worst." As miraculous as Europe's economic revolution was, "comparable development in Asia would have been supermiraculous."

Is that true? Does a look at Asian culture and history reveal indefinite inertia, suggesting that the industrial revolution was a fluke? Or does it show us, rather, that the supposedly inscrutable Orient is actually quite like the Occident—prone to harness new technologies and follow them to deeper and vaster social complexity?

THOSE ZEALOUS MUSLIMS

Landes spent part of his magnum opus *The Wealth and Poverty of Nations* trying to figure out why the westernmost of Oriental cultures, the Islamic civilization of the Middle Ages, had not been destined for industrial greatness. His answer, in part: short time horizons. Whereas Europe's pragmatic medieval Christians coolly pursued "continuing, sustainable profit," the rampaging Muslims were propelled by "fighting zeal" and paused "only for an occasional digestion of conquest and booty."

It's true that many western Europeans pursued profit smartly. In the previous chapter, we saw how some basic elements of capitalism coalesced in Europe during the late Middle Ages—notably the justly celebrated *contratto di commenda,* used to pool capital for trade.

But the idea of the *commenda* may well have come from the Islamic world. Before the *commenda* appeared in Italy in the tenth century, the very same tool, under another name, was used by Muslims as they turned Baghdad and Basra into centers of world commerce, trading goods ranging from paper and ink to panther skins and ostriches. As early as the late eighth century, texts of the Hanafite school, one of four Islamic legal traditions, discuss the *commenda*—and the business partnership, another capital-pooling tool.[†] (At about the same time, checks drafted in Baghdad could be cashed in Morocco, a convenience not offered by European banks until centuries later.) Over the years, Hanafite scholars would again and again defend the legal infrastructure of finance on grounds of "the need of trade" or "the attainment of profit." In this light, Landes's simple dichotomy—that European Christians were moved by sustainable "profit," whereas those zealous Muslims were just "doing God's work"—begins to blur.

Indeed, one of Muhammad's great accomplishments, and one key to Islam's potency, was making the larger world safe for commerce. In the early seventh century, before he started preaching in Mecca, the town's main commercial lubricant was its sacred shrine, the Kabah; violence was forbidden in its vicinity, so otherwise contentious Arab tribes could meet and trade. Muhammad and his successors, metaphorically speaking, expanded that sacred realm across much of the known world. For him—as for other great leaders before and since—waging war turned out to be a way of waging peace.

Of course, during the early Islamic expansion, the war part predominated. In that sense Landes's cartoonish sketch of the Muslim mind has a kind of time-bound truth. But as the Middle Ages progressed, and the Islamic empire grew and crystallized, stretching from Spain across North Africa to Pakistan, its formative mind-set faded. With the trust barrier between distant lands now eroded by a common religion, and communication barriers penetrated by the spread of the Arabic language, this huge swath became a low-friction zone for commerce.[†] Taxation replaced booty as the empire's financial base.

The Muslims, as people are wont to do, retained and refined information technologies that further reduced the friction, including some early algorithms of capitalism, ranging from the *commenda* to basic accounting. (Speaking of algorithms: the word "algorithm" comes from the name of the ninth-century Islamic astronomer and mathematician al-Khwārizmī, who also popularized the term *al jabr*, or *algebra*.) Though tracing the path of medieval memes is tricky, some of these algorithms seem to have reached Europe in time to help usher in the High Middle Ages. It is probably no coincidence that the hotbed of medieval Europe's inchoate capitalism was Italy, with its Mediterranean exposure to Islamic culture.

CHINA'S CAPITALIST TOOLS

Meanwhile, at the other end of Asia, the Chinese were also adept at greasing the wheels of commerce. By the ninth century, if not earlier, tea merchants were using "flying money"—the rough equivalent of traveler's checks—that spared them the risk and burden of lugging copper coins. Eventually, the merchants who issued the checks realized that they could invest some of the deposited money. Thus did the idea of banking dawn on China before it dawned on Europe.

There were other ways to raise capital in China during the European Middle Ages. As many as sixty merchants might together finance the construction of a fleet of ships, then own them collectively. People of more modest means invested in trade expeditions via merchants they knew.

Shipping was vital not just to overseas trade but to commerce within China, thanks to a thick network of rivers and canals that featured fancy locks for handling inclines. Marco Polo, hailing from thirteenth-century Venice, was no stranger to boats, but even he was floored by the traffic on the Yangtze River around the city of I-ching. "I give you my word that I have seen in this city fully five thousand ships at once, all afloat on this river." Actually, Marco Polo's word and a dollar fifty will get you a ride on the subway; he is notorious for exaggeration. But even a more sober source, the historian Jacques Gernet, says that China's internal network of waterways, 50,000 kilometers long, was "traversed by the biggest and most various collection of boats that the world had ever seen."

Farmers shipped fruits and vegetables to urban markets, harvested lumber for shipbuilding, made salable tools, pressed oils for medicines and hair creams. By the late Middle Ages, Chinese peasants, writes the historian Mark Elvin, were "adaptable, rational, profit-oriented, petty entrepreneurs." They don't seem to have found their alleged "magical traditionalism" stultifying. With or without a Protestant work ethic, they worked.

In China, as in Europe, merchants sensed their common interest and formed associations. They never lobbied for commercial freedom as effectively as their western counterparts; Chinese cities didn't become self-governing. But the Sung government, which assumed power in the tenth century, did grasp the value of freer markets, and changed its modus operandi. Rather than control prices and get its share of the pie via requisitioned labor, it let goods flow freely and got its share via sales taxes. Like European leaders and Islamic leaders, Chinese leaders saw the downside of a too-heavy hand.

In China during the Middle Ages, as usual with market economies, the virtues of size showed themselves. There were silk factories with 500 looms and iron factories employing thousands. Near the close of the eleventh century, China was producing 150,000 tons of iron a year, an output that Europe as a whole wouldn't match until 1700.

One driver of economic, technological, and scientific advance was printing. China invented paper and had both wood block printing and movable type before they showed up in the west. They were used largely to spread practical knowledge. Hence such books as *Pictures and Poems on Husbandry and Weaving* and *Mathematics for Daily Use.* Some books came from private publishers, but many had an official air, such as the five-volume *Remedies from the Board of Harmonious Pharmaceutics.* Book titles of the age suggest the dawning of a scientific mind-set and science's natural drift toward specialty: *Treatise on Citrus Fruit,* say, or *Manual of Crabs.*

Whether China of the Middle Ages showed much real impetus toward modern science is still debated. Some say no, insisting that its expertise reached no further than technology. In truth, though, science and technology are inseparable, in two senses. First, because an understanding of the laws of the universe is always at least *implicit* in technology. (Francis Bacon said, "Nature to be commanded must be obeyed.") Second, because, beyond a certain level of technology, the understanding tends to become explicit. Arguably the most profound scientific truth is the second law of thermodynamics, which notes the inexorability of universal chaos and thus (as we'll see in part II) defines the current against which organic evolution and cultural evolution both swim. The earliest statement of the second law came in the nineteenth century from the Frenchman Sadi Carnot, who described himself as a "constructor of steam engines," the vocation that indeed had led him to see the gist of the second law.

By the same token, China's invention of the magnetic compass brought the articulation of laws of polarity and magnetic induction long before these things were discussed in Europe. China of the four-teenth century, Elvin believes, was on the "threshold of systematic experimental investigation of nature."

With or without formal science, China was the technological cen-ter of the world. The Chinese invented gunpowder, and by 1232 an iron bomb known as "Heaven-shaking thunder" was deciding the outcome of battles. By the early 1300s, a water-powered, thirty-two-spindle spinning machine could produce 130 pounds of thread a day, making it "several times cheaper than the women workers it replaces," as one observer noted. It was as advanced as any such machine that Europe would see for more than 300 years.

All told, China's technological base during the Middle Ages was a harbinger of modernity. Printing! The magnetic compass! Bombs! Hair cream!

THE BRINK OF GREATNESS

There was a time, early this century, when medieval Asia's technolog-ical feats were scarcely acknowledged by western historians, and so posed no threat to the standard view of the Orient as constitutionally sedate. But since mid-century, thanks partly to Joseph Needham's landmark *Science and Civilization in China,* such extreme Eurocentri-cism has waned. Today the only reason to argue with Jacques Gernet's verdict—that "the two great civilizations of the 11th to 13th centuries were incontestably those of China and Islam"—would be to quibble over whether Islam, which fell into disarray in the tenth century, really deserves equal billing. Mark Elvin, marveling at the Chinese water-powered spinning machine, has written, "if the line of advance which it represented had been followed a little further then medieval China would have had a true industrial revolution in the production of textiles over four hundred years before the West."

Faced with the spectacle of a world-dominant China in the late Middle Ages, Europe's more persistent cheerleaders have turned it to their advantage. Now that we know how close China came to having its own industrial revolution, its failure to actually have one is all the more inexcusable! So China, once deemed an earnest but dim-witted student, is reclassified as a bright underachiever, but still gets a failing grade. "The mystery lies in China's failure to realize its potential," Landes declares.

Actually, if you look at China after its "brink of greatness" period, the failure to industrialize doesn't look all that mysterious. It is just another example of the caprice of history—the way political decisions and other flukes can alter the course of events for decades or even centuries without reversing the basic direction of cultural evolution as played out globally over the millennia. Before examining the particular roll of the dice that seems to have spelled future stagnation for China, let's briefly examine its precursor: the barbarian onslaught that shook China in the late Middle Ages. This incursion illustrates some of our favorite barbarian themes.

NEW, IMPROVED BARBARIANS

In this case the barbarians were the world-famous Mongols, as in Genghis Khan. As is often the case, the barbarian assault was a tribute to the victim. It was on the strength of Chinese memes that the barbarians encroached on China, conquering the whole nation in the late thirteenth century. The Mongols combined their expert horsemanship with Chinese iron technology and Chinese principles of administration and siege warfare.

The conquest, though disruptive, wasn't fatal. The Mongols, like so many barbarians, preferred inheriting an empire to destroying one. Indeed, once the initial chaos subsided, Mongol rule was in some ways a shot in the arm. The conquest brought political unity to China. In the now-expanded body of low-friction exchange—with a single currency, and a newly extended main artery, the Grand Canal—commerce flourished.

It even flourished beyond China's borders. The Mongols had pushed their western frontier all the way to the Caspian Sea by the time of Genghis Khan's death in 1227, and they later extended it into Turkey and eastern Europe. In the process they knocked off the Abbasid caliphate, the second of the great Islamic dynasties. The expansion was epically bloody. But, like the Muslims before them, the Mongols realized that, once the pillaging is over and you've got an empire to run, peace is a wonderful thing. They kept trade routes safe, and in return for thus lowering the communication and trust barriers, they exacted what has been compared to the modern value-added tax, at around 5 percent.

This was a bargain. Compared to the series of power brokers through which Jewish and Muslim caravans previously passed, the Mongol thoroughfare offered "less risk and lower protective rent," according to the social scientist Janet Abu-Lughod. By the end of the thirteenth century, with the Mongols having brought China into direct contact with Europe, there existed a "world system that . . . had made prosperity pandemic."

The Mongol Empire's transcontinental highway, paired with seagoing trade routes to the south, thus carried Eurasia's invisible brain, and its invisible hand, to new evolutionary heights. They were still primitive organs by modern standards; goods and ideas traveled between east and west without east and west having a clear idea of each other. A global village it wasn't. Still, mutual awareness was now higher than in ancient times, when the Chinese thought cotton coming from the west was fleece from "water sheep," and Romans thought their imported silk grew on silk trees. In the mid-fourteenth century, a papal envoy to China even managed to convince some Europeans that Asia did not, in fact, contain whole nations of monsters (although, he conceded, "there may be an individual monster here and there").

Alas, at around this time, economic decline beset both China and Europe. By Abu-Lughod's reckoning, this is no coincidence, but rather a sign of interdependence. Mutual loss, after all, is the seamy underside of that bright promise of non-zero-sumness, mutual gain; stagnation, like prosperity, can become pandemic.

The fourteenth-century downturn was accentuated by what may have been a more literal Eurasian pandemic. William McNeill has suggested that the bubonic plague began in the interior of China and moved over Mongol trade routes to the Black Sea and then finally, via Mediterranean shipping, spread across Europe. Whether or not the black death did move transcontinentally, it certainly could have, and there lies its value as a metaphor that captures a basic trend of history. As economic and social integration grow in depth and scope, the welfare—the health—of ever more distant peoples becomes correlated. The web of non-zero-sumness expands.

To be sure, many scholars would question Abu-Lughod's claim that the economic fates of Europeans and Chinese had by the fourteenth century grown tightly interwoven. Still, with the web of non-zero-sumness growing larger and thicker, linkage this firm was bound to arrive sooner or later. Even if the eastern and western downturns of the 1300s weren't related, the eastern and western downturns during the great twentieth-century depression were.

This is one irony of globalization. The impetus behind it is strong largely because individual states see that their long-term interest lies in plugging into the system. But when the system hits a downturn, they would be better off if somehow they could magically become less plugged in—temporarily, at least. Modern China survived the early ravages of the 1997 Asian financial crisis in good shape partly because its currency was not easily converted into other currencies. Even so, China stuck to its plans for convertibility, because in the long run you're better off plugged in.

China, of all nations, should know. Its epic mistake—the mistake that got it labeled an underachiever—was unplugging from the system beginning in the late fourteenth century. The consequences of this, a single political decision, have ever since been taken as final proof of some deep anti-modern streak in the Asian character.

CHINA CHICKENS OUT

The unplugging came during the Ming dynasty, which reigned from 1368 to 1644. When the Ming vanquished the Mongols, it was in theory a glorious moment for China: renewed native control after a century of barbarian rule. But since these particular barbarians had been a gateway to global commerce, the truth was more complicated. The Mongols still controlled trade routes in central Asia, but were now less enthusiastic about using them to channel commerce toward China. Besides, the Ming rulers weren't big on free trade. Ever fearful of incursion, they constricted overland commerce in iron, arms, even textiles, lest these things empower barbarian neighbors yet again. As if to provide future historians with a nice symbol for a siege mentality, Ming emperors built the Great Wall of China.

The lack of overland trade needn't, by itself, have been devastating. There was always the sea. But Ming rulers proved touchy about contact not just with barbarians, but with aliens of a manifestly refined variety. Foreign perfume, among other things, was banned at the end of the fourteenth century, and in various other ways, at various times, the Ming made seagoing trade tough.

The irony of this isolationism is that during the early Ming period the Chinese were king of the sea. In 1405 the emperor dispatched a fleet of 317 vessels—nearly twice the size the Spanish Armada would reach—to explore trade routes along southern Eurasia. The voyage (manned, one historian notes, by "a can-do group of eunuchs") was the first of seven expeditions, which over the next three decades would reach India, Africa, and the Persian Gulf. The odysseys were chronicled in books whose titles suggest that the Chinese were feeling their oats: *Marvels Discovered by the Boat Bound for the Galaxy*, for example, and *Treatise on the Barbarian Kingdoms of the Western Oceans*. Yet in 1433, the Ming retreated from big-time sailing, eventually banning the construction of large ships.

The reasons for China's withdrawal are much debated. An anti-eunuch faction of bureaucrats played a role, but most observers agree that something larger was at work, too. The historian John King Fairbank concluded simply that "anti-commercialism and xenophobia won, and China retired from the world scene." Others discern a more rational motive. The historian Peter Perdue sees China consciously shifting resources from one vast project—ocean voyages that had shown little profit—to another vast project: building the Great Wall to keep barbarians at bay. In this view, the move was neither mindlessly xenophobic (centuries of barbarian rampages were no figment of China's imagination) nor anti-modern (the Great Wall was high tech back then).

Whatever the cause for China's withdrawal, the timing was bad. For centuries, China had been a big exporter of good ideas, and western Europe a big importer. Now, just as Europe's social brain was really humming, China opted out of the exchange.

The timing was bad in a second sense, too. Over the next half-century, European nations would embrace sailing big-time, and find the New World. Some scholars believe this stroke of fortune explains why the industrial revolution happened where it did. Europe stumbled onto a trove of precious metals and vast farmlands and hacienda-ready farmers, all just waiting for exploitation. The ensuing enrichment helped finance a burst of technological progress that pushed Europe into the lead during a crucial phase of technological evolution. To believe otherwise—in particular, to think the industrial revolution reflected some intrinsic European brilliance that has only since penetrated the dimmer parts of the world—is to exhibit "a serious malady of the mind" also known as Eurocentrism, writes the geographer J. M. Blaut, a leading proponent of this theory. If Blaut is right, then the rise of the west to world dominance during the eighteenth, nineteenth, and twentieth centuries is essentially the result of a lucky break—geographic proximity to an untapped hemisphere.

To back up his claim, Blaut argues that, as of 1491, Europe was not ahead of the rest of the world in any obvious, across-the-board way. Indeed, the world was already awash in incipient capitalism, or "protocapitalism." Cities "strongly oriented to manufacturing and trade" stretched "around all of the coasts of western Europe, the Mediterranean, East Africa, and South, Southeast, and East Asia."

This is indeed evidence that, had Europe not led the world into industrialization, someone else would have—even supposedly solipsistic pre-modern China, given enough time. Still, that's a long way from saying that Europe's only leg up in the world was a few big ships and a straight shot to the West Indies. There is one other asset Europe had that neither China nor any other empire had: the absence of empire.

Large, unified polities are two-edged swords. On the one hand, they offer big, low-friction zones for trade—an especially valuable thing in ancient times, when marauders often lurked beyond state bounds. But this day-to-day benefit coexists with a long-run liability: imperial governments have often resisted changes that are key to continuing viability amid technological flux. We've already seen this logic at work in medieval Europe, where feudal fragmentation, for all its day-to-day downsides, had the upside of encouraging experimentation with economic and political algorithms.

At the end of the Middle Ages, as monolithic China turned inward, Europe was crystallizing into a land of nation-states, and so its contentious dynamism persisted. By their nature, Europeans were just as capable of formulating self-defeating policies as Ming emperors were. It's just that, in a more immediately competitive environment, someone else was bound to try a better policy, so bad policies came back to haunt you sooner. And once somebody *did* try a good idea, it could spread to competing polities fast by emulation.

Columbus himself illustrates the point. He sought Portuguese financing, but the king of Portugal turned him down—rather as the Chinese government had decided half a century earlier that long westward voyages weren't worthwhile. But there was one difference: Portugal, unlike China, had lots of neighbors that were a stone's throw away. Columbus went to Spain, got support, and came back from the New World triumphant. Within a few years Portugal was playing the discover-America game, too. The "sail westward" meme, having proved its value, proliferated.

As consequential as this meme was, the more important thing, for this book's purposes, is how memes in general exploited the political landscape of Europe. In this hothouse of interstate competition, technologies of energy, of materials, of information—including algorithms of capitalism and of political governance—were bound to keep sprouting and spreading. For example, patent rights, which helped make initiative worthwhile, were granted in Venice in 1474 and diffused to much of Europe by the middle of the next century.

In this light, Europe's eventual triumph is not just *consistent* with a directional theory of cultural evolution; the theory virtually *predicts* such a triumph. After all, the speed of any evolutionary process depends heavily on two factors: how fast potentially fruitful novelties arise, and how fast manifestly fruitful novelties spread. Europe in the fifteenth century, teeming with competitive but mutually communicative polities, scored higher in both categories than any other part of Eurasia. In such a setting the "sail westward" meme—and profitable memes in general—were (a) likely to take root in one polity or another; and (b) having proved their worth, likely to spread.

Can this simple, almost superficial contrast—China's centralized control versus Europe's fragmentation—really explain a difference in economic development that has often been attributed to a deep streak of mystical traditionalism in Asian culture? One way to find out is to look at a part of pre-modern Asia that shared much of China's cultural heritage but lacked its monolithic structure: Japan.

Like China (and via China), Japan was deeply influenced by Buddhism. But unlike Ming China, Japan saw a collapse of centralized rule in the late fifteenth century. The shogun's power seeped out to feudal lords, creating an amorphously competitive environment reminiscent of late medieval Europe. Sure enough, there ensued a familiar pattern: markets flourished, towns expanded (some started commodities exchanges), and the political power of merchants grew. Jesuits who came to Japan in the sixteenth century compared the city of Sakai to Europe's medieval "free cities." Edwin O. Reischauer has described this period as one of "extraordinary cultural innovation, institutional development, and even economic growth" in spite of an atmosphere of "great political confusion and almost constant warfare." (Or, perhaps, *because* of the confusion and warfare?) By the late sixteenth century, "the Japanese, who only a few centuries earlier had been a backward people on the edge of the civilized world, had grown to the point where they were able to compete on terms of equality with the Chinese and also with the Europeans. . . ."

All told, if the key to the "European Miracle" lies in geography, it is not so much Europe's and China's relative proximity to America as it is Europe's and China's *political* geographies. Europe comprised lots of independent laboratories for testing memes, while China possessed political unity—an asset, to be sure, in matters of everyday commerce, but a handicap in any long-run race for technological preeminence.[†]

WELCOME TO THE NEIGHBORHOOD!

A number of scholars have acknowledged that Europe's broken political landscape played a role in its rapid advance. For some of them, such as David Landes, this is among the reasons to doubt that China, left to its own devices, would ever have reached the industrial age.

They are missing a key point. This European advantage—being a neighborhood of competitive laboratories—was an advantage of degree only. All nations have *some* relatively robust neighbors within *some* proximity. China had Japan, among others. That's why no government can countenance stagnation forever without facing the consequences. Even the much-maligned Ming dynasty periodically felt the need to flirt anew with international trade (which it had never quite stifled anyway, thanks to the enterprise of Chinese and Japanese smugglers). And, though technological advance slowed to a crawl during much of the Ming and Manchu periods, it didn't stop—and the economy continued to grow.

Not only do all states have some competitors within their neighborhoods; *the number of those competitors grows inexorably*. The reason is that, as the means of transport and communication advance, the size of a "neighborhood" grows. That is what China and Japan had begun to learn by the sixteenth century, but were taught with special force during the nineteenth century, when westerners in gunships showed up and demanded access to Asian markets: Europe and Asia were now in the same neighborhood.

Such jarring encounters can incite a nativist reaction. At the turn of the twentieth century, China's Boxer Rebellion provided a fine metaphor for the illusions that nourish such reactions; it was inspired by a cult whose rituals were thought to render members impervious to western bullets. This thesis was abandoned in the face of painful evidence, as was the larger thesis of imperviousness to western influence. Witness China—and the rest of Asia—today.

Of course, in the view of Landes and other champions of Europe's greatness, to do what China is doing today—cloning western technologies and economic principles—is cheating. The question, as they want to pose it, is whether China would have industrialized on its own—no outside help allowed.

It would be tempting to answer yes if the question didn't hover so close to meaninglessness. Could France have industrialized "on its own"—without using steam engines made in Britain? Could Britain have developed the steam engines had not a Frenchman earlier shown that steam could move a piston? Could either France or Britain have reached the verge of steam power without first absorbing capitalist algorithms from Italy, which in turn seems to have gotten some of them from Islamic civilization? The answer to these questions isn't no; it's yes, *but*—but it would have taken them longer to reach these milestones, because they would have been laboring under the handicap that afflicted China.

The point is just that western economic historians have stacked the deck. They habitually compare progress in China to that in Europe, although China is just one polity and Europe is a synergistic cauldron of them. If China is going to be pitted against Europe in the game of hypothetical economic development, shouldn't it at least be allowed to team up with Japan?

Apparently not. Landes speculates that the Japanese, but not the Chinese, would "sooner or later have made their own" industrial revolution, even without European contact. Let's play out this thought experiment. If the Japanese had indeed done so, then China would have adopted Japan's industrial technology—either willingly or under coercion of one sort or another. Indeed, high-tech Japanese aggression was eventually a spur to Chinese modernization. But in Landes's calculation, apparently, China doesn't get credit for such derivative development. Then how, in his calculation, does Japan get credit for being in a position to industrialize in the first place? It got there largely with technology from China, after all—beginning with writing.

For that matter, how does Europe get credit for being in a position to industrialize in the first place? In the early seventeenth century, amid the faint stirrings of the industrial revolution, Francis Bacon singled out three technologies that had "changed the appearance and state of the whole world": printing, gunpowder, and the magnetic compass. We now know that all three were first invented in China, and it may be that all three diffused to Europe. (And paper became affordable in Europe thanks to waterwheel paper mills, which seem to have been an Islamic contribution, first appearing in Baghdad.)

Throughout history, cultural evolution has transcended political bounds. Ideas have swished back and forth across continents as the centers of innovation shifted. The question of what individual nations can do "on their own" is more or less pointless—and the fact that economic historians often ask it of Asian nations, but rarely of European ones, doesn't do much to shore up its utility.

Once you view China and Japan as part of a larger east Asian brain, some noted examples of China's forgetfulness during the listless Ming period lose their force. It's true, for example, that China's encyclopedic *Exploitation of the Works of Nature,* written in 1637, was destroyed (perhaps because of the author's political views). But by then there were Japanese editions, which survived. Once again, the world makes backup copies.

ZEN AND THE ART OF COMMERCIAL EXPLOITATION

A Chinese magnum opus on how to "exploit" nature is rather at odds with the stereotype that past Asian cultures, unlike past western ones, abided in harmony with their habitats. According to this view, as summarized by one proponent, the western belief that nature exists "to be manipulated and enjoyed" insofar as technology permits is "exceptional in human history, and is mainly derived from the anthropocentric philosophies of Judeo-Christian religions." Tell that to the Chinese who during the T'ang dynasty of the early Middle Ages wiped out the forests of North China to create some lebensraum.

The idea that westerners exploit nature while easterners commune with it is akin to a larger fallacy that has long haunted the study of economic history: religious determinism. In the standard version, western religion—whether the "Judeo-Christian ethic" or the narrower "Protestant work ethic"—explains why full-blown capitalism and industrialization first appeared in the west. While Christians are putting in a good day's work, Buddhists sit under trees.

It's true that Buddhist doctrine, as laid out by the Buddha, doesn't sound like fuel for material acquisition. Then again, the teachings of Jesus Christ aren't a capitalist manifesto, either. But religious doctrines evolve. China's first pawn shops were run by Buddhist monks. In seventeenth-century Japan, a Buddhist monk advised that "All occupations are Buddhist practice; through work we are able to attain Buddhahood"—an utterance that has rightly drawn comparison with the Protestant work ethic. Meanwhile, over in the cities of Mughal India, purportedly otherworldly Hindus were, as the historian Paul Kennedy has put it, "excellent examples of Weber's Protestant ethic."

Confucianism, unlike Buddhism, treated the profit motive with suspicion well into the Middle Ages. Then again, as Kennedy has noted, in that regard it is reminiscent of the pope's condemnations of usury during the Middle Ages. But this papal doctrine adapted to commercial exigencies, and so would Confucianism. In the thirteenth century, when Eurasia was spanned by a commercial web of unprecedented density, commerce was abetted by Confucians and for that matter by Christians, Muslims, Buddhists, and Zoroastrians—a transcontinental patchwork of spiritual traditions, all with one thing in common: their adherents were human beings, and thus liked mutually profitable exchange. Economic ecumenicalism.

Religions don't always adjust to the dictates of economic growth. In the short run, their attitudes toward technology—including, sometimes, a professed abhorrence of it—can matter greatly. But in the long run—over centuries, not decades—religions either make their peace with encroaching economic and technological reality or fade into obscurity. The Old Order Amish are admirable in their principled refusal to adopt modern technology, but they are not the wave of the future.

MEANWHILE, BACK IN ISLAM . . .

By the eve of the industrial revolution, prospects that the Islamic states of western and southern Asia might host it were dim. The Ottoman Empire had flourished for a time—not just on the strength of conquest, but also by restoring security to international trade routes and charging for the service. But the regime grew oppressive. The printing press was banned lest dissidence arise, and justice was warped by bribery. In short: the empire not only failed to lower the two great barriers to non-zero-sum interaction—the communications barrier and the trust barrier—but actually raised them. Some Ottoman rulers compounded matters by resisting cultural input from the west, and by the time this policy was clearly reversed, in the nineteenth century, it was too late.

Mughal India, born in the early sixteenth century, showed promise for a time, with robust commerce and an advanced banking system. And one of its earliest rulers, Akbar, faced with the prospect of governing lots of Hindus, announced that respect for all religions was now the will of Allah. But the later empire saw institutionalized discrimination, the razing of Hindu temples, and a self-indulgent ruling class. When a local prince yawned, according to one contemporary observer, "all present must snap their fingers to discourage flies."

Both of these Islamic states, having failed to nourish economic and technological growth, paid the price. The Mughal empire expired in the eighteenth century, the Ottoman in the twentieth. Why did they fail to thrive? Theories abound, but, obviously, we see some familiar culprits: parasitic governance and an oppression that left much non-zero-sumness untapped. India, with its caste system, is a famously vivid example.

Maybe, then, the world should be thankful that neither of these empires survived to become a much-emulated model. When regimes that ban printing presses and mandate bias are given the thumbs down, we can only compliment history on its judgment.

At the same time, it's worth noting that these civilizations gave much to the world's river of memes. We've already noted a few of Islam's great gifts; and both Turkey and India, well before Islamic governance, made their own bequests. Around four millennia ago, Turkey—Anatolia, back then—seems to have given Eurasia the idea of smelting iron, the substance that would finally give form to the industrial revolution. And India has been the epicenter of great innovation in two ethereal realms: spiritual and mathematical thought.

In the spiritual realm, India gave us Buddhism, the first major religion to stress tolerance and nonviolence, the only major religion to spread far and wide without conquest, and arguably the major religion whose founding doctrines (unembellished by later additions) most readily survive the modifying force of modern science. In mathematics, India gave Europe, among other things, the concept of zero and the decimal number system, including the numerals that are misleadingly called "Arabic." (If you don't think they were a big advance, try doing some multiplication using Roman numerals.)

India in the late twentieth century began to reclaim its mathematical legacy, not just with deep contributions to theory, but in such practical matters as software design. This is a reminder—as are the "Japanese miracle" and later economic leaps in Asia—that leadership in cultural evolution is fleeting. All large expanses of Eurasia have had their days in the front of the pack, and their days closer to the rear. The story isn't over yet.

If we relax our focus—ask not *which* of the Eurasian cultures is leading the pack, but whether the front of the pack is advancing, whether the cutting edge of Eurasian culture, and thus of social complexity, is moving forward—the answer is that there have been few if any times since the initial closing of the "Eurasian ecumene" in the first century A.D. when the answer wasn't yes. The peoples and states come and go, rise and fall, but the memes keep flowing upward.

There are phases in cultural evolution that are by their nature big leaps; a technology, or a constellation of them, proves so explosive that its lucky hosts suddenly seem light-years ahead of other cultures. Those laggard cultures, indeed, look so pathetic that it is tempting to ask whether there isn't some qualitative difference at work, some special something that rendered the "leading" culture, and it alone, capable of crossing the technological threshold.

Surely such flattering interpretations accompanied encounters ten millennia ago between agricultural societies and hunter-gatherer societies, or, five millennia ago, between literate, state-level societies and illiterate chiefdoms. In those cases we have hard evidence that the interpretation is wrong, the thesis of exceptionalism a self-indulgent illusion: we now know that both farming (and chiefdoms) and writing (and state-level societies) appeared independently, multiple times.

In the case of the industrial revolution, evidence of this sort is unattainable. A revolution at this technological level occurs in an age when a global brain is taking shape, and news can travel around the planet in months. So any subsequent episodes of industrialization—such as Japan's less than a century after Europe's—are necessariliy derivative. Thus the thesis of European exceptionalism can never be conclusively disproved. Still, it is hard to take that thesis seriously if you are aware of the tens of millennia of cultural evolution that led to the industrial revolution; aware of the diffuse power evinced by that evolution at every turn; aware that what creates great technological change isn't so much great cultures as the greatness of culture itself.

One key to culture's greatness is its indifference to local politics. China during the Ming period fails to fulfill the promise of the Sung—but that's okay: there's always Japan; there's also England, France, Italy, India, Egypt, and so on. All these societies had their ups and downs—because of the vagaries of history, the luck of the geographic draw, the greatness or abjectness of political leaders. But through it all, probability overwhelmingly favored progress. If stagnation or regression beset one place, there were always other places.

Isaiah Berlin once wrote a book called *Historical Inevitability* that was devoted to debunking virtually all grand theories of history, certainly including directional theories. Among its errors was imputing to such theories a strict determinism—a belief that every detail of the future could in principle be predicted, that every detail of the present exists by necessity. To read his book, you would think that all who see pattern in history believe, as he put it, that "everything that we do and suffer is part of a fixed pattern."

The truth is more nearly the opposite. The key to the pattern of history isn't the fixedness of everything that people do. The key is the pattern's long-run *imperviousness* to the *lack* of fixedness. A Ming ruler, perhaps out of sheer caprice, rescinds China's oceanic voyages, and the most sophisticated nation on earth turns inward—yet the big picture remains unchanged: globalization and the information age, with all their political import, are in the cards.

Chapter Thirteen

MODERN TIMES

As if to offer proof that God has chosen us to accomplish a special mission, there was invented in our land a marvelous new and subtle art, the art of printing.

—*Johann Sleidan (1542)*

Religion in 1600 presents a grim face. . . . Prickly, defensive, gladiatorial debates were conducted endlessly between theologians who seldom if ever met and who wrote in prose intelligible only to their own kind. Lutheran, Calvinist, and Catholic armies were poised within a few decades to lacerate each other on the battlefields of central Europe.

—*Euan Cameron*

In the eighteenth century, Voltaire described the Holy Roman Empire as "neither holy, nor Roman, nor an empire." This was not the first time the empire had failed to live up to its billing. Since the day in A.D. 800 when the pope crowned Charlemagne, emperors had sometimes differed with popes over who was in charge, and the ensuing dustups typically left neither party looking very holy. And, holiness aside, the emperor's authority was often less than imperial, thanks to uppity feudal lords. By Voltaire's time, what clout the emperor did possess was confined to German lands. In the early years of the following century, the empire expired altogether.

This raises a question. Why, after Europe's recovery from the disarray of the early Middle Ages, did empire never lastingly return? After all, the prosperity of the late Middle Ages restored the logic behind large-scale government. With commerce once again a long-distance affair, commensurately broad governance could ease and protect it. Besides, with war as popular as ever, big states should in theory have swallowed little states ad nauseam. Also favoring large polities was that eternal force of history, the egos of rulers. Yet all imperial designs—the Hapsburgs', Napoleon's—ultimately failed. And it wasn't just *western* Europe that seems to have grown resistant to lasting conquest. Southeast Europe meanwhile freed itself from the long-redoubtable Ottoman Empire. More recently, Russia's imperial dominance of eastern Europe lasted a mere half-century.

Obviously, as the previous chapter suggested, the absence of lasting empire in recent centuries has something to do with Europe's being a crazy quilt of different languages. But that can't be the whole story. The eastern half of the Roman Empire had been quite multilingual, but it endured imperial rule for centuries. By Voltaire's day, something had changed, making linguistic boundaries more formidable. The political model that had prevailed across much of Eurasia in A.D. 100—large, multilingual empires—was becoming an endangered species in Europe. It would also falter in the Near East and, eventually, in much of Asia. Why?

To try to account for such a complex and momentous change with a single explanation would be foolhardy. But let's try anyway: the printing press. At least, let's examine the possibility that the press, more than any other factor, rendered vast multilingual empires unwieldy to the point of being unworkable.

There are other reasons to spend a chapter dwelling on the press. Not for nothing have some historians used Gutenberg's first movable-type press, circa 1450, as the official starting point of Europe's "modern era," that half-millennium of brisk change that got us where we are today. The printing press helped overhaul religious thought and ushered in both the scientific and industrial revolutions. In so doing, it hastened the coming of other world-changing information technologies—telegraph, telephone, computer, Internet. In 1450, most Europeans would have laughed at the notion of a single, intricately woven global civilization (and perhaps at the notion of a globe). Yet already they possessed the basic machinery for creating this world.

The press did more than pave the way for the information technologies that today are revolutionizing life; it foreshadowed them. In its specific, sometimes paradoxical effects, the print revolution parallels the latest phase of the microelectronics revolution. Indeed, there is no better historical preparation for thinking about how the Internet will reshape political and social life than seeing how the printing press reshaped them. The late modern era—today—is in many ways the early modern era, only more so.

TECHNOLOGIES OF PROTEST

When history books note the importance of the printing press, they often emphasize its role as a purveyor of pure, clear knowledge. By stressing the enlightening aspects of the press, they are buying into the view of human progress that prevailed in the eighteenth century (known, not coincidentally, as an enlightenment-era view). In this view, the source of history's directionality is intellectual advance—scientific, technical, political, moral. Over time, people build better machines, better governments, better societies, better moral codes; they rationally discern the good and rationally achieve it. Condorcet, the iconic enlightenment-era progressivist, envisioned a time when elitism and prejudice would dissolve in a sea of virtue and wisdom. (Now *that* would be progress.)

Obviously, intellectual, even moral, progress does happen—and the printing press did its share to further them. But, as we've seen, information technologies are instruments not just of enlightenment but of power. In ancient times, writing had bestowed power; when literacy spread, power spread beyond a tiny elite. With the printing press, and its proliferation, came another episode in the decentralization of power. And, though this power sometimes rested on enlightenment, there was no necessary connection between the two.

One way to see this distinction is to look at the press's role in the Protestant Reformation. There are two basic stories about how the printing press fostered the Reformation. The first is that it brought Bibles within reach of laypeople, allowing them to get their religious instruction from the source and thus form their own opinions about church doctrine, with no coaching from the pope. This story is especially popular among Protestants, and there is some truth to it.

But the more generally important story is the one hidden in the word "Protestant." The printing press lubricated protest. It did so by lowering the cost of reaching and mobilizing a large audience. Before the invention of printing, publishing en masse had been hard unless you could afford the upkeep on, say, a few dozen monasteries full of scribes. (For a student in Lombardy during the fifteenth century, just before the coming of movable type, the price of a law book was more than a year's living costs.) Now, with printing cheap, an eloquent agitator with a catchy idea could occupy center stage.

Martin Luther, a theologian of modest prominence, affixed his critique of Catholic doctrine to the door of Wittenberg's All Saints Church on October 31, 1517, and within weeks three separate editions were rolling off the presses in three cities. A sixteenth-century writer observed: "It almost appeared as if the angels themselves had been their messengers and brought them before the eyes of all the people." Luther expressed shock at the sudden currency of his thought and agreed that the new technology had the earmarks of divinity; printing was "God's highest and extremest act of grace, whereby the business of the Gospel is driven forward."

Of course, the pope had a different view on whether the Gospel, rightly perceived, was being driven forward. And here we see the problem with stressing the "enlightening" aspects of the press. Given the subjective nature of theological judgment, it will forever be debatable whether the press, in promulgating Luther's theses, was furthering human knowledge. What is not debatable is that the press was sending signals that aroused and helped organize a particular community of interest.

The same distinction holds today. When lobbyists use a more recent technological advance—computerized mass mailing—to target a narrow interest group, the mailings may or may not contain truth; often, in fact, they exaggerate the threat that this or that policy poses to the audience. Nonetheless, they succeed in mobilizing the audience, getting it to cough up donations, or to fax senators, or whatever. These mass mailings—fact or argument, true or false—are signals that give energy and cohesion and thus power to a community of interest that might otherwise be amorphous and powerless.

An information technology constitutes, among other things, a nervous system for social organisms—organizations of clergy, say, or organizations of heretics. The better the nervous system, the more agile the organism. For centuries before Luther, as one scholar has observed, the church hierarchy had "easily won every war against heresy in western Europe because it always had better internal lines of communication than its challengers." The press changed that, chipping away at the pope's spiritual authority.

By the same logic, the press chipped away at secular authority. In fact, the two forms of rebellion sometimes fused. In 1524, German peasants revolted, demanding an end to serfdom. Some rebel leaders had been inspired by Luther's teachings, including Lutheran pamphlets that held up the earnest, hardworking peasant as symbol for the simple purity of ideal Christian life. The peasants also emulated Luther's use of the press, publishing a list of twelve grievances.

As it happened, their hero let them down, siding with the ruling class. To argue against serfdom, Luther wrote, was "dead against the Gospel." After all, "Did not Abraham and other patriarchs and prophets have slaves?"

Still, try as Luther might to confine his radicalism to theology, the cleavages within Christendom that the Reformation revealed would time and again turn out to coincide with political fault lines. The "wars of religion" that racked Europe in the late sixteenth and early seventeenth centuries were also wars of politics. In the Netherlands, Calvinists fought to loosen the yoke of their distant and oppressive Catholic ruler Philip II, the Hapsburg king of Spain. In various German states, Protestants struggled for states' rights against the Holy Roman Emperor. By the mid-seventeenth century, the Netherlands was free from Hapsburg control, and the Holy Roman Empire was effectively dead. A primary cause was the centrifugal force of the printing press. The press mobilized religious dissent and political dissent, and often the two worked in synergy.

Still, to call the press a wholly fragmenting, decentralizing force would be to oversimplify. Instruments of efficient communication are tools for mobilizing groups that have something in common—a political aspiration, a religious belief, a language, whatever. If the commonality implies opposition to a central authority, as it did for Luther's followers, the result can be fragmentation, or at least a diffusion of power. But if the group's common bond stretches across existing boundaries, bridging prior chasms, the effect can be to glue fragments together, to aggregate power.

A good example is that mixed blessing of the modern age, the shared national sentiment that—especially in its more intense, self-conscious forms—is known as nationalism.[†] This sinewy sentiment, if used deftly by a politician, can erode the power of local rulers, expanding authority. The result—a centrally governed and culturally coherent region bound by a sense of shared heritage, shared interest, and shared destiny—is the nation-state. We take nation-states for granted today, but they didn't always exist. To understand the printing press's role in their evolution, we need to first understand the forces that were encouraging this evolution well before the press arrived on the European scene.

PESKY NOBLES

The roots of the European nation-state can be seen at least as far back as the twelfth century, in the ongoing economic recovery from the "dark ages." Though towns had gained a measure of freedom from rural dominance, liberating commerce locally, the landscape was still strewn with obstacles to long-distance exchange. The main problem was the motliness of feudal government. Laws and regulations differed from place to place, jurisdictions overlapped, and disputes, even battles, cropped up between neighboring lords, or between town and country.

In this atmosphere, a monarch could win gratitude by performing the public service of harmonizing law and settling disputes. Indeed, by bringing trust and predictability to the system, he could unlock enough economic non-zero-sumness to pay his salary in the form of taxes. Describing the general drift toward centralization in the late medieval and early modern period, E. L. Jones has written, "In return for a lion's share of the small productive surplus, the ruler supplied justice." As early as the end of the twelfth century, this implicit social contract between monarchs and locals had laid the rudiments of England's modern justice system and had brought new concord and homogeneity to France, paving the way for the peaceful, stable prosperity of the thirteenth century.

In essence, what was happening was the oldest story in the book: as commerce expands, governance follows, clearing the way for smoother and stronger commerce. The fourteenth century would bring various disruptions—notably the plague—but in the fifteenth century this trend toward national political organization would resume.

History textbooks often make this second phase in the centralization of rule sound abrupt—the coming of the "New Monarchs." Indeed, the change is sometimes attributed largely to a single technological development. With the big cast-iron cannons of the fifteenth century, "great men could no longer brave the challenges of their rulers from behind the walls of their castles," as one single-volume world history puts it.

Any story repeated by as many historians as the cannon story must hold some truth, but as usually told it obscures as much as it reveals. In particular, it downplays the question of why kings could afford so many cannons, and so many troops to accompany them. The answer is that kings taxed the townspeople. And why were the townspeople willing to pay taxes that would be used to blow away nobles? Because the nobles were among the aforementioned obstacles to commerce; they engaged in marauding and petty warfare that complicated the life of people who wanted to stick with the business of business. As the historians R. R. Palmer and Joel Colton put it: "The middle class was willing to pay taxes in return for peace." By waging war against the reactionary nobility, the monarchs were waging peace—one of their obligations under their implied contract with taxpaying urban merchants.

This is the real story behind the centralization of national rule in the fifteenth and sixteenth centuries: commerce demanded nation-wide harmony, and subsidized the extinction of impediments to it. The particular instruments of extinction—the cannons—are largely beside the point. If they hadn't existed, the merchants' money would have gone to buy some other form of military might that would have spelled equally certain doom for the pesky nobles. The trouble with the textbook "cannon story" is that the textbooks never ask: Wait a second—why couldn't the nobles have just bought enough cannons to blow away the *king*'s castle? A large part of the answer is that the king had commerce—non-zero-sumness—on his side, and the nobles didn't. In the great non-zero-sum games of history, if you're part of the problem, you'll likely be a victim of the solution.

NATIONAL SINEWS

The press reinforced the drive toward national rule in two ways. First, it unified the cultural base of large swaths of land, standardizing custom and mythology and, above all, language. In the late Middle Ages, the scholar Adam Watson has written, "one dialect shaded almost imperceptibly into the next, the Romance languages from the Low Countries to Portugal and Sicily and the Germanic from Holland to Vienna." The press changed that, tamping down dialectical differences, creating large blocks of mutual intelligibility—"unified fields of exchange and communication," as the political scientist Benedict Anderson has called them.

Second, the press began to foster a kind of day-to-day national consciousness. By the early 1500s, single-topic "news pamphlets" were harmonizing English sentiment, reporting on battles, disasters, celebrations. In the ensuing centuries, as journals and true newspapers evolved, the printing press would give more and more fiber to national feeling. Whole states would become, in Anderson's terminology, "imagined communities."

The symbiotic development of printing and nationhood differed from place to place. For the French and the English, it was in the context of an already distinct and growing national organization that the press made its mark. Among central European peoples—the Italians, the Germans—national rule wouldn't arrive until the nineteenth century, so the press, and strong national sentiment, didn't just consolidate the nation-state, but paved the way for it.

Even here, within central Europe, different nations followed somewhat different paths. The German states, their leaders reluctant to surrender local sovereignty, were united by Bismarck via war and intimidation. Italian unification, while not wholly peaceful, was closer to a voluntary conglomeration. But whatever the route, the printing press figured crucially—via newspapers (which had proliferated wildly during the eighteenth century); polemical journals (such as Joseph Mazzini's *Young Italy*); popular books (*Grimm's Fairy Tales* created a national German folklore); and weighty tomes (especially German ones romanticizing the idea of a unique national character, a *Volksgeist*—an idea that found receptive audiences elsewhere in Europe).

In no nation did the printing press *cause* national governance. Amid economic recovery and expansion, the logical scope of governance was growing with or without the press. But the press strengthened the logic and gave nationhood a particular cast—a coherence, a basis in shared language, culture, and feeling. And this nearly tangible unity, in turn, made nations naturally formidable units, resistant to conquest. In a sense, the printing press was less important in carrying centralized governance up to the national level than in stopping it there—keeping it from rising higher, to the level of empire.

We've already noted one early example of this dynamic, when the Calvinists of the Netherlands—the "United Provinces"—successfully resisted Hapsburg dominance in the late sixteenth and early seven-

teenth centuries. This would hardly be the last example, either for the
Hapsburgs or for other aspiring imperialists. In the early 1800s, in the
wake of the French Revolution, Napoleon tried to use the now-firm
French nationalism as a base for empire-building. It worked for a
while, but ultimately he foundered on, among other things, nationalist resistance in places such as Spain and the German states.

In addition to the technological obstacle to European imperialism
posed by the press, there were (as sometimes happens in history!) historical factors. During the Hapsburgs' bid for greatness, the nearby
Ottomans, fearing a rival empire, aided recalcitrant states. For that
matter, even if we confine our gaze to the realm of technology, the
press was not the only thing that bolstered nation-states. The burgeoning roads of the late Middle Ages were conduits, however informal,
for news and other data. They would become better ones as national
postal services evolved during the fifteenth and sixteenth centuries.

Still, even taking account of these and the many other factors that
helped invigorate the nation-state, the printing press stands alone.
More than any other single thing, it accounts for the basic, oft-noted
irony of the modern age in western Europe: political power migrated
both upward and downward. Previously, European governance and
cultural identity had tended to be local, a vestige of feudalism—and
when lines of allegiance or authority did go further in scope, as with
the pope or the Holy Roman Emperor, they were often at the other
end of the spectrum, spanning the continent. But with the modern
age, tiny polities became less and less economical, and vast monoliths
became less and less tenable. The nation-state, with a cultural and
political integrity crystallized by the press, emerged at the expense of
both.

THE LOGIC OF PARADOX

This simultaneous upward and downward migration of power is
sometimes called a "paradox," but on close inspection, it isn't truly
contradictory. Whether dividing or uniting, the printing press was
often elevating coherence.

The empires it helped break up were in some sense[†] arbitrary
empires—ungainly expanses encompassing sharply different cultures
and languages. Meanwhile, when the press helped fuse small polities,
the borders it erased were in many cases artificial and dysfunctional—
impediments to economic efficiency. These impediments were espe-

cially costly when affinities of culture and language would otherwise have allowed smooth concourse by keeping the two barriers to non-zero-sumness—the trust barrier and the communication barrier—quite low. Witness the economic potential unleashed in Germany after Bismarck united it.

These two effects of nation-state formation—breaking up the arbitrarily united and merging the dysfunctionally divided—sometimes happened in one fell swoop. The Italian states of Lombardy and Venetia, incongruously part of the Austrian empire for much of the 1800s, broke off and fused with their closer relatives as part of Italy's emergence in the second half of the century.

The process of nation-state formation has taken awhile to play itself out. Only toward the end of the twentieth century does the idea of the vast multinational empire seem finally to be giving up the ghost around the world. But note how, even back in the nineteenth century, empires increasingly confined their exploitation to areas far from the influence of the press. Western European nations managed colonial empires consisting of pre-industrial and mostly illiterate peoples on various continents. The Russian and Ottoman empires, too, subjugated the illiterate. And as the Ottoman Empire began its long nineteenth-century disintegration, those lands that managed to carve out autonomy or outright independence were often places with the most exposure to the print revolution, such as Greece and Serbia. (A single Serb, Vuk Karadžić, developed a Serb alphabet, published a Serbian grammar book, translated the New Testament, and compiled *Popular Songs and Epics of the Serbs,* paving the way for a Serbian nationalism that, for better or worse, would prove durable.)

The nation-state, having coalesced with the press's help, was still subject to some of its centrifugal force; with pamphleteering cheaper and cheaper, malcontents could always be a headache. So, just as the pope had his Index of Prohibited Books, secular rulers, from the early modern era onward, tried to control the press. In the late sixteenth century, Britain's Star Chamber confined printing rights to two universities and twenty-one London print shops, hoping to rein in the "great enormities and abuses" caused by "diverse, contentious and disorderly persons professing the art or mystery of printing or selling of books." In France, before the Bastille was stormed in 1789, more than eight hundred authors, printers, and book sellers had been imprisoned there.

Of course, none of this worked. Both Britain and France became enduringly pluralistic. This story is a common one in western Europe and across a still-growing portion of the world. Political freedom, notwithstanding its setbacks, seems to be the basic direction in which the world has been headed for centuries now. And, at the risk of oversimplifying, the main reason is ever cheaper and more powerful information technology, as represented by the printing press. By carrying the cost of mass publication to lower and lower levels, the press allowed less and less privileged groups to mobilize against repression.

PRINTING AND PLURALISM

The connection between freedom and the press is not trivially simple; it's not that printing always gives the masses overwhelming firepower against tyranny. In the hands of government, the press is an instrument of propaganda. (Napoleon, never shy about sharing his views, seized the newspaper *Journal des débats* and renamed it *Le Journal de l'Empire*.) Indeed, a sufficiently extreme regime can silence non-government presses, as Stalinist Russia showed.

No, the reason printing implied eventual political freedom is not that it can't be silenced, but, rather, the high cost of silencing it. Both technological innovation and everyday capitalism are collaborative enterprises, dependent on the free and fast transmission of data. During the eighteenth century, when the world saw an explosion of newspapers, they often arose, Anderson has noted, "essentially as appendages of the market"—read for their commodity prices, their schedules of ship arrivals and departures. Leaving presses free enough to serve economic purposes while stifling any political dissent they might lubricate is, to say the least, a challenge. Modern leaders have come to realize what feudal potentates of the late Middle Ages came to realize: you need to grant some freedom to get wealth.

But in modern times, the implied freedom was greater. The lower the cost of transmitting information, the more broadly and intricately productive a society's invisible brain can be, both in the short-run sense of everyday economics and the long-run sense of technological advance. But realizing this potential means empowering a commensurately broad swath of the population to serve as nodes in that brain: fostering literacy, for example, and giving people some leeway in what they read and write. As various scholars have noted, it may be no coincidence that Britain, which after the "Glorious Revolution" of

1688 led Europe in liberty (and which had a daily newspaper seventy years before France did), is where the industrial revolution reached critical mass; or that the Netherlands, which rivaled England in its antipathy toward despotism and in the vibrance of its press, had paved the way for industrialization during its "Golden Age" in the sixteenth and early seventeenth centuries.[†]

The economic logic of freedom is sufficiently subtle to have eluded a number of European rulers in the early modern age. Indeed, on balance, in the centuries after the printing press was invented, European government grew *more* despotic. But the trend was doomed. The economists J. Bradford De Long and Andrei Shleifer have shown that, as a rule, the more despotic the governments, and the tighter their reins on the press, the less prosperous their polities were. Even during absolutist Spain's so-called Golden Age, Spanish cities saw declining wealth and population.

Kant, writing in the late eighteenth century, observed, "If the citizen is deterred from seeking his personal welfare in any way he chooses which is consistent with the freedom of others, the vitality of business in general and hence also the strength of the whole are held in check. For this reason, restrictions placed upon personal activities are increasingly relaxed." What he didn't note is how information technology, in the form of the printing press, had intensified this logic—much less how information technology might further intensify it in coming centuries.

In the middle of the nineteenth century, Europe's reactionary forces made an epic stand against the currents of cultural evolution. They were responding to a wave of revolt that in 1848 swept mainland Europe, from France in the west to the Slavic east of the Hungarian empire. Though demands varied from place to place, they generally centered on political freedom and autonomy for subjugated national groups—two key implications of the printing press. The press's role is further underscored by the revolt's nickname—"revolution of the intellectuals"; it was spearheaded by writers and editors and students, and indeed got a chilly reception from many illiterate peasants.

After all the shouting was over, and revolutionary gains had succumbed to counterrevolution, Europe, though freer here and there, looked disappointingly like the status quo ante. Still, for the attentive despot, the handwriting was on the wall. Czar Alexander II, on launching a campaign for economic development in the 1850s, relaxed censorship, and newspapers and journals flourished. In the same decade the Ottoman Empire launched the same kind of reform, abolishing torture, establishing equality before the law, letting western liberal ideas issue from the press. (As is often the case, enlightenment had been sponsored partly by war. The Russian and Ottoman empires had been humiliated in the Crimean War—the Russians by losing to western powers, the Ottomans by desperately relying on them.)

In neither empire did a miracle ensue. It took repeated convulsions for them to finally disintegrate and for their cores, Turkey and Russia, to gain a measure of political liberty. The process continues, and is still fitful. Around the world, reforms will suffer relapse. But the premise of reformers is less in doubt than ever. Totalitarianism is dead, and authoritarianism isn't looking too healthy. Human nature hasn't changed; there are still plenty of people who would like to be masters of the universe. But increasingly they realize that, thanks to technology, they can't be. The days when vast centralized governments could control the flow of written information—and still prosper—are gone. What ended them is a series of information technologies that lowered the cost of communication and (thus) raised the sheerly economic value of freely flowing information. First among these technologies is the printing press. It created tons of potential synergy, but set strict rules for realizing it: people could play their new non-zero-sum games well only if fairly free.

PLURALISM AND PARASITISM

Lest we lapse into sunny optimism, a few words on the pitfalls of political pluralism are in order. These pitfalls, scarcely visible when you adopt the "information as enlightenment" perspective, become clearer when you see information as influence.

Pluralism—the sharing of power by diverse groups—has the considerable upside of complicating efforts by ruling elites to seize the lion's share of the non-zero-sum surplus. In a society's ongoing (and often implicit) negotiations over how to divvy up that surplus, growing pluralism means that bargaining power is more broadly distributed. This is truly a trend worth celebrating—especially if you recall the starker concentrations of power that (so far as we can tell) were common in chiefdoms and ancient states.

But, even as pluralism thus prevents abuses of power by a ruling elite, it enables the newly empowered to themselves become abusive. Consider the late medieval migration of power from the rural nobility to the bourgeoisie. Among the now-ascendant groups were merchant and craft guilds. Sometimes their power went to good social use. The guilds' selective membership policy, which confined credentials to qualified and honest practitioners, shored up trust in their vocation and in commerce generally—a fine thing. But to some extent the guilds kept admission tight just to suppress competition. And this wasn't the only maneuver they used to keep their jobs artificially cushy. When the printing press showed up, and threatened the jobs of French copyists, the copyists mobilized to thwart it. This was good for the copyists—for a while at least—and bad for book buyers, papermakers, and would-be printers.

In other words, this interest group had generated positive sums *internally,* organizing to pursue common goals. But among those goals was to harm the larger society; externally the group's effect was to dampen overall non-zero-sum gain.

One function of government is to keep interest groups from thus sapping the public good. To some extent, the monarchs who pioneered centralized government in the late Middle Ages did that. They got the guilds to accept innovation, offering the compensation of expanded and smoothed commerce. Still, during the modern age, this challenge of governing was bound to stiffen, as the printing press and other information technologies reinforced pluralism, dividing power among a larger and larger array of groups that could now mobilize effectively. Pluralism beats despotism, but it isn't nirvana; by making parasitism more of an equal-opportunity endeavor, it complicates the challenge of keeping the society's overall non-zero-sum gains robust.

WHAT ABOUT CHINA?

One obvious question: If indeed the printing press tended to encourage prosperity and the diffusion of power, what about China? China invented printing, after all, and was using it centuries before Gutenberg set up shop. So why was it Europe that led the way both to modern industrial prosperity and modern liberal democracy? Why didn't printing have the economic and political effects in China that it had in Europe?

The first half of that last question is easy: Printing *did,* broadly, have the economic effects in China that it had in Europe. The printing of books swept China during the Sung era, a millennium ago. Then, as later in Europe, one result was to spread technical knowledge, raising both the productive use of existing technology and the chances of further innovation. The ensuing vibrance has gotten Sung China compared to Europe during the Renaissance. The age when Chinese printing matured was an age of prosperity and know-how unrivaled around the world. Leaders in information technology tend to be leaders, period.

But what about the political fallout? Why didn't Chinese printing bring the same seismic changes as European printing? For one thing, it was a different kind of printing. Though the Chinese invented movable type, they didn't much use it. Their script—partly phonetic but still heavily ideographic—helps explain why. Each printer needed to keep thousands of characters in stock, as opposed to the hundred or so that would fill the European printer's job case. Imagine trying to compose a page of text from a palette that large, and you'll see why Chinese printers generally stuck with monolithic, hand-carved woodblocks. Though master carvers worked swiftly, printing costs still couldn't plummet in China as they would in Europe in the centuries after Gutenberg. The printing of timely, small-circulation pamphlets, in particular, was not so practical.

Still, shouldn't there have been at least *some* pyrotechnics? The flooding of China with books during the Sung era suggests that printing had gotten much easier, and literacy more common. So why doesn't Chinese history have anything even remotely resembling a Martin Luther?

Actually, itinerant preachers did roam the land and distribute eccentric theological tracts. But, as Peter Perdue has noted, prevailing Chinese religion was less rigid, thus less fragile, than Catholic dogma. Though Confucianism was the official state doctrine, featured on civil service exams, the government didn't mind people worshiping their own deities so long as worship didn't upset the social order.

What's more, China was less ripe for nationalist rebellion than Europe. The fabric of economic interdependence was strong by the age of printing. The Grand Canal had connected China's two great rivers—the Yellow and the Yangtze—since the seventh century. And the fact that Chinese script was intelligible to speakers of all dialects further bound the nation.

But perhaps the biggest reason that China wasn't torn asunder by printing is that its ruling class, intentionally or not, made a kind of peace with the medium by institutionalizing pluralism. In the age of the press, the government opened itself to popular feedback. Complaints and suggestions were assessed by three separate agencies, whose autonomy from the emperor's meddling was painstakingly ensured. There was special emphasis on giving the growing literate class a sense of participation. Civil service exams became more meritocratic than ever, shrinking (though certainly not ending) the role of nepotism and cronyism. A smart, diligent, literate Chinese boy could through education wind up a respected bureaucrat.

This newly just system, along with the ongoing expansion of literacy, eroded the ancient status system in which birth had been destiny. This erosion had other sources, including a market-based prosperity that slowly elevated nouveau riche merchants in the scheme of things. Still, as the historian Charles O. Hucker has observed, "the role of printing in the social leveling process can hardly be overemphasized."

The cynical view is that the Chinese government was simply co-opting the literate classes; it dangled the bauble of prestige before them and then, as they pursued it, brainwashed them with a Confucian education that glorified duty. But even in thus "controlling" the effects of printing—even in channeling members of the expanding literate classes into the state apparatus—the Chinese government had conceded, in some measure, printing's diffusion of power. For service in government meant influence on government. During the Sung era the civil service grew in size and acquired unprecedented power at the expense of court eunuchs and the royal family. "The emperors themselves played only a secondary role, leaving the limelight to their ministers," notes Jacques Gernet.

Obviously, this was nothing like modern representative democracy. Then again, the diffusion of power in early modern Europe was nothing like modern representative democracy. But in both China and Europe the dawn of printing saw an expansion in the number of people who could influence government. Only in China the influence was more orderly.

The system didn't work forever. In fact, it *couldn't* work forever. As the number of literate, aspiring civil servants grew, a declining portion had their aspirations realized. Besides, after the Sung era, China's rulers—the Mongols, the Ming, the Manchu—evinced less consistent interest in a well-ventilated government, staffed by the ablest applicants and structurally open to dissent. Take the first Ming emperor, for example. One faction argued that he should spend more time in the north, near the levers of government. In reply, he had 146 adherents of this view beaten, and 11 died. The same emperor had tens of thousands of people beheaded, and he decapitated the civil service itself, eliminating the offices of prime minister and secretariat.

Later Ming emperors weren't so flamboyantly despotic, but they tended to keep power concentrated in the throne, never restoring the pluralism that a strong civil service had brought to the Sung era. And many were either incompetent or distracted by such fervently pursued pastimes as cricket breeding. It may be that China's failure to fulfill the promise of the Sung era lay as much in the general mediocrity and heavy-handedness of Ming rule as in the famous Ming decision to end oceanic exploration.

No law of history says governments can't retreat from pluralism; repression and wanton self-aggrandizement are always options. But a law of history *does* seem to say that these courses tend, in the long run, to sap economic vibrance. Or, more precisely: the law says that any economic vibrance will in the long run pale by comparison with that in more liberal societies, all other things being equal. And China after the Sung era roughly complies with that law.[†]

THE INDUSTRIAL REVOLUTION

The economic historian Joel Mokyr has attributed Europe's industrial revolution to "chains of inspiration"—one idea leading to another; even seemingly distinct breakthroughs, such as Watt's steam engine, turn out to be the fruit of wide collaboration, sometimes among people who never met. But "chains of inspiration" could be taken to explain the whole history of technical advance, which, after all, has always been driven by a collective social brain. What was new about the industrial revolution was how short the links were in time even when they were long in space; Europe was becoming a very large yet fast brain. A primary reason was the printing press, a new means of forging links in the chains of inspiration—a new technology for making new technologies.

There were various other technologies and metatechnologies conducing to technological evolution. One was England's state-of-the-art governance of intellectual property, dating back to its patent law of

1624, which spurred the full disclosure of technical details that the printing press could then spread via books and journals.

We think of industrial technologies as things of matter and energy, of iron and coal. And this is at some level true; certainly a key consequence of the industrial revolution was the oft-cited one of replacing animate sources of energy—people and other animals—with inanimate sources. Still, even during the industrial revolution, long before the "information age," processing matter and energy was intertwined with processing information. Consider the locomotive. Its engine was not only a cutting-edge processor of energy; it had a governor—a feedback loop—and thus processed data about its own state. What's more, on the social level, the locomotive often served the processing of information as well as of matter and energy; trains carried not just steel and coal, but mail and newspapers. Only with the coming of the telegraph, in the 1830s, would communication technology begin to diverge from transportation technology.

The locomotive, along with other rapid data carriers, accented the truth highlighted by the printing press: the more easily data can move, the larger and denser a social brain can be. The vast, fast collaboration allowed by information technologies slowly turned the multinational technical community into an almost unified consciousness. Increasingly, good ideas were "in the air" across the industrialized world.

Witness how often the same basic technological breakthrough was made independently by different people in different places at roughly the same time. And witness—as testament to the impetus behind easing communication—how often these independent breakthroughs were in information technology itself: the telegraph (Charles Wheatstone and Samuel F. B. Morse, 1837); color photography (Charles Cros and Louis Ducos du Hauron, 1868); the phonograph (Charles Cros—again!—and Thomas Edison, 1877); the telephone (Elisha Gray and Alexander Graham Bell, 1876)—and so on, all the way up to the microchip (Jack Kilby and Robert Noyce, 1958–1959).

These independent epiphanies speak of the overwhelming likelihood of the various breakthroughs in modern information technology. And each such breakthrough—by further easing the transmission of data, whether by sound, print, or image—only raised the chances of further breakthroughs. Via endless positive feedback, the technological infrastructure for a global brain was, in a sense, building itself.

And this infrastructure, of course, did double duty. It sponsored not just long-range inventing, but the workaday business of animating the economy; it carried the signals that allocated goods and services. As the process got faster—as, for example, locomotives, governed by telegraph signals, carried nicely illustrated Sears catalogues to small American towns and then carried order forms back to the big city—everyday economics came to resemble a kind of superorganic metabolism. And the superorganism increasingly assumed planetary scope.

ONE WORLD?

Historians differ over when it becomes fair to speak of a single world economy. (As early as 1500, even though Europe's stance toward the New World was essentially parasitic? In 1800, by which time some residents of the New World—albeit mostly immigrants from the Old World—were engaged in more equitable exchange with Europe? Or not until a century later, when trade levels approached those of today?) But this disagreement, really, is the point: progress toward a global web of interdependence was so long and relentless that pinpointing a threshold is hard.

This progress lay not just in the growth of international exchange, but in its changing nature. In ancient times, with transport crude, slow, and pricey, most trade involved goods with a high ratio of value to mass: jewelry, tapestry, and other exotic, superfluous things. ("Splendid and trifling," as Gibbon said of trade in the first century A.D.) But over the centuries, as transportation grew more cheap and routine, trade in bulky essentials grew practical. Even in the Roman Empire, hauling wheat long distances over water had made economic sense. In modern times, hauling such things far overland began to make sense.

As regions specialized in the things they did best, and imported the rest, mutual reliance for such basics as food, clothing, and tools could ensue. Non-zero-sumness flourished not just within states, but between them.

This economic fact had political consequences. War had long been by some measures a negative-sum game, since even the "victor" suffered. Now, with economic interdependence among nations growing, war became still more disruptive, more negative-sum—and peace more widely welcomed. Within three weeks of the end of hostilities between the United States and Britain in 1814, coffee prices in Saudi

Arabia dropped 30 percent. And this was *before* the steamship and the railroad train.

With the benefits of peace growing, stability became more and more a conscious goal of foreign policy.[†] Not coincidentally, it was the intensely commercial Italian states of the early modern period that had pioneered the idea of having a resident ambassador in another state. By the early nineteenth century, with the industrial revolution turning even large nation-states into intimate neighbors, Europe's great powers had made ambassadors a permanent fixture.

Today we take for granted the corollary notion that formalized civility should be the normal state of relations among polities. But the notion is far from venerable. During the Middle Ages, to the extent that there had been anything worthy of the name "international law," it consisted of the following premises, as one scholar has summarized them: "the high seas were no-man's-land, where anyone might do as he pleased"; in the absence of agreements to the contrary, rulers were "entitled to treat foreigners at their absolute discretion"; and war was "the basic state of international relations."

That this was no way to run a planet gradually dawned on humanity, and got especially forceful articulation in 1625, when Hugo Grotius, in his treatise *Law of War and Peace,* noted that the waging of war tended to be self-defeating—to produce negative sums, in other words. Grotius's point has slowly won wide acceptance. The world's failure thus far to act on this knowledge, and vanquish war, should not obscure the real achievements of the modern age: internationally accepted rules governing conduct on the seas, treatment of foreigners, and many other things.

Like so much else in the modern age, this trend has drawn strength from the printing press. The intellectual historian Bruce Mazlish has noted that, even as the press energized nationalism, it helped create a world literature; great works were translated and published abroad. These works included Grotius's *Law of War and Peace* and various other tomes and essays in the sciences and humanities. Intellectuals, along with merchants, increasingly became a transnational class, creating a supranational consciousness that nourished the evolution of international law and international ethics. The press, having helped break up empires, helped weave filaments among the fragments.

Once again, the printing press was exhibiting its paradoxical properties, as both solvent and glue. And once again, the paradox is resolved by viewing the press as a congealer of common interest, a tool by which people with shared goals reap positive sums through concerted action. As we'll see in the next two chapters, the same paradox, and the same resolution, applies to later information technologies.

Chapter Fourteen

AND HERE WE ARE

> Web to weave, and corn to grind;
> Things are in the saddle,
> And ride mankind.
> —*Ralph Waldo Emerson*

The philosopher Karl Popper felt that "the belief in historical destiny is sheer superstition." Besides, he added, even if there were a destiny, it would be unknowable. "There can be no prediction of the course of human history by scientific or any other rational methods."

Popper's basic argument was simple. History is heavily influenced by the growth of knowledge. And we can't predict the future growth of knowledge. After all, if we could say today what new things we'll know tomorrow, then we'd already know them today, and they wouldn't be new tomorrow. Right?

Right. If we knew today how to build an affordable desktop computer that is fifty times more powerful than current desktop computers, we'd already have it on our desks. On the other hand, does anyone doubt that eventually we *will* have such computers on our desks? For purposes of prediction, isn't the fact that we'll have them more important than the question of exactly how we'll make them?

But wait. There's always a chance, according to Popper, that science and technology will come to a halt. This could be arranged "by closing down or controlling laboratories for research, by suppressing or controlling scientific periodicals and other means of discussion . . . by suppressing books, the printing press, writing, and, in the end, speaking." Well, maybe. But, even assuming that a government had the power to do that, it would soon find itself governing a not very powerful nation. To stop technical progress is to reserve a place in the dustbin of history.

That helps explain why several trends span all of human history: improvement in the transport and processing of matter, improvement in the transport and processing of energy, improvement in the transport and processing of information. We know that these trends will continue, even though we don't know the technical details that will sustain them. Or, at least, we know with, say, 99.99 percent confidence that these trends will continue. That's good enough for me.

Of course, predicting the persistence of technical trends is a long way from predicting their social consequences. When we move from the former to the latter, our confidence drops below 99.99 percent. Still, it doesn't get anywhere near zero. There are trends in social and political structure that more or less follow from trends in technology.

Finding predictive value in trends makes me guilty of "historicism." According to Popper, historicism involves the belief that insight into the future can be had by discovering "the 'rhythms' or the 'patterns,' the 'laws' or the 'trends' that underlie the evolution of history." Popper considered historicism not just misguided, but dangerous. For example, people persecuted under Hitler were victims "of the historicist superstitions of the Third Reich." Popper failed to note that by his definition the Enlightenment-era thinkers who founded American democracy, with their progressivist view of history and their attendant sense of mission, harbored "historicist superstitions" as well.

Whether the world of tomorrow will indeed be a logical outgrowth of today's trends is, of course, a question we can't settle today. The (fleeting) beauty of any predictions made in the next two chapters is their temporary unfalsifiability. But one thing we *can* do today is see whether the world of today is a logical outgrowth of yesterday's trends. If it is, then maybe extrapolating from trends isn't quite the muddled and heinous endeavor that Popper alleged it to be.

Consider seven basic features of the contemporary world that have gotten much attention from social and political analysts. All of these features are genuinely important—even, in some cases, as important as the analysts claim. But none of these features is new. All, indeed, are grounded in very old, very basic dynamics of cultural evolution. Their past stubbornness is valid reason to expect their future persistence.

Not-so-new feature #1: The Declining Relevance of Distance. A book published in 1997 was titled *The Death of Distance*. But of course, distance isn't quite *dead*. The promotional blurb on the back of another recent book put the matter more soberly: "The constraints of geography are shrinking and the world is becoming a single place." True. True now; true a century ago, as steamships plied the waterways, telegraph lines crossed oceans, and railroads criss-crossed continents; true five centuries ago, when Columbus crossed the Atlantic; true two millennia ago, when the Silk Road took shape. Ever since boats were paddled and trails were blazed, distance has become less and less an obstacle to contact. As transportation and communication get smoother and cheaper, long-distance trade and collaboration make more sense.

Not-so-new feature #2: The "Ideas" Economy. According to the futurist George Gilder, the second half of the twentieth century saw the dawning of the "microcosmic" era, which brought "the ascendancy of information and mind in contemporary technology, and hence in economics as well." Thus, the value of computer hardware "resides in the ideas rather than in their material embodiment." Gilder has a point: you won't recover much of the original cost of a new computer if you sell it as scrap metal. Then again, that's also true—if a bit less so—for a new Buick. And it was true—if less so still—for a 1939 Buick. And for the cotton gin, and so on. The growing "designedness" of things is a by-product of technological evolution generally, not of the microelectronics revolution.

Actually, the "ideas" that go into designed goods are only one of the reasons their value so exceeds the value of their raw materials. The other factor is labor—the meticulous assembly that subtle design often entails. Together, these factors encourage long-distance trade by giving practical items a property once confined to gems, silks, and other exotics—a high ratio of value to mass. Along with falling costs of transport, the rising ratio of value to mass has been a basic elevator of trade in recent centuries.

Not-so-new feature #3: The New, Weightless Economy. Various people have made two observations: (1) information, in its modern incarnations, doesn't weigh much; (2) more and more of what we sell and trade is information. Both claims are true, and it's also true, as claimed, that they are connected. One reason trade in information is brisk is that sending $10,000 worth of software across the Atlantic doesn't cost as much as sending, say, $10,000 worth of pig iron. But this logic is simply the latest, if a quite dramatic, example of the above-noted long-term trend: higher and higher ratios of value to mass.[†] Purely informational products are products in which raw materials count for very, very little, and design counts for very, very much.

Not-so-new feature #4: Liberation by Microchip. Suddenly, we are told, information technology is on the side of freedom. This observation gained currency in the final years of the twentieth century, quickly attaining the status of boilerplate in American presidential speeches. Before leaving to visit China in 1998, Bill Clinton observed: "In this global information age, when economic success is built on ideas, personal freedom is . . . essential to the greatness of any modern nation."

It's true that the latest information technology is forcing historically oppressive nations to grant more freedom if they want more prosperity. But, as we've seen, the printing press had much the same effect. And earlier, in Europe's late Middle Ages, an information metatechnology—incipient algorithms of capitalism—made new local freedoms the price for prosperity. Even back in ancient times, the coming of that potent information technology, money, argued for a certain personal liberty, as free markets now made more sense than before. So too with ancient writing. Had those Assyrian traders in the late second millennium B.C. been denied the skills and freedom to write their own contracts, Assyria would have been the poorer for it.

The basic trend is this: new information technologies open up new vistas of non-zero-sumness. But typically the transmutation of non-zero-sumness into positive sums depends on granting broad access to those technologies, along with the freedom to use them well. And, over the long run, polities that fail to respect this liberating logic tend to get punished with relative poverty. Far from being new, this is to some extent the story of history.

One thing that *is* new is how vividly and swiftly the polities get punished. Political leaders now see their competition up close, in real time. And, with technological change coming faster than ever, stagnation breeds calamity in decades, not centuries.

Another new thing is the *extent* of the decentralization of power that is now essential to prosperity. Printing presses, though more cost-

effective than monasteries full of scribbling monks, were still a fairly
pricey form of mass communication. (As the old maxim has it: Never
get into an argument with someone who buys ink by the barrel.)
Today, though, a kind of printing press is standard desktop equipment
in advanced economies. And it is more powerful than the old kind. At
almost no cost, it can make any diatribe available to computer users all
around the planet.

Even governments not enthusiastic about this power, such as
China's, have had a hard time curtailing it without hobbling their
economies. The reason—the link between prosperity and nationally
diffuse data processing—was spelled out by one reform-minded Chinese thinker, Hu Weixi, in 1998. He observed that China's 1.2 billion
people "are not only a 'labor force'; they are also the world's largest
thought warehouse and brain. We can thus use the magic weapon of
freedom of thought to achieve success." (As of 1998, Chinese thinkers
of Hu Weixi's stripe were having forums on the right-wing economist
Friedrich von Hayek, a hero of libertarians and one of the first economists to view an economy as an information-processing system. Long
before the fall of communism, Hayek identified its oft-overlooked
weakness: not only did it fail to offer an incentive to work hard; it
forced signals connecting supply and demand to travel a tortuous path
that invited distortion.)

Seeing history as the unfolding of human liberty is one symptom of
being a "Whig historian." According to Herbert Butterfield, who
coined this pejorative, a Whig historian is someone who tends "to
emphasize certain principles of progress in the past and to produce a
story which is the ratification if not the glorification of the present."
But the fact is that things can work vice versa—the present can ratify a
particular story about the past. When Butterfield wrote, in 1931, various historians—the classic nineteenth-century Whig historians—had
depicted history as the unfolding of freedom. But they hadn't suggested a plausible mechanism to explain why exactly freedom tended to
triumph later in history rather than earlier. The present age—the information age—has suggested the mechanism by letting us watch it work
in real time. Every year, advances in information technology are making Stalinist economics less tenable (and, by some accounts, are making
centralized organizational hierarchies in general less tenable).[†] From
this fast-forward vantage point, we can look back and see the same basic
dynamic at work, albeit more slowly, in centuries, even millennia, past.

Not-so-new feature #5: Narrowcasting. Suddenly we have moved from the age of "broadcasting," featuring a few TV networks offering mainstream fare, to the age of "narrowcasting," featuring lots of more specialized channels. This is truly a new development—new, at least, for this particular medium. But an analogous thing happened with print as the costs of publishing dropped in the fifteenth century and continued to drop thereafter. As publishing got cheaper, publishers could serve narrower and narrower slices of the populace. The first two magazines in the United States, published in the mid-eighteenth century, were called *American Magazine* and *General Magazine,* much as early broadcast networks had names like the American Broadcasting Company. By the end of the nineteenth century, there were magazines such as *Popular Science*—rough equivalents of the Discovery Channel.

This parallel between media is often lost in the course of cosmic media theorizing. The emphasis, instead, is placed on the supposedly determinative *differences* among media. Marshall McLuhan, for example, said that "the medium is the message"—that different media have different intrinsic, culture-shaping properties. Thus the phonetic alphabet was said to be a "hot" medium (though ideographic script, oddly, was "cool"), and TV was said to be "cool" (though movies were "hot").

What did McLuhan mean by "cool" and "hot"? Deciphering McLuhan's prose is not always a cost-effective way to spend time. Besides, the merit of his particular taxonomy is not the main point here. The main point is to deflate such typologies in general—to question the notion that media shape culture and history mainly by virtue of their peculiar sensory properties. Whatever the differences among video, audio, and the written word (and there *are* important differences), their essential potential is the same: they are instruments of communication, persuasion, coordination. All can ease the mobilization of groups with a common goal. You could in principle have started the Protestant Reformation via TV. (Luther *was* a preacher, after all, like Billy Graham.) But TV didn't exist.

Moreover, when TV *was* invented, the core similarities between video and the written word were concealed by the fact that, at the time, the two media had different economic properties. Since the invention of writing in ancient times, a series of innovations—paper, ink, mass-produced paper, the printing press, better printing presses, better mail service—had made this means of sending signals quite cheap. But the cost of producing and widely distributing video—whether movies or TV shows—was still high. In the case of TV, the preciousness of video was further heightened by the finiteness of the broadcast spectrum—a finiteness that led to government control of the medium via broadcast licensing. All in all, splinter groups need not apply. (Ozzie and Harriet did not have countercultural tendencies.) But cable TV would change things a little, and as the World Wide Web goes broad bandwidth, things are changing a lot, driving the cost of telecasting down near zero.

This factor—the *economics* of a given medium, not its purported temperature—is the key to its social effects, and thus a key to history's basic direction. Because in the long run, all information technologies follow the same economic pattern: using them gets cheaper and cheaper; narrowcasting is the fate of all media.

The excessive focus on the *type* of medium, and the underemphasis of its economics, is understandable. At any given time, after all, the different types of media, being in different evolutionary stages, may have radically different economics, and thus seem qualitatively different. Given the cost fifty years ago of video production and transmission, it was natural to think of TV as inherently a mass medium, controlled by big shots, in contrast to such low-cost media as paper. (In Orwell's *1984,* it was Big Brother who communicated by TV, while subversives had to scurry around with handwritten messages.)

So too with computers. Half a century ago, when they were monstrous and monstrously expensive, they seemed just the tool for top-down government. Indeed, some communists hoped they would do for command economies what money had done for market economies. Maybe a building full of computers in downtown Moscow could orchestrate the Soviet economy! By the early 1960s, cybernetics was an officially hot topic in the Soviet Union.

But, meanwhile, microelectronic progress was favoring the market; cheap computing power would soon spread across society. Video, too, would become a tool of the masses, via the camcorder. And now these two media—video and computer—are evolving in synergy. As video production costs plummet (thanks to desktop editing), the Internet is driving transmission costs way, way down. Now even subversives can have TV shows.

What matters about the Internet, then, is not whether it is "hot" or "cold." It can be either, transporting phonetic script or ideographic script, TV shows, movies, or radio. What matters is that the net is rapidly lowering the cost of all these media. Though declining costs of communication are an age-old story, the suddenness of the current drop is bound to be jolting.

All of this may help explain why the long-run political picture (growth in freedom and pluralism over the millennia) looks different from the short-run picture (vacillation between despotism and pluralism over the decades and even centuries). When new information technologies first appear—writing, audio, video, computers—they are often very elite-friendly; for economic and other reasons, control over dissemination is not widespread. So technological evolution periodically favors the centralization of power—specifically, when whole new *kinds* of information technology, such as writing or audio or video or the computer, first appear.

Further muddying the long-term pattern is the fact that, in a given place and time, whether a technology conduces to despotism or pluralism or neither can depend on context and contingency. Radio and TV, pre-Internet, could be used in a mildly pluralistic way, as in the United States, or in a totalitarian way, as in Nazi Germany and the Soviet Union.

Still, as we've seen, in the long run governments have little choice. Even China, an authoritarian (and once-totalitarian) nation, had seen by the late 1990s that it needed the Internet. And its attempts to filter the net—glean the economic benefits without enduring the political consequences—had hardly enjoyed unalloyed success; the nation was more porous to outside information than at any point since the communist revolution. Conceivably, the regime could reverse this trend—but the price would be a dismal economic future. Ultimately, around the world, the combined logic of technology and politics points toward the cheaper, more profuse, more pluralistic use of all media.

Not-so-new feature #6: Jihad vs. McWorld. This phrase, popularized by the political scientist Benjamin Barber's book of the same name, refers to a paradox. On the one hand, the world is growing more "tribal," breaking up by ethnicity or religion or language, as when Yugoslavia dissolves into factionalism; or when French-speaking Canadians push the secession of Quebec; or when fundamentalist Muslims oppose secularists in Turkey. On the other hand, there is the globalization of economics and culture, as represented by a McDonald's in Moscow, say, or MTV airing in Europe. Thus, says Barber, the planet is "falling apart" yet "coming together," and the tension between the two forces has sent the world "spinning out of control." Or, in the formulation of an earlier but less famous analysis, there is at once fragmentation and integration—"fragmegration."

What is really happening here is, at its core, neither new nor deeply contradictory. It is the story of the modern age, going back to the printing press. Information technologies, by lowering the cost of data transport, are making it easier for entities with a common interest to coordinate.

Sometimes the result is "McWorld." TV broadcasters in various countries share an interest with their viewers and with the owners of MTV and the musicians MTV features—an interest in sustaining the phenomenon of MTV. In a somewhat different way, information technologies (and transportation technologies and other technologies) give McDonald's owners and managers and workers around the world a common interest in sustaining the phenomenon of McDonald's— an interest they share with their customers.

Meanwhile, the other half of Barber's paradoxical dichotomy— "Jihad," the "fragmenting" force of "tribalism"—is also lubricated by the declining information costs of coalescence. In that regard, modern tribalism is like such earlier, print-fueled Jihads as the Reformation or the surging Serbian nationalism of the nineteenth century. The first big boost for Quebec's independence movement came when Canadian broadcasting began to get narrow—in 1952 with the first French Canadian TV channel. Today there are various Francophone cable networks. They not only serve as mouthpieces for pro-sovereignty politicos, but also weaken bilingualism, which in earlier days drew strength from American broadcasts that seeped across the border, into the brains of teenagers who had nothing better to watch. (Of course, politics has influenced the timing of all this. The end of the Cold War weakened various national bonds, including those in Yugoslavia and in balkanized African states whose superpower sponsors had once muted division. But in effect what happened was simply the thawing of a natural, long-term process that the Cold War had frozen—the continued crystallization of nationalist coherence through the ongoing decline of information costs.)

Thus Jihad and McWorld—both sides of "fragmegration"—have a

common impetus. Improvements in information technology ease interaction among people who benefit by interacting. And not only are the two halves of this paradox thus bound by a common origin; their effects turn out to be not as contradictory, nor even as different, as you might think. Quebec's tribalism, for example, is a two-way street. The satellites and underground cables that divide Canada with Francophone narrowcasting also mean that TV channels from France can be broadcast in Quebec. Is this transoceanic export of culture an example of McWorld, like other such exports, or of Jihad, since it helps balkanize Canada?

Again, analogues can be found in centuries past. Nineteenth-century nationalism unified while dividing, at once bringing Italy together and breaking the Austrian empire apart. And, during the Reformation, the same printing press that split the European church actually *strengthened* its international sinews. Previously, Catholic ritual had differed from region to region; mutations crept into liturgical texts when copied by local scribes, and the pope tolerated these quirks. Now, via the press, liturgies were standardized—McLiturgies. For that matter, the press gave the new Protestant doctrines more coherent international reach as well. As the scholar Elizabeth Eisenstein has observed, "the main lines of cleavage [between Catholic and Protestant] had been extended across continents and carried overseas along with Bibles and breviaries." Again the question: Jihad or McWorld? Fragme or gration?

Barber's sense that the world is "spinning out of control" is understandable. As Marx saw, when technologies change fundamentally, the economic, social, and political relationships premised on them must sooner or later change as well. The ensuing adjustments can be wrenching. Still, chaos is not the natural culmination of basic historical forces. What basic historical forces are doing is driving the system toward a new equilibrium, in which social structures will be compatible with technology. The big question is how chaotic the *transition* will be—a question that is for the human species to decide. We'll explore this question in the next two chapters.

Not-so-new feature #7: The Twilight of Sovereignty. This is the title of one of many books about forces that seem to be eroding the nation-state's control over its destiny. In a way, the basic idea is a corollary of not-so-new-feature #6. McWorld—globalization—is the great crippler of national governments. A nation's central bank has waning influence, thanks to globalized financial markets that suck currency out of the country or pump it in. And those same financial markets, paired with robust trade, mean that economic instability on one side of the world can prove contagious, leaving nations prey to forces beyond their grasp.

True, but not altogether new. A century or so ago, trade levels, as a fraction of economic output, were almost as high as they are now, and money rushed across borders, if not nearly as fast as today. The dominance of Britain's pound sterling had so fused some financial markets that, as a popular saying had it, "When England sneezes, Argentina catches pneumonia." Contagion then, as now, worked in both directions. In 1890, Britain faced a financial crisis after falling grain prices rendered Argentina unable to make loan payments to the Baring Brothers bank. Nor was international contagion anywhere near new even then. Half a millennium earlier, Britain's reckless international borrowing to finance the Hundred Years' War wound up ruining banks in Florence.

There's no denying that present-day financial interdependence is unprecedented. But the difference is as much quantitative as qualitative.

A DEEPER UNITY

Speaking of things that are not so new: There is nothing new in pointing out that many things thought new are in fact not. Expert debunkers have already assaulted, for example, the notion that economic globalization is a creation of the twentieth century. But the point of this exercise goes beyond cliché demolition. The point is that trends can be of predictive value. If a century or two ago you had looked around and observed, "In the long run, disseminating information tends to get cheaper, and so does transporting products, so that the world becomes a smaller place," you would indeed have observed trends that could be validly extrapolated into the future, notwithstanding Karl Popper's views on the intractable epistemological problems associated with such a maneuver. There is a deep unity between past and present.

In fact, the unity goes one level deeper. All seven of the above not-so-new features boil down to the growth of non-zero-sumness; they either cause it, or are caused by it, or *are* it. Thus:

(1) To say that distance has become less and less economically relevant is to say that relationships can be non-zero-sum across larger and larger distances.

(2 & 3) To say that the ratio of value to mass has risen is to note one of the causes of this declining relevance of distance, this expanding web of non-zero-sumness.

(4) To say that new information technologies ultimately encourage liberty is to say that oppressed groups—dissidents in an authoritarian state, say—can more easily unite in pursuit of freedom, realizing positive political sums among themselves. It is also to say that the allure of the positive economic sums promised by new information technologies is what ultimately leads governments to grant broad, liberating access to them.

(5) To say that casting gets narrower is to say that smaller and smaller communities of common interest can pool their resources to realize positive sums. Thus literate gun buffs, in effect, pool their resources to sponsor *Guns and Ammo* magazine, and the Golf Channel is underwritten by duffers.

(6) Both sides of the Jihad vs. McWorld paradox are non-zero-sumness in action. Quebec's "tribalism" consists of people seeking a common goal that they can gain only in concert. Bonds of mutual benefit also unite McDonald's far-flung web of businesspersons and customers.

(7) Even the "twilight of sovereignty," which sounds so lacking in the cheery properties often associated with non-zero-sumness, is fairly brimming with the stuff. Consider the most famous sovereignty sapper, the globalized economy. Obviously, the flow of goods and services across borders, and the piles of money pushed across those borders by financiers and currency traders, are the result of zillions of non-zero-sum interactions each day. But perhaps more interesting are the larger non-zero-sum games that result from the disruptive effects of this commerce. With economic downturn more contagious than ever, nations see common cause in forestalling regional collapse, and in dampening turbulence.

PROGRESS AT LAST

So what does all this tell us about the present and future? What are we to make of all these non-zero-sum trends? Given that the trends were at work before this century, it is tempting to answer with this lukewarm prediction: the future will consist of more of the same. And it will, at least in the sense that all of the above trends are likely to continue. On the other hand, even long-standing trends move slowly at times and faster other times, and this may be one of the other times. With the coming of TV, the computer, the microcomputer, and allied technologies, this century has seen breakthroughs in information technology that rival all past such breakthroughs, even the inventions of writing, money, and the printing press. Given the centrality of information technology to non-zero-sumness, and the centrality of non-zero-sumness to social structure, is it possible that we are passing through a true threshold, a change as basic as the transitions from hunter-gatherer village to chiefdom, from chiefdom to ancient state?

And if so, what lies on the other side? Is it the promised land? After eons of oppression can liberty at last blossom worldwide? After millennia of mindless strife can there be peace? Or do the more virulent forms of tribalism, and the more volatile aspects of a globalization, foreshadow a future of turbulence and chaos? These are questions we'll address in the next two chapters. Until we do so—and maybe even after—singing "We Are the World" will be premature.

Still, we might nonetheless pause for a celebration of sorts. For the basic arrow of history—the conversion of non-zero-sumness into sums that are on balance positive, and the attendant creation of yet more non-zero-sumness—does seem to have entailed a certain kind of moral progress. By this I don't mean that savages are savage, barbarians barbaric, and civilized peoples civilized. Indeed, it would be hard to argue that there was net moral gain between the hunter-gatherer and ancient-state phases of cultural evolution. The Egyptians had slaves—which virtually no known hunter-gatherer societies had—and their soldiers returned from wars of conquest proudly brandishing the severed penises of slain foes.

Even early "western" civilization didn't bring great moral improvement. Classical Athens is famous for its enlightenment—its art, its science, its egalitarian ethos. But when conquering other cities the Athenians had a habit of executing all male citizens.

Yet progress was in the cards. The reason is that—for the Greeks as for everyone else—broadening interdependence was in the cards; and interdependence has a way of breeding respect, or at least tolerance.

In the case of the Greeks, the interdependence was partly economic, but mainly military. As various Greek states found common cause in fighting off Persians, Athenians began to concede the essential humanity of non-Athenian Greeks.

That this advance in moral philosophy extended no further than the expanding web of interdependence is evident in the counsel that, according to Plutarch, Aristotle gave to Alexander the Great: "to have regard for the Greeks as for friends and kindred, but to conduct himself toward other peoples as though they were plants or animals." Even today, for that matter, respect for people's basic humanity—that is, viewing people as people, worthy of decent treatment—may not extend much further than practical considerations dictate. But practical considerations dictate a larger moral sweep now, because interdependence has grown further. You simply cannot do business with people while executing all their male citizens, and increasingly we do business with people everywhere. The growth of non-zero-sumness, a growth driven by technological change but rooted more fundamentally in human nature itself, has in this one basic and profound way improved the conduct of humans. In fully modern societies, people now acknowledge, in principle, at least, that other peoples are people, too.

As we've seen, in the process of expanding, non-zero-sumness has brought not only more respect for more people, but more liberty for more people. The point isn't just—as such thinkers as Adam Smith have been saying since the eighteenth century—that free markets are best operated by free minds. The point is that the ongoing evolution of information technology heightens this synergy, underscores it, makes it something rulers can less and less afford to ignore.[†]

The world remains in many ways a horribly immoral place by almost anyone's standard. Still, the standards we apply now are much tougher than the standards of old. Now we ask not only that people not be literally enslaved, but that they be paid a decent wage and work under sanitary conditions.[†] Now we ask not only that dissidents not be beheaded en masse, but that they be able to say whatever they want to whomever they want. It is good that we thus agitate for further progress, and all signs are that this agitation goes with the flow of history. Still, it is hard, after pondering the full sweep of history, to resist the conclusion that—in some important ways, at least—the world now stands at its moral zenith to date.

Chapter Fifteen

NEW WORLD ORDER

An Asean-10, as it is sometimes called, will be a dynamic, free-market area of 500 million people by early next century. Already, ASEAN wields enormous diplomatic clout, either driving or codriving APEC, ARF and ASEM.

—*Wall Street Journal*

In 1500 B.C., there were around 600,000 autonomous polities on the planet. Today, after many mergers and acquisitions, there are 193 autonomous polities. At this rate, the planet should have a single government any day now.

World government? Traditionally, the idea of world government has been embraced mainly by left-wing peaceniks. In other corners, it draws various kinds of disdain, two of which are of special interest. One school finds the notion hopelessly unrealistic—and files those left-wing peaceniks under the label "woolly-minded one-worlders." The second school finds the notion plausible but terrifying—and speaks ominously of a coming "New World Order."

Roughly speaking, what divides these two schools is which half of "Jihad vs. McWorld" they extrapolate from. Many members of the first school look to the future and see ever-more-virulent tribalism: civil war, cross-border ethnic strife, and terrorism—all empowered by new and deadly technologies, and all in the explosive context of overpopulation and environmental stress. Thus, according to the writer Robert Kaplan, the "grid of nation-states is going to be replaced by a jagged-glass pattern of city-states, shanty-states, nebulous and anarchic regionalisms." Private armies and drug cartels will flourish and "criminal anarchy" will emerge as "the real 'strategic' danger."

In scenario number two, the problem isn't chaos, but rather a spooky kind of order. The order emanates partly from the multinational corporations and globetrotting financiers who animate McWorld. They swear allegiance not to any nation, but to profit alone, and they've implanted their values in such supranational bodies as the International Monetary Fund and the World Trade Organization, whose tendrils threaten to slowly engulf and then smother national self-determination. In this view, the alphabet soup of supranational organizations—IMF, WTO, UN, NAFTA, and so on—is the harbinger of a coming planetary authority, the sovereignty-crushing New World Order. In some apocalyptic visions on the fringes of Christian fundamentalism, such institutions literally represent the Antichrist.

One difference between these two schools—between the people who fear chaos and the people who fear order—is that the latter are more likely to be insane. They have a tendency, for example, to mistake unassuming dark-colored helicopters for attack aircraft sent by the United Nations.

Ordinarily, sane people are a more reliable guide to the future than crazy people. But here the opposite may be the case. If history is even a roughly accurate guide, much power now concentrated at the level of the nation-state will indeed migrate to international institutions. World *government*—a single, centralized, planetary authority—may or may not arrive, but something firm enough to warrant the name world *governance* is in the cards. World governance, you might say, is human destiny, the natural outgrowth of the millennia-old expansion of non-zero-sumness among human beings.

This isn't to say that the "chaos" theorists are wholly wrong. Indeed, there are at least two senses in which history will vindicate them.

First, the "fragmentation" and "tribalism" that are part of the chaos scenario are indeed growing, and will keep growing. In fact, that massive net decline in the number of polities on earth—from 600,000 to 193—masks a recent reversal. Over the past century, the number of polities has actually *grown*. But, as we'll see, this and other manifestations of "tribalism" are not just reconcilable with world governance; oddly, they are inseparable from it, integral to it.

Second, the chaos at the heart of the chaos scenario is not a figment of anyone's imagination. But, as we'll also see, it is precisely this chaos that is helping to drive the world to the final level of political organization, the global level.

THE LOGIC OF UNITY

The reason to expect the eventual triumph of global governance lies in three observations. (1) Governance has always tended to expand to the geographic scope necessary to solve emerging non-zero-sum problems that markets and moral codes can't alone solve. (2) These days many emerging non-zero-sum problems are supranational, involving many, sometimes all, nations. (3) The forces behind this growing scope of non-zero-sumness are technological and, for plain reasons, bound to intensify.

Consider the long, slow economic recovery of the European Middle Ages, when commerce began to flow beyond the bounds of localities. Growing trade among German cities was threatened by pirates and brigands. As we've seen, the merchants in these cities solved the problem by forming the Hanseatic League—not a full-fledged national government, but still a form of governance. Meanwhile, elsewhere in Europe, it was kings who smoothed the path for commerce, quelling trouble, harmonizing law, creating the nation-state. At work here was a general principle: as commerce expands, enmeshing more and more people, it gives them a shared interest in protecting it from friction and disruption.

Today, as commerce swamps the borders of nation-states, there are many sources of potential friction and disruption. One is trade disputes, which carry the threat of trade wars. Another, put on display during the Asian financial crisis of the late 1990s, is lack of "transparency," of good financial data about ostensibly healthy nations—a shortcoming that can lead finally to a panicked and ruinous exit of foreign investment.

Both of these problems raise the "trust" barrier to non-zero-sum gain. And trust barriers generally get breached by governance in one sense or another. Among the emerging institutions of world governance that address these particular trust problems are the World Trade Organization and the International Monetary Fund. The WTO rules on trade disputes (much as national courts resolve quarrels between companies, though with less binding authority, at least for now). The International Monetary Fund lends money to troubled nations to prevent panics (the rough global analogue of a nation's bank deposit insurance) and tries in exchange to get sound management and clean bookkeeping—transparency.

It is easy to laugh at nationalist fanatics who see these bodies as embryonic threats to free nations everywhere, who rail against the "secret tribunals" of the World Trade Organization. Chances are, after all, that your home has never been searched by an officer wearing a badge that says "IMF" or "WTO."

Still, if what authority these bodies do have exists in response to increasingly non-zero-sum relations among nations, and if technology is very likely to sustain this trend, then it doesn't take a wild-eyed visionary to imagine more and more authority migrating to the global level. Indeed, the Asian financial crisis, by reminding the world of its interdependence, brought calls for stronger international authority from such non-wild-eyed people as the chairman of the U.S. Federal Reserve Board. As the *New York Times* summarized the post-Asia conventional wisdom, "only the most dogmatic free market ideologues think the increasingly integrated global economy can get by without a super-national organization to serve as financial watchdog, scold, mediator and lender of last resort."

Watchdog, scold, mediator, lender—these don't sound very forceful. But to be a "lender of last resort"—to lend when the private sector refuses to—is to subsidize, and with subsidy comes power. Indeed, much of the American government's power over states consists not of legalized coercion, but of strings attached to subsidies. Mainstream plans for post-Asia reforms of the IMF followed this model of government control—a stronger role as lender of last resort, along with tighter strings: more financial disclosure, more fiscal discipline, closer regulation of banking.

As the Asian crisis subsided, the more ambitious of these plans began to fade. Even so, by the spring of 1999, the IMF had voted to set up a permanent "contingent credit line" for emergency use by countries that had met IMF standards of transparency and austerity. Since countries certified as eligible for these bailouts will presumably have an easier time attracting foreign capital, there will be a firm incentive to pass muster. Thus the "contingent credit line," however tenuous it sounds, may evolve into a systematic extension of IMF power.

And bear in mind that supranational lending institutions, even when their influence is more ad hoc, can seem plenty coercive to nations badly in need of a loan. Here is a sentence from a 1998 issue of the Asian *Wall Street Journal:* "The World Bank plans to oppose a South Korean government plan to set up a fund to buy equity stakes in troubled companies, probably dooming the proposal." To the South Koreans, this leverage no doubt felt strikingly like supranational authority.

Also bear in mind that loose organizations which solve non-zero-sum problems have a history of foreshadowing firmer authority. In the nineteenth century, various Italian states harmonized their tariff schedules, deepening economic integration and easing the eventual march toward national political unity. Who knows what could eventually become of the World Trade Organization?

This is in one sense a silly comparison. Economics was hardly the only force at work in nineteenth-century Italy. That other great mixer of non-zero-sum cement, hostility toward an alien power (Austria), played a role. Clearly, such hostility won't be a factor in the evolution of world governance, assuming any extraterrestrials out there continue to keep their distance.

Indeed, for political conglomeration to take place without war either in the backdrop, as a threat, or front and center, as a reality, would mark a contrast not just with Italian and German history, but with history, period—and, for that matter, with prehistory. As we've seen, when social organization moved to the supravillage level of the chiefdom, fighting very often, if not always, figured in. Sometimes the fighting was local, with one village chief conquering the others and thus becoming founder of a chiefdom. But when the merger happened less coercively—as one would hope in the case of world governance—an enemy seems to have loomed outside the chiefdom.

The United States, it is true, moved from a loose confederation to a true nation during peacetime. But it was war that had pushed the states over the threshold to confederation, and the specter of future war figured in the arguments for centralization laid out in the *Federalist Papers.*

In short, if space aliens don't attack the planet, yet it nonetheless moves toward firm supranational governance, the transition will be without known precedent. Still, there are two big reasons not to rule out such an aberration.

GETTING PULLED TOGETHER

First, though it's true that few if any big, voluntary political consoli-
dations have taken place in the prolonged absence of war and the
threat of war, it's also true that, so far as we know, there's never been
a prolonged absence of war and the threat of war. Who knows how
powerful economic logic alone might prove if it didn't keep getting
a boost from collective antipathy? If common enemies didn't keep
pushing people together, maybe people would have time to relax and
get pulled together—more slowly, but no less surely.

The European Union may now be illustrating the point. It began as
a zone of liberalized trade, in some ways like the North American
Free Trade Agreement. But look at how much power is centrally
wielded now. In 1996, amid fears of Mad Cow Disease, the EU
banned Britain from exporting beef—not just to the other nations in
the Union, but to the whole world! Among the other things the EU
has done matter-of-factly: prohibit member states from importing
Iranian pistachio nuts (which seemed to contain high levels of a natural carcinogen); and tell member states they could—and, indeed,
must—permit the sale of the impotency drug Viagra.

If a nation-state can't decide where to export its beef, which pistachio nuts are acceptable, and what remedies are available to impotent
citizens, then it is not fully sovereign in the old-fashioned sense of the
word. How did this happen? How is it that one minute you're a mere
free-trade zone, and the next minute—or, at least, the next half-
century—you're banning pistachio nuts? How did the European Coal
and Steel Community morph into the European Community and
then into the European Union? Actually, the transition is surprisingly
logical: one non-zero-sum game leads to another, which leads to
another, and so on. The logic is worth fleshing out, in part because
the World Trade Organization shows signs of following in some of
the EU's footsteps.

First, nations trade with one another (non-zero-sum game number
one). Then they see further gains in mutually lowering tariffs (game
number two). Then they decide all would benefit by dampening
quarrels over what qualifies as a violation of this agreement, so they
set up a way to settle disputes (game number three). Crossing this last
threshold—forming an inchoate judiciary—is what turned the Gen-

eral Agreement on Tariffs and Trade into the World Trade Organization.

But adjudication entails tricky questions. For example: How do you handle covert trade barriers? When a country makes it illegal to import shrimp caught in nets that kill sea turtles, is that really an environmental law, as advertised, or protectionism in disguise? Both the EU and the WTO weigh in on such matters. Thus did the WTO wind up telling the United States to quit barring shrimp imports from several Asian nations. In theory, the U.S. could have ignored the ruling. All the WTO would have done in response is to approve retaliatory tariffs by aggrieved countries—tariffs those countries could levy even without the WTO's blessing. Still, the U.S. has benefited from so many WTO rulings that it has a stake in preserving respect for them. This, really, is the whole idea behind governance: the voluntary submission of individual players to an authority that, by solving non-zero-sum problems, can give out in benefits more than it exacts in costs. This net benefit is why WTO rulings may well become more binding, whether through sheer custom or through new law.

As supranational governance greases transborder commerce, more questions arise. If each nation has different food-testing and food-labeling laws, won't that make life more complicated and expensive for pan-European food producers, and indeed for their customers? Moreover, if each nation has separate food laws, mightn't they morph into covert trade barriers? Mightn't French farmers support laws that deem Portuguese food processing substandard, or that define "cheese" so stringently as to keep some imports from carrying that label? All told, then, wouldn't uniform food laws make things cheaper and simpler for most people? Yes, says the EU (game number four) and thus acquires another function. (And yes, said the United States early in the twentieth century; it was growing interstate commerce that carried such regulatory functions from the state to the federal level.)

As Europe has already shown, the chain of non-zero-sum games needn't end with regulatory and judicial power. With transnational commerce growing, all those national currencies became a bother. There was the cost of currency exchange and nagging uncertainty about exchange rates. Wouldn't most Europeans benefit from a single currency? Yes, the EU decided. But one currency meant one central bank. So each nation lost its autonomous central bank and, at a more symbolic level, its currency (or, as they aptly say in Britain, its sovereign). At this point—with nations having surrendered their control of monetary policy—the line between a loose association of nations and an outright confederacy has arguably been crossed.

A "to-be-sure" is in order. To be sure, European integration was, from the beginning, not a strictly economic project. It was conceived as a way of "waging peace"—reuniting a war-torn continent with economic bonds that would make war unthinkable. (The plan seems to have worked.) Still, there was also a purely economic logic behind fusion, which is why European corporations pushed for it and the stock market applauded it. Indeed, just as Europe was unifying its currencies, *The Economist* published an article called "One World, One Currency," noting the analogously powerful economic logic behind global monetary union. The article stressed the near-term political impossibility of such a goal, and, moreover, some economists doubt its sheerly economic wisdom. Still, as exchange rates gyrated in the aftermath of the Asian crisis, there was serious talk in both Argentina and Mexico about adopting the U.S. dollar as official currency.

All in all, the moral of the EU story is: Presto! International trade, via self-regenerating non-zero-sumness, can expand the scope of governance. No extraterrestrials necessary! As technology continues to shorten economic distance, the logical scope of supranational governance could conceivably become the whole planet. This may be hard to imagine now, given the cultural and linguistic diversity of the world and the simmering hostility among some of its peoples. But remember: If ninety, even sixty, years ago, you had predicted that someday France and Germany would have the same currency, the reply would have been: "Oh, really? Which nation will have conquered which?"

GETTING PUSHED TOGETHER

Even if murderous extraterrestrials aren't a strict prerequisite for global governance, they would be a big time saver. Even if sheerly economic forces could in theory get people to put aside their petty differences (a big if), nothing does the trick quite like the common threat of death.

As it happens, the end of the second millennium has brought the rough equivalent of hostile extraterrestrials—not a single impending assault, but lots of new threats that, together, add up to a big planetary security problem. They range from terrorists (with their menu of

increasingly spooky weapons) to a new breed of transnational criminals (many of whom will commit their crimes in that inherently
transnational realm, cyberspace) to environmental problems (global
warming, ozone depletion, and lots of merely regional but still supranational issues) to health problems (epidemics that exploit modern
thoroughfares). None of these things has quite the galvanizing effect of
space invaders, but they are all scary, and they all imply supranational
governance in one sense or another. They threaten many nations with
common perils that are best overcome through cooperation.

The timing is convenient. With economic organization reaching
the global level, and governance showing faint signs of doing the
same, that great historical congealer of governance—an external
enemy—disappears by definition. Meanwhile, a whole slew of non-
zero-sum problems arise that are rather like an external enemy; they
push people together, to escape common calamity, rather than pulling
them together for common gain.

You can already see some of the new structures of governance
emerging. They are tenuous, but in their very weakness future
strength is visible. Consider the Chemical Weapons Convention. The
CWC gives an international body unprecedented power to spring
surprise searches on any member state at the behest of any other
member state. In 1997, when the U.S. Senate debated the treaty
before ratifying it, critics warned about a loss of American sovereignty. In reply, defenders of the convention found themselves in the
odd position of stressing its underlying weakness: if the inspectorate
tries to search your garage, the American government can stall, and if
the search seems unconstitutional, can thwart it. Presumably true.
Still, if other countries can do the same, then much of the CWC's
value to the U.S.—the ability to demand inspections in foreign countries—goes down the drain. And if the U.S. wants to realize that
value—if it wants other nations to surrender a bit more sovereignty—
then *it* will have to surrender a bit more sovereignty. That's the way
these games work.

The question, then, is whether the value of a stronger inspectorate
will ever warrant the price of lost sovereignty. And the answer is
almost certainly yes. The spread of technological information—
including how to make weapons—has always been unstoppable in the
long run, and new media, notably the Internet, are making the long
run very short.

Actually, chemical weapons are the least of the problem. On the Richter scale of weapons of mass destruction, they barely register. It is biological and nuclear weapons that ensure that by 2020 any well-funded terrorist group will have the know-how to kill 50,000 people in any given city.

Biological weapons are much easier to make and to hide than nuclear weapons. The encroachment on sovereignty that they will call for is, to current sensibilities, shocking. All kinds of industrial and medical facilities will have to be monitored. The personal possession of some equipment will probably have to be banned, and surprise inspections will be necessary. This prospect—that some supranational agency could demand to search your basement or your kitchen freezer—would now strike most Americans as unthinkable, even if local police were in tow to guard against abuse.

But trauma has a way of making the unthinkable widely thought. In the middle of World War II, the historian Arnold Toynbee met with a number of notables in Princeton, New Jersey, to discuss the postwar world. By the end of the meeting, Toynbee had convinced John Foster Dulles—a temperamentally conservative man who would later become secretary of state in a Republican administration—that world government was essential. Dulles signed on to the group's conclusion that "as Christians we must proclaim the moral consequences of the factual interdependence to which the world has come. The world has become a community, and its constituent members no longer have the moral right to exercise 'sovereignty' or 'independence' which is now no more than a legal right to act without regard to the harm which is done to others."

And World War II, bear in mind, was not as scary for the average American as biological weapons will be. There was no chance in 1942 that whole American cities would be decimated without warning. Once this threat becomes real, appreciable sacrifices of sovereignty are among the less extreme solutions that will get trotted out. (And among the more benign. Persecuting particular groups, such as Muslims, may seem far-fetched now, but recall the internment of Japanese Americans during World War II. Wouldn't supranational governance be preferable?)

Lots of non-zero-sum problems show the same logic: basic technological trends make their growth all but certain, and their eventual solution will likely entail real accretions of supranational governance.

Global laws on the prescription of antibiotics? Sure, if their too-casual use creates strains of super-bacteria that can cross oceans on any airplane. Limits on each nation's saltwater fish harvest, complete with random inspections of any fishing boat and stiff penalties? Sure, as the oceans thin out. An International Monetary Fund whose regulatory powers approach those that national governments now possess? Quite possibly, in the wake of a global depression.

And then, of course, there is that notorious sovereignty-sapper, cyberspace, which empowers offshore tax evaders, offshore info-terrorists, offshore copyright violators, offshore libelers. (And some-times the culprit may be on-shore, using an offshore computer.) Nations will find it harder and harder to enforce more and more laws unless they coordinate law enforcement and, in some cases, the laws themselves.

Viewed separately, the various layers of supranational governance may not seem momentous. But they add up. We can't predict for sure which problems will find strong supranational solution, but a broader prediction is possible: as technological evolution keeps doing what it has always done—expand and deepen non-zero-sumness—much supranational governance will be in order.

In the end, then, the "chaos" and "order" scenarios find a kind of reconciliation. Whole new species of chaos will indeed arise, but—assuming we respond to them wisely—they will drive the world to a new level of political organization that is capable of preserving order. What the "chaos" theorists fail to see is that chaos is just a non-zero-sum problem, something people are good at solving.

GOOD TRIBALISM

Some chaos theorists also suffer from a skewed view of modern "trib-alism." They fret about its disruptive effects without realizing that, for every newly empowered, destablizing "tribe," there are scores more "tribes" emerging that are harmless if not benign—tribes that will help bring order to the new world.

The first step toward seeing why is to remember that a prime ener-gizer of tribalism is communications technology. For Irish who are serious about their Irishness, there is the Gaelic Channel (*Teilifís na Gaelige*). And virtually all tribes—from Quebec's separatists to north-ern Italy's separatists to France's Corsican separatists to Spain's (and France's) Basque separatists—are using the Internet for orchestration.

The same information technology that serves these old-fashioned "tribes"—bound by language or religion or cultural history—also serves "tribes" of a newer sort, bound by common interests ranging from the political to the recreational. This trend long predates the Internet and cable TV. Beginning in the early 1960s, computerized mass-mail so lowered the costs of organizing that such once-obscure groups as the American Association of Retired Persons became political potentates.

Since then desktop publishing and other microcomputer technologies have brought the threshold for organizing down so low that few common interests are too obscure to warrant mobilization. Nebraska's Original Betty Club, confined to women named Betty who live in Nebraska, seeks to boost esteem for a name that, they lament, has fallen out of fashion. The "Christian Boys and Men Titanic Society" is "dedicated to the notion of saving women and children first." The "National Association for the Advancement of Fat Acceptance" is self-explanatory.

As long-distance contact gets cheaper and easier, more and more of these "tribes" are transnational. The European Headache Federation, according to the *Wall Street Journal,* was founded in the 1990s to "elevate the status of the headache," a mission that entails publishing the "Headache Yellow Pages."

Many transnational groups are of greater geographic scope than the European Headache Federation, and of greater moment. Environmental groups, using "EcoNet," "GreenNet," and other cyberspace channels, were among the first to congeal globally via the Internet. Adherents of the chaos scenario, such as Robert Kaplan, commonly stress the roiling effect of both tribalism and environmental degradation. But the very technology that drives some of the tribalism is helping to solve environmental problems; it is empowering supranational enviro-tribes—technically, "non-governmental organizations," or NGOs—that are shaping policy at the supranational level, where it often belongs. They have had real influence on accords ranging from global warming to overfishing, and have made their presence felt in corporate headquarters the world around. To protest logging by Mitsubishi, the Rainforest Action Network let concerned world citizens use its Web site to fax the car maker, generating so much paper (a tree's worth?) that Mitsubishi changed its fax number. (Meanwhile, Mitsubishi also had a brush with a supranational feminism tribe. Japanese women's groups picketed a shareholders' meeting over sexual harassment at a Mitsubishi plant in Illinois.)

Supranational "tribes" are now agitating in such once-national policy realms as factory working conditions. In the process, they illustrate that governance needn't always involve governments. Consider the success of the International Labor Rights Fund, the Lawyers Committee for Human Rights, and other nongovernmental organizations in negotiating a code governing wages, workweeks, and child labor in clothing factories. Nike, Liz Claiborne, L. L. Bean, and other clothiers agreed to the code so their products could sport labels attesting to humane working conditions. Nike, meanwhile, as part of the Federation of Sporting Goods Industries, was also negotiating with such NGOs as Unicef and Oxfam Christian Aid over conditions in sporting goods factories. And the South Asian Coalition on Child Servitude, working with religious, labor, and consumer groups, got some rug makers in India to allow surprise factory inspections so that their rugs could carry the "Rugmark" label, signifying that adults making the local minimum wage had woven them.

This is not government as we've come to know it; no elected officials, no tax dollars involved. But, to the extent that it works, it is governance, in somewhat the sense that the merging of merchants into the Hanseatic League was. And it is governance aimed at some of the problems in the chaos-theory laundry list, such as the cultural dislocation that comes from the rapid industrialization of traditional societies.

Lest we get misty-eyed over the beneficence of the International Labor Rights Fund and like-minded souls, let's pause for a dose of cynicism. American labor unions didn't spend so much time lamenting working conditions abroad before workers abroad started taking jobs from American workers. Now union leaders worry deeply about those conditions—whenever they keep wages so low that American union members can't compete.

Then again, lobbying has always been self-interested. That's governance as usual. More interesting is what's new about the lobbying: it's transnational. Karl Marx isn't these days considered a great prophet, and his suggestion that "workers of the world unite" doesn't seem to have been widely heeded. But economic logic is moving that prophecy toward a modest kind of fulfillment.

After all, even as American workers try to make some Asians jobless—such as children with very low wages—these Americans see eye to eye with other Asian workers, such as the ones who, after transnational regulation, will get the (now higher-paying) jobs instead of the children. And when American unions press for labor accords in trade agreements—when they want NAFTA to ensure Mexican workers the right to organize—they are singing in tune with most Mexican workers. In general, workers in high-wage and low-wage countries have a common interest in elevating pay in low-wage countries (so long as large numbers of workers in low-wage countries aren't priced out of the market altogether, a threat that doesn't loom large).

As it happens, NAFTA included no meaningful labor accords; it passed the U.S. Congress on a center-right coalition, so its structure is conservative. But this is no necessary feature of trade blocs. The European Union gets very involved in workplace regulation—and, in general, proves that a trade bloc can embody leftish values.

Will the World Trade Organization ever go the way of the European Union? Will qualifying for membership someday entail more than low tariffs? Child-labor laws? Workplace-safety laws? Right-to-organize laws? And what about environmental law? Might adherence to global environmental treaties become part of staying a member in good standing of the WTO?

None of this is unthinkable, especially if the American left—and the left in other nations—make such changes a condition for accepting WTO amendments that big business supports. We may well someday see workers of the world truly uniting, like environmentalists of the world. The old political debates will continue—How to balance economic equality with efficiency?—but they will be mediated by supranational bodies of governance, whether regional or global.

Indeed, 1999 saw a glimmer of such a day during wrangling over who the next director of the World Trade Organization would be. As the *New York Times* noted, the United States—under a left-of-center government beholden to labor and environmental groups—favored the candidate more likely to "advocate environmental and labor issues." France did too, even though its farmers feared he would take a harsh view of agricultural subsidies. A European diplomat said of French politics: "This is a case where the labor forces who fear low-wage competition have won out over the agricultural interests."

Such interest-group struggles are nothing new. But they didn't use to involve bodies with names like "the World Trade Organization."

In another sign of consequential global governance, hotels near the World Trade Organization headquarters in Geneva have become a hangout for lobbyists—from Aetna, Citibank, the International Federation of Accountants, and so on. Intense lobbying also attended the 1996 agreement among 160 nations—just about the whole world—on a common copyright code, reached under the auspices of the UN's World Intellectual Property Organization. Meanwhile, lobbyists for Consumers International approached the UN's Codex Committee on Food Labelling about getting genetically modified foods so labeled. After failing there, they moved on to the Convention on Biodiversity, a transnational group that includes the EU and Japan.

Note the pattern: Again and again, supranational "tribes"—environmental groups, labor groups, human rights groups, trade groups, multinational corporations—abet order, not chaos. Their narrow but long reach moves law and regulation toward global harmony. These different tribes don't stand in paradoxical contrast to globalization, any more than myriad kinds of cells stand in paradoxical contrast to the larger organism they constitute. In both cases, the fine-grained diversity is integral. The body politic couldn't reach the global level if interest groups didn't get there. To say that social complexity grows in depth and scope is to say that division of labor, including division of political labor, grows more fine-grained yet of broader reach. In that sense, supranational "tribalism" is a natural outgrowth of the whole history of humankind.

BAD TRIBALISM MADE GOOD

And what of the more literal, more disturbing forms of tribalism? What of raw nationalism, or the ethnic balkanization of states? These do seem resurgent—not surprisingly, given the end of the Cold War and the march of technology. But at times they fall under new forms of supranational sway. Twice during the 1990s, arguments between nations over disputed islands wound up not in war but in the World Court. And in 1998 a Rwandan mayor was condemned to life in prison for abetting genocide—sentenced not by victorious enemies, but by a UN tribunal.

Indeed, there is growing reason to believe that, even as tribalism, galvanized by microelectronics, balkanizes nations, it will fit neatly into the New World Order. Quebec's separatists vowed that, if liberated from Canada, their new nation would straightaway join NAFTA and the World Trade Organization. This would be a smart political move—a way for Quebec to bind itself to the United States, Mexico, and the larger world, diluting dependence on its recently divorced spouse. But whatever the motivation, the effect would be to cement Quebec into supranational bodies.

The same logic holds on the other side of the Atlantic. On the eve of European currency unification, polls showed that no nation supported the EU more strongly than Ireland. After all, membership in the EU affords the Irish some economic distance from the resented British. By the same token, if any of Europe's separatist movements ever gain sovereignty—in northern Italy, Basque Spain, Corsica—expect them to have membership forms on the EU's doorstep the next day.

Faced with this spectacle, we will once again find it hard to figure out which side is winning: Jihad or McWorld. Would this be a more tribalized or more globalized world? Or, to get back to the very beginning of this chapter: How do you do the accounting here? If in 2025 Italy splits in two, but both halves are plugged into a network of supranational governance more solid than anything that exists today—if, indeed, they've ceded large chunks of sovereignty by joining it—then has the number of sovereign polities in the world grown or shrunk? There is some point at which supranational governance becomes firm enough so that the things we now call nations are more like provinces. What is that point?

This question helps rectify the anomaly noted above: after the number of polities had fallen for most of human history, it rose during the twentieth century, as empires breathed their last. Now we can see continuity beneath this quirk. For as the number of "sovereign" polities grew, the degree of their sovereignty was waning. These were not, after all, the good old days of the Middle Ages, when states were expected to act as they pleased on the high seas and to treat visitors from other states with caprice. By the nineteenth century, the notion of international law had taken root, and the intangible strength of international norms was growing.

This new support for supranational rules was based largely on growing economic interdependence, which made war more and more a lose-lose game. Indeed, war was not just a lose-lose game, but a lose-lose-lose game—bad for the warriors, but also, increasingly, bad for their neighbors. Kant had seen this logic beginning to unfold back in the late eighteenth century, and had extrapolated from it ambitiously: "The effects which an upheaval in any state produces upon all the others in our continent, where all are so closely linked by trade, are so perceptible that these other states are forced by their own insecurity to offer themselves as arbiters, albeit without legal authority, so that they indirectly prepare the way for a great political body of the future, without precedent in the past."

Not long after Kant dropped this hint about human destiny, Napoleon proved that destiny had not yet arrived. But the havoc he wreaked had its upside. Peace was ratified by the Congress of Vienna in 1814–1815, an unprecedented gathering at which all European states tried to forge a structure for lasting order. As it happened, that structure would be used not just to prevent war, but to stifle the liberal revolts of 1848, so the Congress of Vienna is often recalled as reactionary. But it set a vital precedent. In agreeing to hold periodic follow-up meetings on continental stability, it became a primordial version of the "great political body" that Kant envisioned.

And it seemed to help. Much of the nineteenth century was a peaceful time for Europe. Indeed, as technology drove trade to higher levels, peace seemed to some the wave of the future. In 1910, Norman Angell famously explained, in the widely translated book *The Great Illusion,* that in an interdependent world, war made no sense as an instrument of policy. Just as famously, the world didn't listen.

After World War I, the League of Nations—a more elaborate version of the "great political body"—was formed to keep the peace. It failed. But today, one world war later, economic interdependence makes war even less rational than before. And nuclear weapons further elevate the irrationality.

Encouragingly, people seem to grasp all this non-zero-sum logic. So far, at least, nuclear powers have tended not to have wars with one another. And as for economic interdependence: wars have lately been confined to areas such as Africa and the Balkans, where historical factors have prevented rich, fine-grained interdependence from evolving—"underglobalized" areas, you might call them. In a striking departure from historical pattern, wars are increasingly things that poor nations do; rich nations intervene to stop them, not to win them. Barring a reversal of time-honored trends, poor nations will get richer, making war a remote prospect in an ever-larger part of the planet.

"Great political body" number three—the United Nations—is the most ambitious one yet. God knows it hasn't done quite the job of preserving peace that idealists envisioned. But it has had more glimmers of success than the League of Nations, and has already lived twice as long as the league.[†]

Moreover, expanding non-zero-sumness has in the meanwhile carried the rationale for supranational governance well beyond peacekeeping. Everything from tax policy to accounting standards to environmental policy to policing power is creeping toward the supranational level.[†]

Whether the eventual result will be a single "world government," as woolly-minded one-worlders have long dreamed, is anybody's guess. Another possibility, and perhaps a more benign one, is lots of overlapping bodies—some regional, some global; some economic, some environmental; some comprising national governments, some comprising non-state actors; and so on. An intermediate possibility is that these sorts of bodies would fall loosely under the auspices of the United Nations (as many inchoate bodies now do).

All kinds of scenarios are imaginable. What is hard to imagine is that this migration of governance beyond national bounds will stop before turning nation-states into something rather like provinces. The migration is driven by clear technological trends that are millennia-old and show no signs of abating: advances in technology—information, transportation, military, and so on—that make relations among polities more non-zero-sum. This logic will draw "tribes" into the New World Order even as they fiercely assert their independence from an old order.

Obviously, some "tribes" are less eager to join than others. But it is unlikely that any nation, no matter how radical its origins, will forever resist globalization. Perhaps the most anti-modern regime in the world today is the fundamentalist Islamic government of Afghanistan, the Taliban. Yet in 1999 the Taliban orchestrated the first large western investment in Afghanistan in two decades—a new telephone system that would finally permit direct-dial international calls. As the *Washington Post* reported, "Taliban authorities . . . said they hoped [the new phone system] would help bring in more foreign trade and investment, reestablish links with Afghan professionals who fled overseas, bring the Internet to their educational system and improve Afghanistan's image abroad." This is not a recipe for continued Jihad.

DIFFUSE DESTINIES

In many developed nations, the drift toward world governance is drawing fire. The nation-state, nationalists complain, is sacrificing its sovereignty. This is true; governmental structures—including supranational ones—always lessen the freedom of their constituents. But at the same time, governmental structures expand freedom. If a city's government is functioning well, its citizens gain the freedom to walk the streets with little fear of assault. Part of the deal, though, is that they don't have the freedom to assault other citizens. If you like the idea of government, that means you cherish freedom from assault more than freedom *to* assault.

So it is with supranational governance. Would you like to be reasonably free of the fear of a global depression? Or would you rather preserve your nation's freedom to raise tariffs at will, or to keep its financial institutions opaque to international view? Do you cherish the freedom to live without fear of dying in a biological weapons attack? Or do you prefer the freedom to live without fear of having your freezer searched for anthrax by an international inspectorate in the unlikely event that evidence casts suspicion in your direction?

Or, to turn such questions inside out: Which sort of sovereignty would you rather lose? Sovereignty over your freezer, or sovereignty over your life?

The loss of sovereignty isn't some novelty dreamed up by bureaucrats at the UN or the WTO. If we define sovereignty broadly—as supreme control over your fate—then the loss of sovereignty is a fact of history, one of the most fundamental, stubborn facts in all of history. Indeed, to observe that history has eroded sovereignty time and again is simply to restate the thesis of this book: that history has elevated non-zero-sumness time and again. For to say that you have been cast into a non-zero-sum situation is to say that you have lost unilateral control over your future—that your destiny has to some extent been taken out of your hands and spread among other people, just as part of their destiny now rests in your hands. Both you and they can regain some measure of that control, some portion of that lost sovereignty, but only through cooperation—a cooperation that involves sacrifices of control, of sovereignty, in its own right. The question is never whether you can keep all of your sovereignty; history says you can't; all along it has been the fate of humankind to have its fate increasingly shared. The question is in what form you want to *lose* your sovereignty.

Of course, even answering that question wisely won't bring instant Nirvana. Once we've recognized the necessity for global governance, we still have to get from here to there. And that could be dicey. After all, history—and prehistory—attest that evolving from one distinct level of political organization to another often brings "transitional instabilities"—a polite term for catastrophe. Elman Service observed, for example, that "as a theocratic chiefdom becomes a state, it is normally in a state of convulsion, if not of full civil war."

Can we manage to move from the national to the supranational level of organization without convulsion? And is there a price for doing so? Would preserving order amid great flux entail large sacrifices of privacy and civil liberty? These are the subjects of the next chapter.

Chapter Sixteen

DEGREES OF FREEDOM

No one can be perfectly happy till all are happy.
—*Herbert Spencer*

In general, history has shown a healthy indifference to the strengths and weaknesses of particular political leaders. Blunders and oppressions in any one part of the world have tended not to be broadly ruinous; there have always been whole continents full of other polities where more enlightened policies could prove their mettle. The basic path of cultural evolution, toward broader and deeper social complexity, has been safe from the ravages of evil and incompetence.

Until recently, it was unthinkable that anyone would end this benign fragmentation. Genghis Khan couldn't possibly have ruled the whole planet, even with his high-tech courier service. Napoleon, living in pre-telegraphic days, would also have found world conquest hard to attain, much less sustain.

But Hitler, in an age of telephones and airplanes, could conceivably have conquered the world. True, stagnation might have ensued, as he suppressed information technologies to head off revolt. But rulers can bear stagnation so long as there are no vibrant societies to vanquish or embarrass them.

By World War II, then, "saving the world" had ceased to be hyperbole. Winston Churchill, in rallying resistance to Hitler, was performing a feat whose greatness wouldn't have been possible centuries earlier. It is in this light, the light at the end of the second millennium, that leaders have come to matter in a larger sense than ever before.

Yet the hypothetical prospect of world conquest isn't the main thing now putting a premium on leadership. Even assuming that governance moves to the global level peacefully and democratically, inspired guidance will matter in a new way. After all, there's only one globe. However many alternatives there are for reforming the IMF after a crisis, they can be tested only in serial, not in parallel. And if the first trial fails spectacularly, that's really bad news.[†]

What's more, the potential badness of bad news has risen. With more souls in the world every century, the sheer weight of potential suffering has reached an all-time high. Hitler and Stalin made this point, and the coming of thousands of nuclear weapons has underscored it.

Of course, if you're a person of sufficiently large vision, you can always shrug this worry off. Even if we wipe out all human beings, some species will survive, such as the famously radiation-resistant cockroach. And if biological evolution is directional (see part II), then maybe there will eventually be a species smart enough to reignite cultural evolution, impelling social organization, once again, toward the planetary level. So global concord will get a second chance!

Personally, I don't feel a strong enough kinship with cockroaches to find much solace in this scenario. In fact, there are mishaps well short of nuclear annihilation that I'd just as soon avoid. I say we take a stab at figuring out what sorts of things great leaders would do to keep our species not just alive, but in reasonably good shape as it makes the tricky transition to a new social equilibrium.

TIP #1 ON SAVING THE WORLD

There are two basic keys to saving the world. The first is to recognize the inevitable and come to terms with it. Granted, this is not wholly original advice. But, obvious as it sounds, the world has often failed to follow it. World War I began with the Austro-Hungarian empire still refusing to accept that, in the age of print, suppressing nationalism was a loser's game. The Hapsburgs not only clung to their imperial Balkan holdings but tried to expand them, notwithstanding obviously fierce Slavic nationalism. The event that triggered the war epitomizes Hapsburg hubris. The archduke Francis Ferdinand was visiting Bosnia when a nationalist terrorist tossed a bomb toward his car. The bomb bounced away before exploding. Rather than pause and humbly reassess the wisdom of this particular visit, the archduke resumed the motorcade and proceeded with his scheduled activities—and was assassinated later that day.

So what inevitabilities should the modern world recognize? For starters, the one the Hapsburgs ignored. The present information revolution is comparable in consequence to the print revolution and carries the same basic lesson. Denying self-determination to homogenous, determined groups will get harder and harder, be they Kosovars, Corsicans, or Tibetans.[†]

Even apparent successes in repressing the new technologies of nationalism foreshadow long-run failure. Turkey, with its restless population of Kurds, had long tried to thwart the British-based broadcasts of Med TV, the world's only Kurdish-language TV channel, by cracking down on satellite dishes. Then in 1999, Britain, Turkey's ally, revoked Med TV's broadcasting license. The director of Med TV vowed to find a new home, and that shouldn't be hard; all it takes is one nation with a satellite uplink. But more important is that in five or ten or fifteen years, when the Internet's broadband revolution has reached Turkey, you won't even need an uplink to broadcast—just a computer. There will be dozens of Kurdish TV channels, and the only way to block them will be to ban modems—this in an age when having a modem will be like having running water. Call this Inevitability Number One—the inexorable spread of technologies of tribalism.

As we saw in the last chapter, technologically empowered nationalism and micronationalism needn't be a long-run problem. Newly sovereign polities will someday be securely cemented into supranational bodies, right next to older sovereign polities (with "sovereignty," in both cases, being not what it used to be). But getting from here to there is no snap. Secession movements usually inspire resistance—which, in turn, usually empowers militants within the secession. So things could get messy in a lot of places fairly fast—a classic "transitional instability."

Making things even messier is Inevitability Number Two: the growing power, compactness, and accessibility of lethal technologies. This trend dates back to the invention of gunpowder and has now gone beyond chemistry, into nuclear physics and biotechnology. Meanwhile, knowledge of how to harness the new lethal forces is rendered ever more available by ever-subtler information technologies. It seems likely that, for some time to come, more and more people will have the option of committing atrocities of greater and greater severity.

These people will include frustrated nationalists, but also lots of other people—most notably people who in one sense or another suffer from the cultural dislocation that globalization brings. There are Islamic and other religious fundamentalists who find their values threatened by modernization. There are plain old Luddites. There are environmentalists radicalized by the pace of deforestation. All these people, thanks to Inevitability Number Two, constitute a new breed of threat—what the journalist Thomas Friedman calls the "superempowered angry man."

In addition to the superempowered angry man, we face what you might call the quite disgruntled man—maybe not upset enough to bomb the World Trade Center, but still displeased with the drift of things, and, thanks partly to information technology, able to have real political effect. This category includes Americans who lose jobs to low-wage foreigners, and French farmers exposed to high-tech competition from abroad. Their political manifestation is a reactionary nationalism with a nasty, nativist undercurrent that, amid an economic downturn, could get pretty ugly. These disgruntled men (and women) aren't going away. Indeed, the source of their grievance—the globalization of capital and technology broadly—is so basic as to constitute Inevitability Number Three.

So there you have it: three inevitabilities, all part and parcel of globalization, and all with disruptive tendencies, at least in the short term. What to do?

TIP #2 ON SAVING THE WORLD

The early-twentieth-century sociologist William Ogburn attributed many of the world's problems to "cultural lag." Cultural lag happens when material culture (technology, basically) changes so fast that non-material culture (including governance and social norms) has trouble catching up. In short, the disruptive part of culture gets out ahead of what Ogburn called the "adaptive" part of culture. Ogburn's general prescription was to speed up the latter—"make the cultural adjustments as quickly as possible." But there is another option: slow down the former—cut the rate at which material technology is transforming the world; make the inevitable unfold at a more sedate pace.

Of course, it isn't that easy. Globalization doesn't come with a velocity-control knob. And the old-fashioned approach to slowing the spread of material technology—raising tariffs—has a history of inviting retaliation and thus yielding full-blown trade wars (the kind that usher in depressions). But there is a safer approach to slowing globalization down just a tad—a supranational approach.

The idea isn't to create a Bureau of Global Slowdown at the United Nations. The idea is simply to tolerate various supranational efforts that are starting to take shape and that, as they solidify, will naturally have a sedative effect. As first-world and third-world workers unite to raise third-world wages (and thus keep first-world wages from free-falling), industrialists will complain that this dulls the market's edge, slowing progress. Yes, it does—but that's okay. As environmentalists unite to save rain forests, or tax fossil fuels, the same complaint will be heard—and the same answer will apply. In the age of the superempowered angry man, and the quite disgruntled man, the slowing down of deeply unsettling change is a benefit, not just a cost, because anger and disgruntlement are world-class problems.

Note that these two particular forms of slowdown—supranational labor and environmental policies—have the added virtue of directly addressing specific sources of anger and disgruntlement: the rapid exodus of blue-collar jobs from developed nations; and the ecological damage that can radicalize environmentalists and that, more broadly, deepens cultural dislocation in such already polluted places as Mexico City and Bangkok. In a sense, then, these policies address *both* halves of the "cultural lag." Rather than choose between slowing the "material" change and hastening the "adaptive" change, we can slow the material change *by* hastening the adaptive change.[†]

In a way, it's a misnomer to call this a "slowing" of globalization. After all, the things that might do the slowing—supranational labor groups or environmental groups, supranational bodies of governance—are themselves part of globalization. What is really happening is that the further evolution of *political* globalization is slightly slowing the evolution of *economic* globalization.

This approach has a proven track record. In the United States during the early twentieth century, as economic activity migrated from the state level to the national level, the national government grew powerful enough to regulate it. And some of the regulation—labor laws in particular—had the effect of subduing capitalism a bit, dulling its harsher edges. This was, among other things, a preemptive strike against chaos (in the form of Marxist revolution) and a successful one.

We needn't worry much about economic globalization grinding to a halt under the weight of regulation. The political prerequisites for real regulation, such as strong transnational labor coalitions, will form only slowly, and the economic impetus behind globalization is mammoth. In fact, the only thing with much chance of stalling globalization for any length of time is the very chaotic backlash—from the angry and the disgruntled—that a slight slowdown might avert. Strange as it sounds, the best way to keep economic globalization from slowing down a lot may be to slow it down a little.

THE LIVES OF A CELL

There is one other inevitability that will shape future life: the growing ability to document people's day-to-day behavior—via Web-browser footprints, credit card records, and other forms of digital data. Whether this is a good or a bad thing depends on your perspective.

On the plus side, it is a weapon against superempowered angry men. The World Trade Center bombers left damning evidence on, among other things, a bank's computers and their own desktop computer. In future cases, expect to see evidence from EZ Pass highway toll booths and those rapidly multiplying security cameras. (After the bombing, a number of New York landmarks, including Rockefeller Center, installed cameras that capture the face and license plate of every driver entering the parking garage.)

All to the good—except for one thing: those surveillance cameras and electronic toll booths chronicle the movements not just of terrorists, but of you. Of course, we could always subdue this data gathering by passing laws. And you could dodge some of the data gathering by avoiding places with security cameras. But you may feel safer parking in a camera-equipped garage. And you may be happy for toll booths to save records of your passage, if that will help catch any terrorist who should plant a bomb in some highway tunnel. This could well be the core threat to future privacy—not its unwanted invasion by Big Brother, but the voluntary surrender of it in the name of security. And, depending on how insecure we feel, we might decide to grant police—national, maybe international police—lots of leeway in using such data as grounds for search. This may seem implausible now, but, as I've already suggested, a single act of biological terrorism in a big city could redraw the contours of plausibility.

Hence a creepy irony of the coming world: even though information technology's basic drift in recent centuries has been to expand freedom—to bring political pluralism to more and more nations—it can, at another level, shrink freedom. As the world comes to resemble a giant superorganism, with a fiber-optic nervous system, we could come to identify with Winston Smith, who, in Orwell's *1984,* is asked by a totalitarian goon: "Can you not understand, Winston, that the individual is only a cell?" But, unlike Smith, we'll have *chosen* the life of a cell.[†]

The trade-off between liberty and privacy on the one hand, and order and security on the other, is a hardy perennial. What is new are two things: the growing technological ease of invading privacy, and the growing technological ease of disrupting order. These are what threaten to cast the trade-off in new and severe terms.

In this light, slowing down economic globalization looks better than ever. After all, one way to soften the terms of this trade-off—to get more order without giving up liberty—is to reduce the number of people who want to disrupt the order: shrink the supply of angry men.

Indeed, if shrinking this supply buys freedom, it's tempting to ask: Are there any *more* ways to shrink it? Any further tips for reducing the number of aspiring terrorists and other hate-filled people? Any untapped channels of influence?

THINGS GET MUSHY

Pierre Teilhard de Chardin, the mid-twentieth-century prophet of globalization, hinted at one. Having forecast the integration of humankind into a "planetized" whole—a giant "super-organism"—Teilhard addressed the obvious, Orwellian fear: the "lives of a cell" scenario. His message: Fear not. "To say 'love' is to say 'liberty.' There need be no fear of enslavement or atrophy in a world so richly charged with charity."

Well that's a load off my mind! As usual, Teilhard's optimism is so fuzzy and boundless as to erode his credibility. And, as usual, his instinct for big-picture dynamics is acute enough to restore some credibility. He seems to sense that, when you face the trade-off between freedom and order, much of your wiggle room comes from a third variable that lies in the realm of spirit—or, to put it more mundanely, the realm of morality; what the world needs is to expand its supply of good will.

Of course, this goal, though moral, isn't reachable only by moral means. We've just seen some political initiatives that, while not likely to unleash a torrent of good will, could at least cut the supply of *ill* will. There are no doubt other such policies. Still, when one group of people harbors contempt, or even disdain, for another group—fundamentalist Muslims for westerners, say, or westerners for fundamentalist Muslims—the problem is indeed at some basic level moral. What is arguably the biggest challenge of the future—staying safe while staying free—may be a project that is as much spiritual as political.

You may object that this project sounds vague and mushy. Well, obviously! That's why it's called a "spiritual" project instead of, say, a "civil engineering" project. Whether it is a doable project is unknown. But moral leadership *has* occasionally met with success. Neither Gandhi nor Martin Luther King, Jr., accomplished all he had hoped for, but both led wrenching and necessary social transformations with less violence than such transformations had historically entailed. And, while both men were at some level politicians, their political force was inextricable from their spiritual force.

History holds other causes for hope, too. At the end of the Middle Ages, when Europe's governance, impelled largely by economics, moved from the local to the national level, the nation-state became, in some measure, a unit not just of political organization, but of moral organization, featuring at times a certain diffuse good will. Religion played a role—monarchs didn't exactly shy away from Christian symbolism—but also important was the sheer sense of common cultural belonging and common destiny. People thought of themselves as being in the same boat.

Well, in some ways, at least, the whole world is increasingly in the same boat. In that light, was Teilhard's optimism so hopelessly unrealistic? Given inspired leadership, how close *could* the world's peoples come to brotherly love—or, failing that, to the less intense psychic unity that a mild-mannered nationalism brings? We'll return to this question toward the end of the book. For now I'll just note that there was a time, centuries ago, when even nationalist sentiment must have seemed improbable to Europeans, given cultural and linguistic differences and venerable hatreds.

There is one other, very different, and somewhat smaller, sense in which modern problems may find quasi-spiritual solutions. Slowing the rate of economic globalization, hence of cultural dislocation, is not only a political project. The less bent on material acquisition people in affluent nations are, the less breakneck the pace of modernization—and, as a bonus, the less environmental havoc there will be. As another bonus, we may discover something sages have been saying for millennia: endless acquisition isn't the route to fulfillment anyway.

MIXED EMOTIONS

Clearly, the argument of this book has taken a strange twist. Two of them, in fact—two twists that run parallel to each other.

First, I've argued above for subduing the pace of history—this after contending for fifteen chapters that the direction of history is largely good. Sounds odd, I know. But it is history's *long-run* course that is mainly good, and, as John Maynard Keynes pointed out, it is the short run—the time frame in which we live our lives—that concerns us most. In the short run, the "natural" course of history has sometimes brought much unpleasantness.

And "cultural lag" seems to have been a big reason. Some historians trace the virulence of twentieth-century German nationalism all the way back to the nineteenth century, when industrialization swept lands that had just barely left the Middle Ages. Russia, even more than Germany, had to fast-forward from an age of serfs into the industrial revolution—and, in a sense, Russia never recovered, never developed fitting governance. Among the casualties that can arguably be chalked up to this cultural lag: the many millions who died in the Holocaust or at Stalin's hands. Today, catastrophes of this size could transpire even without the sponsorship of a national political leader. That's life in a world featuring unaccounted-for nuclear materials, pervasive biotechnology, and lots of unhappy campers. Hence my concern about the number of unhappy campers.

Which brings us to the second strange twist: the sudden welling up of moral sentiment—my rhapsodizing about the need for good will and about the evils of avarice. The previous chapters had evinced a certain austere admiration for the "unsocial sociability" that Kant saw in the human psyche. After all, this tension within human nature has sustained the largely healthful drift of history. The tireless pursuit of social status, even of conquest, has ultimately elevated the human condition, allowing more and more people to live, on balance, better lives. How can I justify turning my back on the very things that got us where we are today—such spurs to progress as greed and hatred?

Well, at the risk of sounding cold-hearted: they've outlived their usefulness. From the beginning their value was of an ironic sort.

Enmity drove society toward larger expanses of amity. Greed and the lust for status, for power over people, helped drive a technological evolution that granted people more freedom. All along, the darker side of human nature was defensible, if at all, only to the extent that it tended to thus negate its own values system. And, all along, there was the implied prospect that, in the end, if the darker side's downside grew and its upside waned, defending it would get hard.

The end is here. With the world's ecosystem already under stress, and billions of additional people apparently on the way, mindless materialism grows more dubious.[†] With society finally globalized, we don't need war to push political organization (that is, the realm of peace) to broader expanse. And with nuclear and biological weapons at hand, full-fledged war—and for that matter full-fledged terrorism—are less palatable than ever. Hatred just isn't what it used to be.

Even Herbert Spencer—who had a certain respect for enmity's fructifying effects—saw the declining virtue of antipathy. He wrote: "From war has been gained all that it had to give. . . ." The social evolution that "had to be achieved through the conflicts of societies with one another, has already been achieved; and no further benefits are to be looked for." Wars, he observed, had not only ceased to be vital to progress; increasingly, they were the cause of "retrogression." (And this was *before* nuclear weapons.)

War has contained the seeds of its own demise all along. This primal form of zero-sum energy, through the very logic of history that it helped impel, was bound to grow more and more negative-sum until finally its downside was too glaring to ignore. In retrospect, it looks almost like planned obsolescence.

If war can indeed be turned into a relic, then the virtue of greed will recede further. From a given society's standpoint, one big upside of wanton material acquisition has traditionally been the way it drives technological progress—which, after all, helps keep societies strong. In the nineteenth century, Russia and Germany had little choice about modernizing; in those days stasis invited conquest. But if societies no longer face conquest, breakneck technological advance is an offer they can refuse, and frugality a luxury their people can afford.

God knows greed won't vanish. Neither will hatred or chauvinism. Human nature is a stubborn thing. But it isn't beyond control. Even if our core impulses can't be banished, they can be tempered and redirected.

Or, more accurately: some impulses can be used against others. People will always seek social status, and revel in the esteem of their peers, but this very thirst can be used to dampen other thirsts. In defining the kinds of behaviors that do and don't win esteem, communities have great power over how human nature expresses itself. Among the things that can in principle become prerequisites for social status (and, indeed, in some communities already are): not engaging in conspicuous consumption; not saying hateful things about whole national, ethnic, or religious groups, or even about other people.

Franz Boas, though not big on generalizing about history, once stated as "one of the fundamental characteristics of the development of mankind" that "activities which have developed unconsciously are gradually made the subject of reasoning." The example he cited was the maturation of scientific inquiry, but one might also cite the maturation of history itself. As the people of the world come to constitute a single invisible brain, they can purposefully guide their course, consciously seeking the worthy goals they were once blindly, often painfully, driven toward.

Kant, you may recall, was so impressed with the way narrow self-interest had served the greater good that he believed some gratitude was in order: Nature should be "thanked for fostering social incompatibility, enviously competitive vanity, and insatiable desires for possession or even power." Well, all right: Thank you, nature. More specifically, thank you, *human* nature. Now get a grip on yourself.

PART II

II

A

BRIEF

HISTORY

OF

ORGANIC

LIFE

Chapter Seventeen

THE COSMIC CONTEXT

The atom is a pattern, and the molecule is a pattern, and the crystal is a pattern; but the stone, although it is made up of these patterns, is just a mere confusion. . . .

—*Aldous Huxley*

One of the most influential science books of the twentieth century was an ultra-slim volume, written by the physicist Erwin Schrödinger, called *What Is Life?* Published in 1944, it helped inspire James Watson and Francis Crick, among other future eminences, to study life's hereditary material.

So what is the answer? What is life? There are questions I don't purport to answer in this book, and that is one of them. But briefly pondering it will help answer some questions that are central to this book. For example: Why is it one book, instead of two? Aren't organic evolution and human history sufficiently different to demand separate treatment?

Early on, I claimed that the answer is no—that the two processes naturally constitute a single story. This claim of unity has several dimensions. First, the claim is that the two processes have common dynamics; "evolution" isn't just a catchy metaphor for cultural change; at some basic level, cultural evolution and biological evolution have the same machinery. Second, they have the same fuel; the energetic interplay between zero-sum and non-zero-sum forces has been similarly pervasive in the two evolutions. Third, the two processes have parallel directions—long-run growth in non-zero-sumness, and thus in the depth and scope of complexity. Indeed, organic evolution, given long enough, was very likely to produce creatures so complex, and so intelligent, as to be capable of sponsoring cultural evolution—

a cultural evolution that would then naturally extend evolution's general drift toward deeper and vaster complexity.

All of this is what the next three chapters are about—showing that, on a number of grounds, it makes sense to see all of history since the primordial ooze as a single creative thrust. But the first step toward doing that is to get clear—or, at least, clearer—on what was so neat about the primordial ooze in the first place: What is life?

THE SPIRIT OF THE SECOND LAW

Schrödinger saw life against the backdrop of the second law of thermodynamics. The second law (in case your mastery of thermodynamics has decomposed over time) is the one that sounds so depressing: entropy—disorder—grows inexorably; structure decays. The logical culmination of this trend is a day when all molecules are randomly distributed. No planets, no stars—nothing but sameness; the universe, as if it had been run through an unusually large Cuisinart, will be a vast puree.

The process is observable even in smaller spaces, and over a shorter time frame, here on earth. Pour cream in coffee, and the initial distinctions in color, texture, and temperature fade, as does the motion created by the pouring. Generally speaking, Schrödinger observed, systems left alone for very long will become motionless and of uniform temperature; eventually, "the whole system fades away into a dead, inert lump of matter."

What makes life so strange is its seeming exception to this rule. Unlike cups of coffee, organisms preserve distinctions—between kidneys and stomachs, between leaves and stems. "It is by avoiding the rapid decay into the inert state," Schrödinger wrote, "that an organism appears so enigmatic."

What's the trick? Is life defying the second law of thermodynamics? No. The process of living, like all other processes, raises the total amount of entropy in the universe, destroying order and structure. Ever compare a five-course meal with the ensuing excrement? Something has been lost.

Obviously, something has been gained, too. The growth of an organism creates new order and structure. But on balance, says the second law, the organism has to consume more order than it creates. And so it does. The key to staying alive (Write this down!) is to hang on to the order and expel the disorder. As Schrödinger put it, "the essential thing in metabolism is that the organism succeeds in freeing itself from all the entropy it cannot help producing while alive." So order grows locally even as it declines universally.

Still, if life doesn't violate the letter of the law, it violates the spirit of the law. To read the second law, you wouldn't expect these islands of structure to arise and persist. Yet the islands not only persist; they grow, by virtue of a crafty trade policy: import structured things (five-course meals) and export less structured things.

How, exactly, does importing structure help life preserve its own structure? What life is mainly after isn't structure in some generic sense, but rather structured *energy*. After all, everyday experience shows us that if energy is to be of any use, it has to be in structured, concentrated form. The heat in a campfire's embers can be used to boil water, but only because it is concentrated in the embers. Once the second law has done its work, and all the embers' heat has diffused into the air, the energy, though still around, is of no use. The key to exploiting energy is to get near a nice, distinct package of the stuff and then capture some of it as it degrades into more entropic, less useful form. It is this captured energy that life uses to build and replenish its structure, to arrange matter into distinctly ordered form and keep it there, notwithstanding the universal tide of entropy.

Thus, when you eat a hamburger, you are ingesting structured, hence usable, energy. This structured energy was donated by a cow, which got the energy from grass, which in turn got the energy from sunlight, which carried the sun's energy to the Earth in concentrated little packets. At every link in this chain, the total amount of entropy in the universe grew, and the total amount of usable energy dropped, along with the total amount of order. Still, here on planet Earth, more and more usable energy was crammed into small organic spaces, and the total amount of order grew, folded into more and more complex forms.[†]

ENERGY AND CULTURE

Some cultural evolutionists have put great stress on the role of energy technology in cultural evolution. That's not surprising. After all, inherent in cultural evolutionism is the idea that human societies are in some ways like organisms. And, as we've just seen, the very essence of an organism is that it captures and processes energy; it uses this structured energy to (among other things) create structured matter:

fingernails, fur, bones, beaks, brains—all the things whose ongoing existence constitutes defiance of the spirit of the second law. Certainly a cultural evolutionist can be forgiven for asking: Well, then, can the same be said for a human society?

Pretty much. Like individual organisms, societies convert energy into material structures. Even the simplest hunter-gatherer societies take energy in the form of food and use it to build shelters. As societies get more complex, energy technologies change. Farming replaces hunting and gathering as the source of human energy, and human energy gets supplemented by nonhuman energy—oxen, waterwheels, steam engines. But one thing stays the same; energy keeps getting invested in structure, albeit more complex structure: temples, factories, jetliners. Indeed, even leaving aside all these material accomplishments, the people themselves, in a less concrete but real sense, constitute structures. A marching army, a cheering crowd, even a far-flung corporation amount to coordinated, hence orderly, arrangements of bodies. These human structures, like the nonhuman kind, depend on energy. And—also like the nonhuman kind—they tend to get more complex as cultural evolution advances.

Rather as an organism's various structures preserve the organism against entropic forces, at least some of a society's structures protect the society against disintegration. Certainly that's true of the walls of Jericho, perhaps the oldest living monument to agricultural productivity. Even a chiefdom's temples—regardless of whether they were tools of oppression and mass control or of benign coordination—were in some sense integrative, keeping a large society ordered, organically coherent. And the corporations that provide a society with food or clothing, as well as the armies that defend it, are there to stave off atrophy or disruption.

The comparison between society and organism should certainly be kept on a leash. For one thing, energy use is less frivolous, more consistently functional, within an organism than within a society. Still, comparing organisms and societies has its payoffs, and justifies some of the attention that cultural evolutionists have paid to energy technologies.

And yet, during part I of this book, I put most of the emphasis on information technologies. Indeed, I argued that what is commonly called the first energy revolution—the coming of farming—was probably more important as an *information* revolution; the residential density allowed by farming brought a quantum leap in the size and efficiency of the "invisible brain."

Why my seeming preoccupation with information technology? In part, perhaps, because we're living in the information age, and, like most people, I tend to see the past in terms of the present. But there is another reason for treating information with respect. Though both information and energy are fundamental, information is in charge. In human societies, energy (and matter, for that matter) is guided by information—not the other way around.[†]

ENERGY AND INFORMATION

One of the first cultural evolutionists to emphasize this fact was Kent Flannery. Even in fairly simple hunter-gatherer bands, he wrote in 1972, there are often "informal headmen, who collect and distribute knowledge about which groves of edible nuts have been thoroughly harvested, which canyons currently have high concentrations of game, and so on."

At higher levels of social complexity, too, information is running the show. Chiefs give signals that channel human energy into the construction of temples. They give signals that determine how many spears should be made and when the spears should be carried off to war. And they do all this only after receiving signals about the state of things in the world.

Ancient states also processed information—often in bureaucratic form—to process energy and matter. So too with modern markets; the invisible hand, as we've seen, is utterly dependent on an invisible brain, a decentralized system of data processing. All told, Flannery argued, "one of the main trends" in cultural evolution is "a gradual increase in capacity for information processing, storage, and analysis."

Does the comparison between organism and society hold up here, too? Do plants and animals process information to process energy and matter?

Yes, all living things do. Even the lowly *E. coli* bacterium does. When in an energy-poor environment, it synthesizes a molecule called cyclic AMP that then binds to the DNA, prompting the DNA to initiate the construction of a flagellum, a tail that whips around, sending the bacterium swimming off until it finds energy-rich environs. The cyclic AMP molecule is a kind of symbol, whose meaning is rather like the meaning of a hunter-gatherer's report to the headman that the local nut groves are barren. And the DNA that responds to the symbol by triggering the construction of a tail is rather like the headman who assimilates the evidence of scarcity and suggests that the band relocate.

The word "meaning" isn't meant loosely here. Charles S. Peirce, founder of the philosophy known as pragmatism, believed that the "meaning" of a message is the behavior it induces, behavior appropriate to the information the message carries about the state of the environment. In this sense, a cyclic AMP molecule has genuine meaning. It leads to behavior (relocation) appropriate to the information conveyed (a local energy shortage).

In the course of wisely relocating, the *E. coli* belies the old notion that DNA is merely a "blueprint" for construction of the organism. DNA isn't just data; it is a data *processor.* Long after a human being has been "built," the DNA in each cell continues to do what *E. coli's* DNA does: absorb information from its environment (which, in the case of humans, is often other cells) and govern its cell's behavior accordingly. And, for that matter, during the building of the organism, DNA is processing information about the larger environment and shaping the organism accordingly—adjusting a child's skin hue to the intensity of sunlight, for example. From birth until death, DNA is brainier than any blueprint.

Data processing permeates even trivial-seeming organic functions. When grass grows upward instead of downward, its cell surfaces have sensed which direction holds more light and heat, and conveyed data reflecting that fact to the DNA, which then deploys signals governing growth. When a broken bone heals, signals must first relay the fact of the injury, and then other signals guide various construction projects, from protective swelling to bone mending.

Even at the level of individual molecules, building and preserving structure in the face of entropy has to involve a kind of information processing, a kind of discernment. The Nobel Prize–winning molecular biologist Jacques Monod observed that protein molecules have the "ability to 'recognize' other molecules" by their shape—and thus array themselves into orderly arrangements. "At work here is, quite literally, a microscopic discriminative (if not 'cognitive') faculty." This "elective discrimination" makes living things "appear to escape the fate spelled out by the second law of thermodynamics." At all levels of any organism, information guides energy and matter in ways that preserve structure—much as it does in human societies.

There is one further analogy between organisms and societies. It isn't just that in both cases energy is marshaled in a way that sustains and protects structure. And it isn't just that this marshaling is always guided by information. It is that *it is the function of the information to guide the marshaling*. The inter-cellular and intra-cellular messenger systems involved in fingernail or bone construction evolved to guide such construction; they were preserved by natural selection because they helped preserve the core of life's structure, the DNA. And the same is true of the information-processing system that governs the construction of huts and temples and skyscrapers, and the maintenance of corporations and armies.

I don't mean that people are genetically endowed with cerebral programs for building temples and running armies. I mean that, more generally, people are genetically endowed with the proclivity to think about building things and orchestrating social enterprises, and to communicate about these projects; and that these genetically based data-processing proclivities were favored by natural selection because they helped sustain DNA, helped keep people alive. Cultural information, like all previous forms of organic information, was created to preserve and protect genetic information.

Viewed against the backdrop of all of life, then, culture was in one sense nothing new: just another data-processing system invented by natural selection to marshal energy and matter in ways that preserve DNA. But it was the first of these systems that began to take on a life of its own, inaugurating a whole new kind of evolution. Natural selection, after inventing brainier and brainier forms of DNA, long ago invented brains—and then finally, in our species, invented a particularly impressive brain, a brain that could sponsor a whole new kind of natural selection.

MIRACLE GLUE

One of the first cultural evolutionists to emphasize energy technologies was Leslie White. Indeed, some mark the mid-twentieth-century renaissance of interest in cultural evolutionism to the publication in 1943 of his paper "Energy and the Evolution of Culture." White was also among the cultural evolutionists who more or less ignored information technologies. In a way, this is not surprising. His landmark paper came out a year before Schrödinger's book appeared, a decade before DNA was discerned, and decades before science had truly grasped the pervasive role of biological information in upholding the integrity of organisms. Thus his attempts to import insights from biology into the social sciences met with limited success. He observed that the question "What holds systems together?" is "as fundamental to sociology as it is to biology," but he couldn't carry the analysis further. Presumably the key was a "force," but in the social sciences we "have no name for this force unless we call it solidarity," and "what its name, if any, is in biology, we do not know."

Well, now we do. In societies, in organisms, in cells, the magic glue is information. Information is what synchronizes the parts of the whole and keeps them in touch with each other as they collectively resist disruption and decay. Information is what allows life to defy the spirit, though not the letter, of the second law of thermodynamics. Information marshals the energy needed to build and replenish the structures that the entropic currents of time tirelessly erode. And this information isn't some mysterious "force," but, rather, physical stuff: the patterned sound waves that my vocal chords send to your ear, the firing of neurons in a brain, the hormones that regulate blood sugar, the cyclic AMP molecule in a bacterium. Information is a structured form of matter or energy whose generic function is to sustain and protect structure. It is what directs matter and energy to where they are needed, and in so doing brushes entropy aside, so that order can grow locally even as it declines universally—so that life can exist.

Ever since the primordial ooze, information technology has stayed at the center of the story. From DNA to the brain to the Internet, information processors keep spawning bigger information processors, and being subsumed by them. Life keeps violating the spirit of the second law on a grander and grander scale. Only by seeing this epic story in its full sweep—not just the last 30,000 years, but the previous several billion—can we address the question of whether it had, in any sense, an author.

Chapter Eighteen

THE RISE OF BIOLOGICAL NON-ZERO-SUMNESS

A beehive is a collaborative enterprise on far more levels than first appears.

—*Matt Ridley*

You may be under the impression that you have a single set of genes, arrayed along chromosomes in the nucleus of each cell. A common misconception. The nucleus is just one of many little subcellular bodies called organelles. And one kind of organelle—the mitochondrion, which processes energy—has its own genes, passed down separately from the genes in your nucleus. Whereas your nuclear genes were drawn equally from your mother and father, all your mitochondrial genes came from your mother. And if you are male, you have no chance of passing them to future generations.

Why on earth would each cell have two sets of DNA? The answer, now generally acknowledged after decades of resistance, is this: Once upon a time, before there were multicelled organisms, the distant ancestor of your mitochondria was a free-living, self-sufficient cell, something like a simple bacterium; and the distant ancestor of the cells the mitochondria now inhabit—your nucleated cells—was also a free-living, self-sufficient cell. Then the two free-living cells merged; the mitochondrion specialized in processing energy, and the larger cell handled other matters, such as locomotion. The two lived happily ever after in blissful division of labor.

I'm such a romantic—always stressing mutual benefit. Some biologists would say that the story as I've rendered it glosses over ugly details. According to the Nobel laureate Renato Dulbecco, the once-autonomous mitochondria are now "subservient to the needs of the cells in which they reside." According to John Maynard Smith and Eörs Szathmáry, authors of *The Major Transitions in Evolution,* mitochondria are "encapsulated slaves," subject to ruthless "metabolic exploitation."

As we'll see, one can argue with such cynical interpretations of the mitochondria's plight, but for now the point is just that biologists actually talk this way—as if a mitochondrion, a dinky blob presumably lacking sentience, *had* a plight. What do they mean? In what sense could an organelle be a "slave" that gets "exploited"? You would think, to hear this kind of talk, that a mitochondrion, like a person, has interests that are either served or not served. Is there any sense in which that's true?

Yes, in a Darwinian sense. In Darwinian terms, living things are "designed"—by natural selection—to get their genes into subsequent generations. To serve their "interests" is to aid this genetic proliferation. To frustrate their interests—to "exploit" them, for example—is to reduce their genetic legacy.

With this vocabulary in hand, we can apply game theory to biological evolution. When two organic entities can enhance each other's prospects for survival and reproduction, they face a non-zero-sum situation; to the extent that their interests are at odds, the dynamic is zero-sum. In this light we will see that biological evolution, like cultural evolution, can be viewed as the ongoing elaboration of non-zero-sum dynamics. From alpha to omega, from the first primordial chromosome on up to the first human beings, natural selection has smiled on the expansion of non-zero-sumness.

ALLIANCE IN THE PRIMORDIAL SOUP

How did life begin? Beats me. But from the beginning, one of its driving forces was non-zero-sum logic.

Consider the two genes that linked up to form the world's first chromosome. (I'm using "chromosome" loosely, to mean any strand of genes, not just the neatly packaged strands that evolved in the kinds of cells that constitute plants and animals.) What were the genes' functions? Again: I have no idea (except that they must have been able to replicate). But this much is clear: by virtue of now being in the same boat, they had a highly non-zero-sum relationship. By and large, what was conducive to the survival and replication of one was conducive to the survival and replication of the other. So whatever functions they *did* perform—building protective insulation, say—they were best advised to cooperate, to unite in pursuit of their common goal, dividing labor synergistically.

Of course, genes don't take advice. And they don't "cooperate" in the sense that people cooperate—size up the other players and decide that coordination is the most sensible option. In fact, once these two genes had hooked up, there *were* no options. The nature of both genes was now set; either they would by their nature cooperate to accomplish more than they could accomplish alone, or they would by their nature miss this opportunity. If the latter, then their lineage would stand a poorer chance of flourishing. Unless a genetic mutation soon brightened prospects for their progeny, this particular experiment in inter-gene cooperation might well fail. The two genes that invented the world's first chromosome, lacking a legacy, would fall into the dustbin of organic history.

That's life: most experiments fail. Still, maybe the pair of genes that invented the world's *second* chromosome, or its third, or whatever, would interact more constructively. Prospects would be brighter for this lineage, which could be fruitful and multiply.

It is in this blind, stumbling way that genes come to "pursue strategies" without thinking about the goal. Over time, natural selection preserves those genes that just happen to do a good job of playing the game.

And when genes are on the same chromosome, cooperation is generally the name of the game. The maturation of a frog is a construction project of such intricate harmony that by comparison the building of a modern skyscraper seems crude and chaotic. Ditto—only more so—for a bear. To say that more and more complex organisms have evolved over time—as they have—is to say that genes have over time gotten involved in more vast and elaborate non-zero-sum interactions. From bacteria to people, biological evolution moved in fundamentally the same direction that cultural evolution has moved in.

Is the story of life really so simple? Just more and more non-zero-sumness piling up? No. For one thing, there are a few fairly daunting evolutionary thresholds that had to get crossed somewhere between a well-designed bacterium and *Homo sapiens*. But, conveniently, crossing these thresholds tends to depend on harnessing non-zero-sum logic, a task at which natural selection manifestly excels.

HOW CELLS GOT COMPLICATED

One of the greatest thresholds was the coming of the "eukaryotic" cell. For much of the early history of life, the most complex being was the prokaryotic cell, which persists today in such forms as bacteria. Prokaryotes are a bit slovenly. For example, their DNA is bloblike, constituting a chromosome only in the loose sense. The eukaryotic cell, which went on to be the building block for plants and animals, is more tidy and bureaucratic. Its DNA is arrayed neatly along distinct chromosomes and housed in a nucleus, whence it issues commands that are shuttled forth. The eukaryote has much division of labor, thanks to its many organelles. These include the mitochondria, mentioned above; and, in plants, the green bodies called chloroplasts, which handle photosynthesis and which, like the mitochondria, descended from a free-living ancestor that fatefully merged with another cell.

Back when many biologists doubted the autonomous origins of mitochondria and chloroplasts, the foremost proponent of the scenario was the biologist Lynn Margulis. Margulis contends that various other organelles, too, had free-living ancestors. The nucleus itself may even be an example. Most mainstream biologists doubt that Margulis is right, but then again, not so long ago they were doubting her story about mitochondria and chloroplasts.

Why all the mainstream skepticism? Margulis, for one, believes that biologists (who tend to be male) have a bias in favor of competition, and against cooperation, as the formative force in evolution. She might plausibly use the scientists quoted above, talking about "exploitation" and "subservience," as Exhibit A in her indictment. Though they accept her thesis that mitochondria came about by merger, they still insist that mitochondria are brutally subjugated by dominant partners. Or, to put their claim in the language of biology: they accept that two distinct entities came together through "symbiosis"—which just means "living together"—but insist that the symbiosis is parasitic, not "mutualistic." Or, to put the claim in the language of game theory: yes, it all started with a relationship between two sovereign beings, but the central dynamic has been zero-sum, not non-zero-sum.

Are these stereotypically male biologists right? Not demonstrably.

Consider Dulbecco's contention that the "needs" of mitochondria are "subservient" to the needs of the larger eukaryotic cell. It rests entirely on his observation that the actions of the mitochondria are governed mostly by genes in the larger cell's nucleus; the big shots in the nucleus give the orders, and the mitochondrion obediently follows them—the very definition of servitude.

It's true that many instructions governing mitochondria issue from the nucleus. Indeed, as other biologists have noted, some genes that were initially in the mitochondrion, and controlled it from there, seem to have migrated to the nucleus, where they exercise remote control. But, as some of those biologists have also noted, this transfer may have been favored by natural selection because it raised the efficiency of the overall cell. If so, then the transfer brightened prospects for the DNA remaining in the mitochondrion as much as for the DNA in the nucleus—since the overall cell is, after all, the boat in which both kinds of DNA find themselves.

Dulbecco is anthropomorphizing mitochondria. Human beings like autonomy, and often resist control. But there's no evidence that mitochondria have a strong opinion about autonomy one way or the other. The only sense in which they can be said to have "needs" or "interests" is in a Darwinian sense. And in Darwinian terms, they are just as interested in the efficiency of the larger cell as is the nuclear DNA. For both of them, the larger cell is home. The two are locked into a highly non-zero-sum relationship.

The jaundiced view of Maynard Smith and Szathmáry also has anthropomorphic overtones, but of a different sort. To back up their view of mitochondria as slaves, they theorize that long ago, at the beginning of the symbiosis, nuclear cells kept mitochondria around "as humans keep pigs: for controlled exploitation." That is: they would let the mitochondria reproduce in captivity, then eat a few, then let them reproduce some more. In this scenario, the cell's current handling of mitochondria—inserting "tapping proteins" to extract energy, rather than eating the whole mitochondrion—is just a higher-tech version of the original enslavement; it is a "more elaborate metabolic exploitation."

But, even granting that mitochondria started out as the subcellular equivalent of pigs, would this really be exploitation? Don't get me wrong. I'd rather be a person than a pig, and I do believe that pigs get the raw end of the pig-person relationship. But that's because when I think about *human* benefit—and, in a way, even when I think about pig benefit—I think about happiness; and in both species (presumably) a certain amount of freedom furthers happiness. But when we talk about nuclei and organelles, we're talking only about Darwinian benefit, about genetic proliferation. And in *Darwinian* terms, domesticated species do very well, thank you. There is today a lot more pig DNA around than its undomesticated kin, wild boar DNA. In that sense—in the Darwinian sense—getting eaten is the best thing that ever happened to pork. Analogously, the "controlled exploitation" of those proto-organelles may well have boosted their legacy, in which case it wasn't exploitation. Certainly they have lots of descendants today—billions in every sizable animal on earth.

In harping on intra-cellular non-zero-sumness, I don't mean to say that the interests of organelle DNA and nuclear DNA entirely coincide. Though the two spend most of their life in the same boat, they do take separate boats to the next generation. Since a mitochondrion's DNA is passed down only via mothers, its Darwinian interests might be served by biasing reproduction in favor of females, so that daughters were the norm and sons the exception. Even if this sexual imbalance cut down a bit on the reproductive success of the overall organism, the trade-off could still be worthwhile from the mitochondrion's point of view. But the nucleus would take a different view, since it gains nothing by a surplus of daughters.

Hypothetical as this logic may sound, it has actually found incarnation—in plants at least. In various plant species, mitochondria have genes that cause the (male) pollen to abort, biasing reproduction in favor of (female) seeds. That this works against the nuclear DNA's interests is evident in the countermeasures it takes. In some cases, nuclear "restorer" genes have evolved to neutralize the bias by boosting the supply of pollen.

This is a reminder that non-zero-sum relationships almost always have their natural tension, their purely zero-sum dimension. It is also a reminder that biologists such as Dulbecco are not wrong to say that tension can exist between mitochondria and nuclei. But they are sometimes wrong in what they see as evidence of tension, and in leaping to the conclusion that tension is pervasive, when in fact it is a small part of the overall picture.[†]

At the very beginning of the mitochondrion-nucleus relationship, to be sure, zero-sumness was probably the main story line. Apparently the proto-mitochondrion first got inside the cell in an act of exploitation gone awry. Either the big cell tried to eat the little cell and failed to digest it, or the little cell invaded the big cell and failed to kill it. But these ignoble origins of a non-zero-sum relationship shouldn't surprise us. The general theme of this book is that non-zero-sumness tends, via both biological and cultural evolution, to *emerge*. And it has often emerged among entities—villages, cities, states—whose relationship had once been overwhelmingly zero-sum. Even if the relationship between mitochondria and the larger cell was initially full of strife and bitter recrimination, today they are locked into an essentially non-zero-sum game, and both play the game well, to their mutual benefit.

The underlying reason that non-zero-sum games wind up being played well is the same in biological evolution as in cultural evolution. Whether you are a bunch of genes or a bunch of memes, if you're all in the same boat you'll tend to perish unless you are conducive to productive coordination. For genes, the boat tends to be a cell or a multicelled organism or occasionally, as we'll see shortly, a looser grouping, such as a family; for memes, the boat is often a larger social group—a village, a chiefdom, a state, a religious denomination, Boy Scouts of America, whatever. Genetic evolution thus tends to create smoothly integrated organisms, and cultural evolution tends to create smoothly integrated groups of organisms.

E PLURIBUS UNUM

In crossing the next great threshold after the eukaryotic cell—the chasm between single cells and multicellular life—natural selection again followed non-zero-sum logic, but this time the logic had a different foundation.

Many single-celled forms of life reproduce without sex—clonally. A cell just splits in two, creating a carbon copy of itself. One consequence is that adjacent cells often have exactly the same Darwinian interest, a fact that makes their merger into a single multicelled organism more feasible. If this makes immediate sense to you, congratulations. If it doesn't, don't feel bad; evolutionary biologists didn't quite fathom the logic until the 1960s, more than a century after *On the Origin of Species,* when William Hamilton authored the theory of kin selection.

What is in an organic entity's "Darwinian interest," remember, is what aids the transmission of its genetic information. So if a bunch of cells have exactly the same genetic information, their Darwinian interests are by definition identical. Suppose, for example, that two cells face starvation, but cell A can somehow save cell B by committing suicide. The net effect of cell A's "sacrifice" is to raise the chances that *its own* genetic information will reach the next generation—since, after all, its own genetic information is being carried by cell B. By Darwinian math, a cell has as large a stake in a clone's welfare as in its own welfare.

Of course, cells don't do math, or conscious calculation of any other sort. So how might a cell come to pursue its Darwinian interest in this circuitous manner—via "kin-selected altruism"? Imagine a mutant gene that just happens to incline a cell to make some sacrifice on behalf of nearby cells. After a few rounds of cell division, a cell containing this gene is surrounded by cells that also contain the gene. So when the occasion arises for this "altruistic" gene to spring into action on behalf of an imperiled nearby cell, the gene is actually rescuing a copy of itself. Sure, the gene stands a chance of perishing in the course of such heroic acts (that's why they call them "altruistic"); but if, on balance, more copies of the gene are saved than lost, the gene can spread through the population (that's why they put "altruistic" in quotes). This gene will outreproduce an alternative, risk-averse gene that would stand idly by while nearby copies of itself perished en masse.

Consider a real-life example of altruism at the cellular level: the cellular slime mold. The slime mold is reminiscent of that old TV commercial about whether Certs is a candy mint or a breath mint. (Both—"it's two, two, two mints in one.") Is a slime mold a society of cells or a single multicelled organism? Both—or, if you prefer, neither; it sits at the boundary between society and organism, never making a firm commitment. Its cells spend lots of time on their own, scooting along the forest floor looking for nutrients, occasionally reproducing by splitting into two. But when food grows scarce, the first cells to feel the shortage emit a kind of alarm call in the form of a chemical called acrasin. Other cells respond to the call, and a transformation ensues. The cells bunch up together, form a tiny slug, and start crawling as one.

Finally, having reached a propitious spot, they set about to create a new generation of slime mold cells. The slug stands up on end and turns into a "fruiting body" that features a sharp division of cellular labor. While some cells—depending on where they happened to be in the sluggish aggregation—become bricks in a sturdy stalk, other cells—again, depending on where they find themselves—become spores, designated for reproduction. The spores rise to the top, for widespread dispersal, and are launched into space to carry the slime mold legacy into posterity. The stalk cells, having spent their last full measure of devotion, now die. They have "sacrificed" their own reproductive prospects for those of their neighbors. But the "sacrifice" isn't real. The stalk cells stand a very good chance of being genetically identical to their next-door neighbors, so they have a strong Darwinian stake in the dispersion of the spores.

Once again, two organic entities—the cells that turned into bricks and the cells that turned into spores—cooperate because they are in the same boat. But in this case the "same boat" doesn't mean "the same physical vehicle," as it does for genes that share a chromosome. After all, for much of the cells' life cycle they are on their own, free to reproduce autonomously; that they "choose" to cooperate in constructing a common vehicle suggests some *prior* non-zero-sum logic, some sense in which they were *already* in the same boat. And they were. They were in the same boat by virtue of having a common Darwinian interest—getting the same set of genetic information sent to the next generation.

The slime mold's aggregation seems so magical that back in the 1930s, the German biologist who first filmed it described it in Bergsonesque terms—as the result of an immaterial "vital force." But in fact the aggregation is orchestrated concretely—by the acrasin that cells emit when it is time to aggregate.

This sending of signals is of course fairly standard procedure in non-zero-sum relationships. In the case of humans the signals may be light waves conveying gestures, or sound waves that carry words, or patterns of ink, or electronic blips. In the case of intra-cellular communication, the signals are often proteins. For example, the nucleus governs the cell's energy processing by having proteins sent to the mitochondrion. But at all levels of organization, when entities coordinate their behavior to mutual benefit, information tends to be processed.

Think back to the *E. coli* bacterium featured in the last chapter, the one that builds a propulsive tail to escape low-carbon environs. This deft maneuver is actually a cooperative venture between two kinds of genes on the bacterial DNA: genes responsible for sensing the carbon shortage, and genes responsible for building the tail. Such cooperation naturally calls for communication between the sensing genes and the building genes. And communication there is, involving a "symbol," the cyclic AMP molecule that, when carbon grows scarce, is sent to genes that initiate the tail-building.[†]

As we saw in part I, communicating—breaching the "information barrier"—is just one of two things that generally have to happen for positive sums to ensue; there is also the "trust" barrier to overcome; the threat of "cheating" must be dampened. At least, that's the case in non-zero-sum games among human beings. But presumably "trust" isn't a big problem among cells. Right?

Wrong—in a certain sense, at least. Biological evolution, like cultural evolution, creates opportunities for cheaters—pirates and scoundrels who would parasitically subvert the greater good if left to their own devices. So natural selection designs technologies of "trust"—anti-cheating mechanisms.

For example, we've seen that the cells in your body get along famously in large part because they are genetically identical. But suppose that, while your cells are dividing after birth, a mutation happens. A new, genetically distinct type of cell is born. Rather than focus on serving the needs of the larger organism, it replicates itself manically. By the time you are old enough to reproduce, there are so many of these mutant cells that they stand a much better than average chance of getting their DNA into the next generation.

In theory, such a cell would be favored by natural selection—at least, in the short run. But in real life, this sort of parasitism couldn't happen. The reason is that back when you were very, very, very young, your "germ line" was "sequestered." That is, the cells that will form your egg or sperm were put aside for safekeeping; try as some mutant skin cell might, it will never get into the next generation, no matter how prolific it is.

Why do animals thus seclude their germ lines? In the view of some biologists, it is precisely to *avoid* this sort of parasitic mutiny; germ-line sequestration is an anti-cheating device—the functional equivalent of a technology of trust.[†] It presumably evolved as organisms that lacked such a technology died out, their coherence compromised by rampant cheating—rather as human cultures are extinguished if they fail to discourage the sorts of parasitism to which humans are prone.

THE CELLULAR SLIME MOLD IN ALL OF US

When our own lineage crossed the blurry line on which the slime mold sits—the line between society and organism—the result probably didn't look much like a slime mold. (For one thing, our incipiently multicellular ancestor probably floated rather than crawled, since it seems to have evolved in the sea. Ever wonder why there's so much salt in your body?) Still, the evolutionary logic behind the harmony among our cells is the same as the logic behind the harmony among slime mold cells: the non-zero-sum logic of kin selection, empowered by the fact that the cells in our bodies are clones of one another. This clonal relationship is the reason that your skin cells issue not the faintest protest while drying up and blowing away. So far as they are concerned, the safety of your sperm or egg cells, as the case may be, is a cause worth dying for.

This logic has encouraged harmony not just within multicellular organisms but among them. The reason is that kin selection can operate, in diluted form, among organic entitities that aren't genetically identical. Full human siblings, for example, while not sharing *all* of their genes, do have lots of genes in common. So some altruism makes Darwinian sense among siblings—it is just a more measured altruism than among clones. You might take a chance on behalf of a sibling—even run into a burning building to save him—but you don't normally act as if his welfare were just as important as your own. The reason is that you are related to siblings by a factor of 50 percent, not 100 percent. In general, according to Darwinian theory, when kin-selected altruism evolves, the degree of altruism will roughly reflect the degree of relatedness.

For example: Ants, thanks to a strange form of reproduction, can be related to siblings by 75 percent, and some biologists think this is the reason for their extraordinary cooperation. Most ants even—like our skin cells—forgo reproduction entirely, and instead perform some mundane task that makes life easier for their kin, such as hanging bloated from the ceiling, waiting to be tapped as a food source in time of need.

We don't know how these bloated, dangling ants feel. Resentful that their genes forced them into this humiliating position? Delighted that they can aid cherished kin? Or do they feel anything at all? But we do know what feelings accompany kin-selected altruism in humans, and one of them is love. Indeed, kin selection is the original source of love. What was almost surely the first form of love in our lineage was a mother's love of offspring, and it is because the offspring assuredly shares the mother's genes that the genes bestowing love, and the altruistic behavior accompanying love, could flourish. So kin selection, in addition to impelling life through key organizational thresholds, carried it across a threshold of no mean spiritual significance as well.

Kin selection isn't the only Darwinian dynamic that binds organisms into webs of non-zero-sumness. There is also reciprocal altruism, which can evolve among non-kin. A vampire bat, on returning from a nightly blood-sucking expedition empty-handed, may accept a donation of regurgitated blood from a close friend—and will return the favor on some future night when fortunes are reversed. Both bats benefit in the long run. Of course, they aren't smart enough to recognize this win-win dynamic. Still, it is non-zero-sum logic that natural selection followed in programming bats to behave as if they *did* understand such things.

The same thing happens in other species: chimpanzees, dolphins, even people. As noted in chapter 2, people, though smart enough to pursue non-zero-sum gain without specific prompting from their genes, do in fact get prompted. Generosity, gratitude, and outrage over ingratitude are genetically based impulses that can steer us into mutually profitable relationships and away from unprofitable ones. There is also a genetically based impulse that sometimes *keeps* us in profitable relationships—affection, an affection that, in the deepest friendships, can approach outright love. If kin selection can be credited with inventing affection, reciprocal altruism gets credit for extending affection beyond kin. And what gets credit for both feats is non-zero-sumness, the logic shared by these two evolutionary dynamics. Non-zero-sumness, in addition to being the reason that organic complexity exists, and the reason communication got invented, is the reason there is love.

The evolution of reciprocal altruism, along with the prior evolution of kin-directed altruism, imbued human societies with structure long before there were corporations (or even markets) or prime ministers (or even Big Men). Biological evolution pushed us a ways up the ladder of social complexity before cultural evolution acquired much momentum. And biological evolution was in turn pushed, you might say, by non-zero-sum logic.

And so it had been from the beginning, since genes on the same chromosome first cooperated. Natural selection is a blind process, but in its fumbling, bumbling way, it has wound up, time and again, happening upon and harnessing the logic of game theory. Just as cultural evolution would follow that logic from hunter-gatherer band to global civilization, genetic evolution had followed it from tiny, isolated strands of DNA to hunter-gatherer bands. This logic organized genes into little primitive cells, little primitive cells into complex, eukaryotic cells, cells into organisms, organisms into societies.

Just how strong was the impetus behind all this? How likely were the amazing feats of non-zero-sum logic? How likely was the biological evolution of complexity, of richer and richer non-zero-sumness? This is a tough question. Certainly, crossing the various thresholds described in this chapter entailed solving more knotty technical problems than I've bothered to enumerate. Still, there is reason to believe, as we'll see in the next chapter, that the answer is "very likely." In the meanwhile, let's briefly marvel at natural selection's unconscious ingenuity, its ability to pursue varied paths to the same end: cooperative integration among organic entities.

But let's not overdo it. The discovery of the role that symbiosis—and mutualism—played in the evolution of eukaryotic cells has led to some dubious rhapsody about the harmony of nature. One author titled a chapter on the symbiotic origin of complex cells, "Is Nature Motherly?" Well, if by "motherly" you mean "gentle," the answer is no. The harmony created by biological evolution thrives on dissonance. It is because natural selection pits fiercely competing models against one another that models exploiting non-zero-sum logic become the norm.

Consider the bottle-nosed dolphin. Male dolphins, programmed to play win-win games, team up with other male dolphins to form a coalition. The coalition then abducts a female dolphin from a competing coalition, forcibly detaining her and taking turns having sex with her. Coalitions even play non-zero-sum games with other coalitions. Coalition A helps coalition B steal a female from coalition C today, and coalition B returns the favor tomorrow.

This is quite a display of cooperation—cooperation on top of cooperation—and is yet another tribute to natural selection's genius. But it isn't a tribute to natural selection's sweetness, and it isn't cause to get all mushy about Mother Nature. In biological evolution, as in cultural evolution, Kant's "unsocial sociability" is a recurring theme

Indeed, "unsociality"—that's a polite way of putting it, actually—is central to natural selection: organisms vie for finite resources, and the losers slip from the pages of history. And the "slipping" assumes such forms as starvation, disease, and getting devoured. Natural selection creates by discarding, and it doesn't discard gently. Tennyson, in a poem that was a favorite of Darwin's wife, Emma, wrote of Mother Nature: "So careful of the type she seems, / So careless of the single life." In biological as in cultural evolution, breathtaking creations come at a horrible cost.

Chapter Nineteen

WHY LIFE IS
SO COMPLEX

While we readily admit that the first organisms were bacteria-like and that the most complex organism of all is our own kind, it is considered bad form to take this as any kind of progression. . . . [One] is flirting with sin if one says a worm is a lower animal and a vertebrate is a higher animal, even though their fossil origins will be found in lower and higher strata.

—*John Tyler Bonner*

Among the technological feats of the twentieth century is the invention of "binary" chemical weapons. Two chemicals, harmless when separate, are toxic if combined. Safe to transport, deadly when deployed.

This was not an original idea. It had appeared millions of years earlier, in the form of the bombardier beetle. In one tank the beetle carries a harmless chemical mix. In another tank resides a catalyst. The beetle adds the catalyst to the mix upon deployment, creating a scalding substance that, via pliable rear-end nozzle, is showered precisely on nearby tormentors.

The bombardier beetle is an example of the evolution of complexity. Clearly, a beetle equipped with two separate munitions compartments and a spray nozzle is more complex than the same beetle lacking these accoutrements.

And this isn't just any old kind of biological complexity. The beetle's arsenal involves *behavioral* complexity. Toxic nozzles aren't much good unless you can aim them and squirt them.

Aiming and squirting—like any impressive behavior—involves information processing, a command-and-control system. In some small measure, then, evolution's elevation of the beetle to bombardier rank involved a growth in intelligence. In other lineages, of course, the evolution of intelligence—of behavioral complexity—has proceeded further. And we have binary chemical weapons, among other things, to show for it.

Was all of this in the cards? I don't mean binary chemical weapons, or bombardier beetles, or human beings or any other particular thing or species on this planet. I mean the evolution of complexity and intelligence. Did basic properties of natural selection make it very likely that someday *some* animal would be smart enough to invent neat gadgets? And figure out that the earth revolves around the sun? And ponder the mind-body problem? Does biological evolution intrinsically favor the growth of biological complexity—including behavioral flexibility, and its underpinning, intelligence? Is this biological "progress" somehow natural?

It has long been unfashionable to answer yes. One big reason is the same big reason that made belief in directional cultural evolution so unfashionable: past political misuse.

Early this century, biological progressivism was dear to the hearts of "social Darwinists," who used it to justify things like racism, imperialism, and laissez-faire indifference to poverty. The logic behind social Darwinism—to the extent that it had a coherent logic—was something like the following: The suffering, even death, of the weak at the hands of the strong is an example of "survival of the fittest." And the "survival of the fittest" has God's blessing. And how do we know that the "survival of the fittest" has God's blessing? Because He built the dynamic into His great creative process, natural selection. And how do we know that natural selection is God's handiwork? Because of its inexorable tendency to create organisms as majestic as ourselves, organisms worthy of admission to heaven! In short, biological progressivism was used to deify nature in all its aspects, and nature, thus deified, was invoked in support of oppression.

The philosophical confusions underlying social Darwinism have been much analyzed, and we needn't repeat the exercise.[†] Nor need we mull the question of whether nature is "good" or "bad"—at least, not until we enter the realm of spiritual speculation several chapters hence. For now we can let matters rest with a simple and obvious point: the fact that an academic proposition can be misused doesn't mean it's not true.

So is it true? Is organic evolution directional? Do basic properties of natural selection pretty much ensure the evolution of complexity, including behavioral complexity, and thus, given long enough, the evolution of great intelligence? Some people say yes (often quietly), and some say no. The most prominent, persistent, and passionate sayer of no is the noted paleontologist Stephen Jay Gould. He has devoted two books to denying "that progress defines the history of life or even exists as a general trend at all." To see progress in evolution, he says, is to indulge a "delusion" grounded in "human arrogance" and desperate "hope"—the hope that we are anointed by nature to sit on its throne, that we are "a predictable result of an inherently progressive process." Gould recommends that we wake up and smell the coffee, confront the harsh prospect that "we are, whatever our glories and accomplishments, a momentary cosmic accident that would never arise again if the tree of life could be replanted from seed and regrown under similar conditions." And he doesn't mean "we" narrowly—not just *Homo sapiens*. If you replayed evolution on this planet, the chances of getting *any* species smart enough to reflect on itself are "extremely small."

RANDOM WALKS

Before seeing what is wrong with Gould's claim, we need to first see that it isn't as sweeping as it sounds. You might think that when he says progress is not a general evolutionary trend, he is saying that evolution doesn't tend to produce more and more complex forms of life over time. But he isn't. He concedes that the outer envelope of organic complexity may tend to rise—that "the most complex creature may increase in elaboration through time." Nor is he saying that the *average* complexity of all species shows no trend. "Life's *mean* complexity may have increased," he allows.

Okay. So in what sense *doesn't* complexity tend to grow via natural selection?

For starters, a few species have gotten less complex through evolution. And many species have gone long periods with little if any growth in complexity. Bacteria showed up billions of years ago, and there's a lot of them still around, evincing no aspiration to climb higher on the tree of life. This point is widely accepted by biologists, as is its upshot: that "orthogenesis"—some sort of mystical inner impetus toward higher complexity, pervading all of life—doesn't exist. Surely Gould is saying more than this?

Yes. He's saying not only that bacteria are pretty simple creatures; he's saying that they outnumber us. Or, as he puts it: "modal" complexity shows no tendency to grow; the level of complexity at which the greatest number of living things resides—the *mode*—has not changed noticeably since at least 2 billion years ago. Back then, most living things were about as complex as a bacterium. One billion years ago, ditto. Now, ditto.

Indeed, not only do bacteria outnumber us; they outweigh us. In fact, they outweigh just about anything, if you add up all the underground bacteria. Also, they can survive under lots of weird conditions. "On any possible, reasonable, or fair criterion, bacteria are—and always have been—the dominant forms of life on earth."

Actually, some people who consider themselves reasonable and fair might opt for an alternative definition of dominance. For example: "ability to blow up the planet," or "ability to figure out how all life forms were created" or "ability to create whole new life forms," or just "ability to put bacteria under a microscope." Gould will have none of this talk. To make the most complex form of life on earth our bellwether for progressive trends would be to exhibit a "myopic focus on extreme values only."

Well, maybe. On the other hand, "extreme values" are the only reason most people care about the question of biological progress to begin with. Was something as complex as us, as behaviorally flexible as us, as smart as us—something with our "extreme values"—likely to evolve? Take away this question, and books about the evolution of biological complexity would be published only by university presses. Indeed, take away this question, and Gould himself probably wouldn't be terribly interested in the subject. As his writing makes clear, his pet peeve with biological progressivism is its past political overtones, notably social Darwinism.

In short, the issue of "extreme values"—the issue of what happens at the outer envelope of biological complexity, where our species resides—is the only reason we're having this discussion to begin with. Gould seems to realize this. For he doesn't, in the end, rest his case on the stagnation of "modal" complexity. He returns at length to the supposedly extraneous issue of extreme values. Having conceded that the outer envelope of complexity may tend to grow, he proceeds to argue that this growth is not truly "directional," but rather "random." This argument is the heart of one of his two book-length ruminations on complexity, *Full House*.

What does he mean by random? Consider a drunken man walking down a sidewalk that runs east-west. Skirting the sidewalk's south side is a brick wall, and on the sidewalk's north side is a curb and a street. Will the drunk eventually veer off the curb, into the street? Probably. Does this mean he has a "northerly directional tendency"? No. He's just as likely to veer south as north. But when he veers south the wall bounces him back to the north. He is taking a "random walk" that just *seems* to have a directional tendency.

If you get enough drunks and give them enough time, one will eventually get all the way to the other side of the street (notwithstanding traffic fatalities involving other, less lucky drunks). That's us: the lucky species that, through millions of years of random motion, happened to get to the far north. But we didn't get there because north is an inherently valuable place to be; indeed, if it weren't for the brick wall, there would be just as many drunks south of the sidewalk as north of it, and the randomness of all our paths would be obvious. That is: if it weren't for the fact that no species can have less than zero complexity, the history of life wouldn't look like a natural progression. Gould writes: "The vaunted progress of life is really random motion away from simple beginnings, not directed impetus toward inherently advantageous complexity."

Again, as with Gould's emphasis on stagnant "modal" complexity, one might ask how much this argument really matters for philosophical purposes. The question behind this whole exercise, remember, is whether the evolution of something as smart and complex as us was very likely. If the combination of a "random walk" and a "wall of zero complexity" leads people to conclude that the answer is yes, then, well, their answer is yes. If, as Gould fears, people are inclined to take a "yes" answer as evidence of higher purpose, they probably aren't going to be too picky about the exact *type* of "yes." God, they will say, works in strange and wondrous ways.

Still, the more factors favoring the evolution of complexity, the more irresistible the "yes" answer is—the more likely evolution was to eventually reach a human level of intelligence. So it's worth seeing if Gould is somehow overlooking "non-random" factors that are conducive to complexity. He is. They fall under the rubric of "positive feedback"—the evolution of complexity strengthens the logic behind the evolution of complexity, which strengthens the logic behind the evolution of complexity . . . and so on.

NON-RANDOM WALKS

Consider, again, the bombardier beetle. Since there was a time when beetles didn't exist, there must have been a time when no animals were specially adapted to kill and eat them. Then beetles came along. Then various animals *did* acquire, by natural selection, the means to kill and eat them. This expansion of behavioral repertoire, itself a growth in complexity, spurred a response: the bombardier beetle's bombardieresque qualities—a counter-growth in complexity. Thus does complexity breed complexity.

One might expect that, given long enough, beetle predators will undergo a counter-counter-growth in complexity, some way to neutralize the beetle's noxious spray. In fact, they already have. Skunks and one species of mice, the biologists James Gould (no relation) and William Keeton have written, "have evolved specialized innate behavior patterns that cause the spray to be discharged harmlessly, and they can then eat the beetles." Until the next round of innovation, at least.

The technical term for this dynamic is the same as the nontechnical term: an "arms race." Over the past two decades, various prominent biologists—Richard Dawkins, John Tyler Bonner—have noted how arms races favor the evolution of complexity. Gould, in the course of a two-volume meditation on the evolution of complexity, doesn't mention the phenomenon.

Finding evidence of arms races in the fossil record is tricky. But one venturesome scientist methodically measured the remnants of various mammalian lineages spanning tens of millions of years and found a suggestive pattern. In North America, the "relative brain size" of carnivorous mammals—brain size corrected for body size—showed a strong tendency to grow over time. And so did the relative brain size of the herbivorous mammals that were their prey. Meanwhile, comparable South American herbivores, which faced no predators, showed almost no growth in relative brain size. Apparently ongoing species-against-species duels are conducive to progress.

Arms races can happen *within* species, not just between them. Ever spend months observing a chimpanzee colony in painstaking detail? A few primatologists have. The male chimps, it turns out, spend lots of time scheming to top each other. They form coalitions that, on attaining political dominance, get special sexual access to ovulating females—at the great Darwinian expense of less successful coalitions. So males with genes conducive to political savviness should on average get the most genes into the next generation, raising the average level of savviness. And the savvier the average chimp, the savvier chimps have to be to excel in the next round. And so on: an arms race in savviness—that is, an arms race in behavioral flexibility. There's little doubt that this dynamic has helped make chimps as smart as they are, and there's no clear reason why the process should stop where it is now.

Meanwhile, female chimps also exhibit political skills, of a somewhat different sort, that raise the survival prospects for their young. Here, too, genes for savviness should in theory not only prevail and fill the gene pool, but, having prevailed, create selective pressure for yet more savviness. It would always pay female chimps to be smarter than average, and, for that very reason, the average would keep rising. Positive feedback.

Natural selection, as described by Gould, has no room for this sort of directional dynamic. "Natural selection talks only about 'adaptation to changing local environments,'" he writes. And "the sequence of local environments in any one place should be effectively random through geological time—the seas come in and the seas go out, the weather gets colder, then hotter, etc. If organisms are tracking local environments by natural selection, then their evolutionary history should be effectively random as well." This would be good logic if environments consisted entirely of seas and air. But in the real world, a living thing's environment consists largely of—*mostly* of—other living things: things it eats, things that eat it, not to mention members of its own species that compete with it and consort with it.[†] And no one—not even Gould—denies that the average complexity of all species constituting this organic environment tends to grow. So the sequence of environments *isn't* "effectively random" over time; there is a trend toward environmental complexity.

And it wouldn't matter if we assumed, along with Gould, that back at the dawn of life the growth in average complexity was wholly random, like the stumbling drunk's path. The fact would remain that, for whatever reason, environmental complexity started to grow. Species, in "tracking" this growth of complexity, can't be described as stumbling around randomly. Their evolutionary change is, by Gould's own definition, directional. And, since they are themselves part of the environment for other species, the process is self-reinforcing. More positive feedback.

A number of evolutionists have argued that if you look at a colony of chimps (our nearest living relatives), you can see some of the social dynamics that pushed our own ape ancestors in our direction, toward greater intelligence. Probably so. In any event, *some* non-random forces seem to have done the pushing. To the extent that we can judge from an imperfect fossil record, the growth in brain size—from *Australopithecus africanus* to *Homo habilis* to *Homo erectus* to early *Homo sapiens* to modern *Homo sapiens*—is brisk, with no signs of backtracking and little in the way of pauses. It looks for all the world like 3 million years of pretty persistent brain expansion.

How does Gould explain this trend? Not readily. The only explanation his worldview would seem to allow is a long series of lucky coin flips—the most serendipitous drunken walk in the history of drinking.[†] Indeed, luck of this caliber might be enough to make you suspect that the coin flips were divinely guided! It's no surprise that the creationist literature contains many approving citations of Gould's work. If his view of natural selection were correct, I would be a creationist, too; natural selection would not be a plausible means of human creation—at least, not of such rapid human creation.

And it isn't just *our* ancestors that, in Gould's scheme, were so lucky. Mammalian lineages broadly exhibit a movement toward braininess. True, individual species can spend a long time without getting noticeably smarter. But examples of mammals—or for that matter multicellular creatures in general—evolving toward *less* braininess are vanishingly rare. What a lucky bunch of drunks animals are!

WHEN BAD THINGS HAPPEN TO GOOD GENES

There is another sense in which luck might figure in evolution, and this sense was the centerpiece of Gould's first book-length assault on biological progressivism, *Wonderful Life.* The book was about the fossils of the Burgess Shale, products of an apparently sudden (as these things go) expansion of biological diversity around 570 million years ago, at the beginning of the Cambrian period. This "Cambrian explosion" is the first well-documented flourishing of multicellular life. Drawing on the work of British paleontologists who had studied the Burgess Shale, Gould used the fossils as a case study in the decisive role of chance, of "contingency."

Gould's argument was simple. Some of the fossils are very weird-looking, and seem to fit into none of the basic categories of animals that now populate the earth. These weird species, or their descendants, apparently went extinct—and through no evident fault of their own. They must have fallen prey to bad luck, some sudden, unpredictable shift in ecology for which their past evolution hadn't prepared them. Had this random shift not happened, had these oddballs proved enduringly prolific, today's tree of life would presumably look very different. Thus can a roll of the cosmic dice fundamentally alter the future of evolution.

Since Gould's book was published, his interpretation of these fossils has been challenged by a number of paleontologists. It now looks as if the Burgess Shale animals aren't nearly as weird as Gould and some other researchers first thought; for the most part the animals fit readily into a standard taxonomic tree, and their descendants are with us still. In the case of a fossil so bizarre-looking that it was named *Hallucigenia,* Gould—following the prevailing interpretation—was looking at it upside down. Those baffling squiggly things on its "back" were legs. And those strangely spiky "legs" *were* spikes—armor, presumably the product of an arms race.

Notwithstanding this revisionism, part of Gould's argument is surely valid. Whether or not the Burgess Shale animals are a case in point, species do go extinct because of cosmic rolls of the dice. A big meteor happens to head toward Earth and then—poof!—no dinosaurs. This sort of sudden and unpredictable emptying of niches undoubtedly shapes future evolution.

To take an example of particular interest: if *our* ancestors had been wiped out through bad luck, then indeed, as Gould has repeatedly proclaimed, human beings would never have evolved. This point—in some ways the central point of *Wonderful Life*—is so unarguable that, so far as I know, it has never been argued against; no sober biologist would claim that there was some kind of inexorability to the evolution of *Homo sapiens* per se—you know, a species five or six feet tall with armpits, bad jokes, and all the rest. The only serious question is whether the evolution of *some* form of highly intelligent life was likely all along—some animal smart enough, for example, to be aware of itself.

Gould skirted this question in *Wonderful Life,* but he later said, in *Full House,* that the answer is no. It might be tempting to agree, if Gould's argument about the drunken stumbling of life were valid. But given the manifest existence of arms races, and the manifest premium those races place on behavioral flexibility, and the ongoing growth in behavioral flexibility since animals showed up on the scene a bit before the Cambrian explosion, the temptation to agree with Gould is quite resistible.

Resistance is rendered even easier when we consider what, at the risk of anthropomorphizing nature, I can only call natural selection's genius. Though a blind process that works by trial and error—and random trial, at that—it has a remarkable knack for invention, for finding and filling empty niches. It has adapted animals to life on land, underground, under water, in trees, in the air.

In the course of this niche-filling, natural selection doesn't just invent remarkable technologies; it keeps reinventing them. Flight and eyesight are two technologies so amazing that they are commonly cited by creationists for their implausibility. Yet flight has arisen through evolution on at least three separate occasions, and eyes have been independently invented *dozens* of times.

Why are eyes such a favorite of natural selection's? Because light is a terrific medium of perception. It moves in straight lines, bounces off solid things, and travels faster than anything in the known universe.

This isn't to say that natural selection is single-minded in its devotion to light. The other familiar senses—smell, sound, touch, taste—are all amply represented in the animal kingdom, and are just the beginning of a long list of organic data-gathering technologies. Indeed, humankind's vaunted twentieth-century advances in sensory technology almost seem like a long exercise in reinventing the wheel. We now have infrared sensors for night vision; rattlesnakes beat us to that one. We use sonar—old hat to bats, and standard equipment on dolphins. Some burglar alarms work by creating electric fields and sensing disturbances in them; so do some fish—the elephant-snout fish of Africa and the banded knifefish of South America (not to be confused with fish that use electricity to stun, such as the 450-volt electric catfish of Africa, or the 650-volt electric eel of South America). Of course, people fathomed the informational value of the earth's magnetic field long before the twentieth century—with compasses—but not as long ago as natural selection did; some bacteria, and various more complex creatures, use the same trick for orientation.

Why is natural selection so attentive to sensory technologies?[†] Because they facilitate adaptively flexible behavior. And what else facilitates adaptively flexible behavior? The ability to *process* all of this sensory data and adjust behavior accordingly. In other words: brains. Not our brains, necessarily, or even "brains" in the technical sense of the term, but rather intelligence as an abstract *property*. It is natural selection's demonstrable affinity for certain *properties*—its tendency to invent them and nurture them independently in myriad species—that renders trivial Gould's truism about how bad luck can wipe out any one species or group of species. At least, it is trivial so long as we're discussing the likelihood of the evolution of the *property* of great intelligence, and not the evolution of a particular intelligent species.

Simon Conway Morris, one of the paleontologists whose Burgess Shale research Gould relied on most heavily in writing *Wonderful Life*, made this general point in a broadside critique of Gould's interpretation of the Shale fossils. "[T]he role of contingency in individual history has little bearing on the likelihood of the emergence of a particular biological property," Morris wrote. The fates of individual species may depend on the luck of the draw. But the properties they embody were in the cards—at least, in the sense that the deck was stacked heavily in their favor.

Consider again the property of eyesight. Obviously, given that natural selection has invented it dozens of times, the chance extinction of even large groups of species possessing it wouldn't much affect the big picture; the *property* of eyesight was destined for invention and reinvention. Can we make the same argument about intelligence? Not exactly, because intelligence is a different kind of property. If we define it broadly enough—as the processing of information to orchestrate adaptively flexible behavior—then it appeared too early, at the very base of the tree of life, to have kept getting reinvented later, in the tree's branches. (Even *E. coli* bacteria "know" enough to find new environs if their surroundings are low in carbon.) Still, *greater* intelligence is something that has been invented billions of times. In all kinds of animal lineages—in mammals, fish, reptiles, insects, birds— there has been extensive growth in behavioral flexibility, and the growth has often come in small increments. Add up all those increments, and what do you have? A pattern.

Gould writes: "Humans are here by the luck of the draw." True. But a human level of intelligence isn't. Given long enough, it was very, very likely to evolve. At least, that's my reading of the evidence. It is also the reading of some eminent evolutionary biologists, such as William D. Hamilton and Edward O. Wilson, though other eminences, such as Ernst Mayr, disagree.

WHAT TOOK SO LONG?

Sometimes people (including Gould) skeptically ask why, if the evolution of intelligence was so likely, it took so long. It's a fair question. Remember the great evolutionary thresholds noted in the previous chapter—from prokaryote to eukaryote, from eukaryote to full-fledged multicelled animals? Sounds like a pretty brisk progression when you say it in one sentence like that, but in fact getting from stage one to stage two may have taken as long as 2 *billion* years. And it seems to have taken at least another 700 million years to get from eukaryotes to the squishy, blobbish multicelled animals that show up in the fossil record before the Cambrian explosion. Then things really get rolling: from squishy blobs to human beings in around 600 million years. Why did organic evolution spend so much time in pre-explosion mode, twiddling its thumbs?

You could ask the same question about cultural evolution. If it was so powerful, why did it take so long to get off the ground? What were all those people in the Middle Paleolithic waiting for? As it happens, the explanation for these two forms of early sluggishness—cultural evolution's and biological evolution's—are, broadly speaking, the same.

Think back in time to the primordial ooze, when the first cells had come into being. Each of them was a potential source of innovation; by genetic mutation, it could come up with a new "idea" for organic design, and that idea, if good, could spread. In a sense, these cells collectively constituted natural selection's "brain."

But mutations—new "ideas"—don't happen all that often, and anyway most mutant ideas aren't good. So it's going to take a whole lot of bacteria to constitute a "brain" big enough to think up good ideas very often. No doubt a good part of life's early life was spent slowly but relentlessly raising the size of the brain. The process is comparable to the tens of thousands of years humans spent slowly increasing their numbers, until finally, around 15,000 years ago, the invisible brain was big enough to generate innovations at a more-than-glacial rate.

Of course, human innovation wasn't *totally* lacking amid the sparse population of the Paleolithic; in fact, it had been slowly accelerating as population slowly grew. And much the same can be said about the early history of organic life. During those 2 billion years before the invention of eukaryotic cells, important work got done. For example, a series of obscure-sounding but important energy technologies appeared: first "autotrophy" then "photophosphorylation" (a way of getting energy from light) and then full-fledged photosynthesis, which incorporated photophosphorylation. So natural selection wasn't just twiddling its thumbs during stage one. In fact, even symbiotic division of labor among cells made an appearance. Gooey mats of wall-to-wall bacteria, it now seems, consisted of two kinds of cells: the ones on top photosynthesized, and the ones on the bottom made a living breaking down the photosynthesizers' waste products by fermentation.

These gooey mats illustrate a key point. The growth of natural selection's "brain" consists not just of rising numbers of organisms, but, at least as crucially, of rising numbers of species. A fermenting bacterium can, by mutation, generate a whole different set of "ideas" than a photosynthesizing bacterium can—new approaches to fermentation, for one. Each new species opens up new "design space," expanding not just the chances of a good idea, but the spectrum of possible ideas.

There's a second sense in which each new species expands design space. Each species is—like the photosynthesizing cell—a potential energy source, just begging natural selection to create a species tailored to exploiting it. In the case of the photosynthesizing cell—exploited by the fermenting cell—the form of exploitation was harmless: the new species simply recycled the waste of the old species. Often the form of exploitation is more predatory. But whatever the form—harmless, predatory, parasitic, mutualistic, whatever—it always seems to arrive sooner or later. Each new species opens up a potential niche, and natural selection excels at filling niches.

Consider the technological opportunities opened by the nectar in flowers. It has led to the hummingbird's long bill and its soft, stationary flight, to the beehive's amazing collective intelligence, and to many other marvels of flower exploitation. And because, from the flower's point of view, these animals usefully transport pollen, flowers have evolved various ways of encouraging the transport. Some flowers briefly entrap beetles, coating them thoroughly in pollen before letting them head for another plant. Scarlet gilia plants in Arizona put out red flowers, which attract hummingbirds, and then shift to an all-white floral display, which attracts moths, after the hummingbirds leave in late summer. Some orchids have come to resemble bees or wasps or flies, thus attracting a male bee or wasp or fly that, in confusedly trying to mate with the flower, picks up its pollen (and sometimes, poignantly, leaves sperm behind).

The diversification of flowers in turn diversifies their symbionts. Different species of hummingbirds have different bill lengths, depending on their favorite flowers, just as different bee species have different tongue lengths.

And so it goes. Growth in the number of species is assured, in the first place, by the expanse and heterogeneity of the earth; a vibrant, spreading species gets split up by mountains or rivers or deserts or meadows or oceans—or sheer distance—and then its fragments adapt to the peculiar contours of the local ecosystem. And, because each new species itself defines a new potential niche for another species, the more species there are, the more there will be. Once again, complexity begets complexity by positive feedback, but in this case it is the complexity of the whole ecosystem that expands. Thus does the size and fertility of natural selection's brain so assuredly grow—slowly at first, but inexorably.

Apparently the brain was getting pretty big and fertile by the time eukaryotes showed up. For the age of eukaryotes seems to have lasted only a third as long as the age of prokaryotes before the epic threshold to multicelled animals was crossed. And, then, around 600 million years ago, came the "explosion" in the diversity of animal life.

Why the "explosion"? Personally, I suspect that the abruptness of Cambrian creativity has been exaggerated by an imperfect fossil record; and that the main story is continued creative acceleration of the sort you'd expect from a process whose inventiveness grows with the number of past inventions. Still, there probably was *some* abruptness, and it's instructive to ask why. Perhaps the most commonly cited reason is the coming of multicellular predation. Much as evidence of human war is abundant by the time cultural evolution takes off, evidence of predation shows up near the beginning of the Cambrian. Fossilized tracks of the now-extinct trilobite can be seen homing in on the tracks of a Cambrian worm—and only the trilobite's tracks emerge from the intersection.

In addition to this anecdotal evidence of predation, there is the sudden trendiness, during the Cambrian, of body armor. Exoskeletons and shells presumably were round two in an arms race, predation having been round one. And, of course, some armaments are more high tech than body armor and sharp teeth. It is fitting that the trilobite, the first predator ever to leave a record of its crime, is also the first animal known to have had eyes. Once you're hunting mobile prey, acute perception helps. Once you're hunted, ditto. And acute perception isn't worth much without rapid data processing.

Biological evolution's gathering momentum, then, had come from two forces that also figured in cultural evolution. First, there was the gradual elevation of the likelihood of innovation, via the growth of a giant "brain." Then there was intensifying competition that raised selective pressure; in both biological and cultural evolution, zero-sumness stimulated non-zero-sumness. Why did the trilobite's genes for legs come to play a non-zero-sum game with subsequently added genes for eyes—and with genes for processing and applying visual data? Because it's a jungle out there.

LEARNING BY DOING

We could say more about why evolution was so slow in the beginning. In particular, we could elaborate on how daunting some of the simple-sounding thresholds to higher organic organization are. The invention of eukaryotic cells, and of multicellular life, entail enough knotty mechanical problems to raise one's respect for natural selection's ingenuity.

At the same time, these problems should caution biological progressivists, such as me, against too easily asserting the inexorability of complexity's growth. Though in some ways organic evolution is a tidier process than human history, in other ways it is more imponderable. And I must admit to feeling slightly less confident in estimating the probability of the biological evolution of a human-caliber brain than in estimating the likelihood—given that brain—of the cultural evolution of a global brain like the one now emerging.

Still, just as the number of independent evolutions of chiefdoms and civilizations valuably corroborates other arguments for cultural evolution's directionality, the number of times that life has passed through particular organizational thresholds carries great weight. Multicellularity was invented over and over again—more than ten times, by some estimates. A proliferation of multicellular life, then, would seem to have been in the cards, with or without the Cambrian "explosion."

The likelihood of the earlier crossing of the eukaryotic threshold is a murkier issue. Of the eukaryotic cell's various organelles—including energy-processing units such as mitochondria and chloroplasts—its defining element is the nucleus. And it's not clear whether the nucleus was invented more than once. But organelles of one kind or another have been "invented"—sometimes in the symbiotic fashion described in the previous chapter, sometimes not—on a number of different occasions. In this sense, at least, the logic behind the complexification of simple cells seems to have been powerful.

In 1951, the British zoologist J. W. S. Pringle wrote a technical paper called "On the parallel between learning and evolution." It is not a bad comparison. Natural selection is not just a process that "invents" new technologies, such as eyes; it implicitly "discovers" properties of the physical world, such as light's reflection. It is this ongoing invention, and implicit discovery, that is an essential, predictable part of evolution by natural selection. The particular species embodying the "learning" are incidental—transient repositories of knowledge, like a textbook that may go out of print even as its contents live on in other books.

No technology has received more attention from natural selection, more refinement via natural selection, than intelligence. Everywhere around us is evidence of the tendency of intelligence to grow through evolution. The most spectacular example, with all due humility, is us. The human brain is the greatest product yet of the larger, throbbing, endlessly inventive "brain" that is the biosphere.

Of course, intelligence is not a one-dimensional thing. We are not just ten times as smart as dinosaurs or a thousand times as smart as bombardier beetles or a million times as smart as bacteria. Our intelligence is qualitatively different from theirs. It has a number of attributes that, together, have created culture—a culture, moreover, that is rich enough to become an evolutionary force in its own right.

In that sense, some of the terminology in this chapter has been a bit loose. I've spoken about the likelihood of evolution's reaching a human "level" of intelligence—as if there were some simple ladder of animal IQ that life climbs, with "self-awareness" at its acme. My excuse, I guess, is that biologists, including Gould, often talk this way—and that, moreover, talking this way is fine for some purposes. But for the purposes of this book's argument, we'll have to now get a little subtler in our conception of intelligence. If you want to know how likely the coming of cultural evolution was, it isn't enough to argue that natural selection tends to create smarter and smarter things. These things have to be smart in several particular ways if full-fledged cultural evolution is to get rolling. Was *this* kind of smartness in the cards?

Chapter Twenty

THE LAST ADAPTATION

As a result of a thousand million years of evolution, the universe is becoming conscious of itself.

—*Julian Huxley*

The first hard evidence of culture in the human lineage comes in the form of crude stone tools fashioned more than 2 million years ago. What made our ancestors start using stone tools? The consensus among archaeologists, according to one of them, is that the habitat was becoming "drier, less forested, and/or more dangerous"; these "environmental pressures for survival" forced early hominids to innovate.

Once again we are face-to-face with the dreaded "equilibrium fallacy," the idea that life changes only when jostled by an outside force. When I first heaped scorn on this idea—in chapter 6, while discussing the origins of agriculture—the complaint was that it warped our view of cultural evolution, making the process look more episodic, less persistent, than it is. But we've also seen—just a few pages ago—how a type of equilibrium fallacy can analogously bias one's view of biological evolution. The assumption that a species will change only in response to a changing habitat—and never because of competitive "arms races" among its members—was shown to abet underappreciation of natural selection's complexifying tendencies. Now we have an example of the equilibrum fallacy that is doubly worthy of wrath, because it combines these two malign effects; it entails a warped view of biological *and* cultural evolution.

For here, more than 2 million years ago, with the first stone tools, our ancestors are in the midst of gene-meme co-evolution. The brain has been growing via biological evolution for some time. But cultural evolution is also in motion. Even before the first stone tools, there were no doubt tools made of less durable stuff, since lost to posterity. And anyway, culture, which today goes well beyond material technology, did so back then, too. Tools aside, inventing or imitating new tricks—for hunting or scavenging or foraging or fighting—would have come in handy. So handy, in fact, that biological evolution would have encouraged this sort of cultural play; to the extent that useful memes paid off in Darwinian terms, aiding genetic proliferation, then natural selection would favor genes for processing memes: genes for innovating, observing, imitating, communicating, learning—genes for culture.

This sort of co-evolution can become a self-feeding process: the brainier that animals get, the better they are at creating and absorbing valuable memes; and the more valuable memes there are floating around, the more Darwinian value there is in apprehending them, so the brainier animals get. In all probability, the first stone-tool-using hominids were already on this co-evolutionary escalator. That would help explain the growth in brain size that had recently distinguished them from the dim-witted *Australopithecus afarensis,* as well as the ensuing eons of rapid growth—all told, a near-tripling of cranial capacity over 3 million years.

Once you're on this sort of escalator, powered by the positive feedback between the two evolutions, there's no obvious reason to stop. If you don't suffer some grave, species-wide misfortune—a meteor collision, say—you're probably headed for big brains and big-time culture. And somewhere along the way, stone tools are pretty sure to get invented. This is not an event that needs a special "explanation." It is just a stage that the escalator passes through (though environmental quirks could of course affect how fast the escalator reaches it).

So, for the destiny-minded observer, the big question isn't: How likely were stone tools? Given the co-evolutionary escalator, stone tools were automatic. The big question, rather, is: How likely was it that the escalator would get cranked up in the first place? My view (surprise!) is: pretty darn likely. Because however dazzling the cultural achievements at the top of the escalator, the various genetically based assets it takes for a species to embark on the escalator in the first place aren't all that exotic.

That isn't to say that our particular ancestors were destined for embarkation. Indeed, our lineage was just flat-out lucky to find itself in possession of the portfolio of key biological assets. But there's a difference between saying it took great luck for you to be the winner and saying it took great luck for there to be a winner. This is the distinction off which lotteries, casinos, and bingo parlors make their money. In the game of evolution, I submit, it was just a matter of time before one species or another raised its hand (or, at least, its grasping appendage) and said, "Bingo."

THE ESCALATOR'S ENGINE

Before running through an inventory of the biological assets that seem to grant admission to the co-evolutionary escalator, let's pause to appreciate the escalator's subsequent momentum; let's see what's wrong with the dreaded equilibrium fallacy in this particular context. Why is it silly to say that early hominids wouldn't bother to improve their subsistence technology unless the environment turned suddenly hostile?

For starters, because the environment was already hostile! When you've got a meager set of tools, no knowledge of fire, prowling predators, and a brain half the size of a human brain, survival can be a real adventure. If crude stone tools help you kill animals, or even chop up animals some other predator killed (and perhaps use the hide as a blanket at night), that's a plus.

But the larger problem with this version of the equilibrium fallacy is that organisms aren't designed to merely survive; they procreate, too. For example, within a "polygynous" species, males compete to mate with as many females as possible. Were our distant ancestors polygynous? Yes. Early hominid males were vastly larger than females. This sharp "sexual dimorphism" is a tell-tale sign of polygyny. It signifies that for generations big strong males have had lots of offspring by lots of females, while little scrawny males, having gotten sand kicked in their face, have had few or none.

This dynamic—"sexual selection"—is one example of the type of intra-species "arms race" discussed in the last chapter. It explains why rams have big horns (rams with little ones got butted out of the sexual arena), why peacocks have gaudy tails (peahens scoffed at dull-tailed males), and why our various primate relatives—gorillas, chimpanzees, baboons—exhibit sexual dimorphism commensurate to their degree of polygyny. And it explains why early hominid males might have eagerly embraced new stone technologies. The more females whose favor they could win, the more progeny they could have.† And one way to win the favor of a female primate is to give her meat—a precious treat amid a diet of fruits and vegetables. Crude as it sounds, various primates have been seen engaging in sex-for-meat swaps. That includes human beings, when observed in their natural habitat, a hunter-gatherer society—though typically excellence at hunting seems to aid a man's Darwinian prospects in subtler ways (such as elevating his social status, which in turn can lead to sex).

In short: it wouldn't take a suddenly worsening climate to interest males in technology that made it easier to kill or butcher animals—especially if the technology conveniently doubled as an aid in killing or intimidating rival males! As zoologists have noted, it may be no coincidence that the human skull gets thicker around the time the hand axe is invented.

For that matter, the genes of females, too, would benefit from cultural innovations that helped nourish them and their offspring. But we're getting ahead of the story. To assume that our ancestors were capable of shaping tools and using them purposefully is to skip over the question of how our species got into the culture business to begin with. What particular biological assets does it take to get onto the co-evolutionary escalator? What exactly did our ancestors have to acquire via genetic evolution before cultural evolution could pick up much momentum?

GETTING A GRIP

Culture is at bottom a way of learning from the learning of others without having to pay the dues they paid. Suppose you are an elephant. Suppose nearby human beings start carrying guns and trying to kill you. Wouldn't it be nice to learn about this threat without risking the discomfort of a mortal gunshot wound?

In southern Africa early this century, a population of more than 100 elephants faced this very situation. Citrus farmers had hired a hunter to annihilate them. They were easy pickings at first, naïvely unafraid. But after watching some comrades die, the elephants grew averse to humans and began emerging from the bush only at night. This isn't too surprising—and it isn't culture, either; it's just firsthand learning. What's surprising is that these tendencies lasted into the next generation, even though the hunting had long since stopped. Apparently young elephants emulate their elders' aversions, thus tapping into the older generation's hard-earned intellectual capital. *This* is culture—the transmission of information from one individual to another by non-genetic means.

Culture in this elementary sense doesn't require all that much in the way of intellectual firepower. Just watch and imitate. Even bird brains can do it. In England, back in the days of milkmen, a bird called a titmouse discovered that by pecking through a milk bottle's aluminum foil cap, it could help itself to the cream at the top. The idea caught on. Before long, no bottle of milk in the British Isles was safe.

Clearly, imitation can be a powerful force. But teaching—active instruction—speeds things up. Since organisms so often live near offspring and other close relatives, actively imparting key knowledge can, by the logic of kin selection, be good for the organism's own genes; so a genetic inclination toward teaching can evolve. When researchers gave tainted fruit to baboons, older males tried it, rejected it, and thereafter threatened young baboons that showed interest in it. Aversion to the fruit spread through the whole troop, though most baboons never tasted it.

These two prerequisites for a human-level culture—learning by imitation, and active teaching—don't by themselves get you anywhere near a human-level culture. What other biological assets helped put our ancestors on the co-evolutionary escalator? One, no doubt, is tool use.

Tool use per se is no human monopoly. Sea otters use rocks to break abalone shells. The Galápagos woodpecker finch, when in the mood for a grubworm, puts a thorn in its beak and pries the worm out of a tree's bark.

I don't envision sea otters bursting onto a path of headlong gene-meme co-evolution anytime soon. There's a limit to the technological level you can reach when you're working with flippers—even if we give natural selection a couple of million years to make the flippers more supple. Indeed, we're the only species with the sort of manual dexterity it takes to, say, build a model airplane. But we didn't start out that way. Millions of years ago, back before our ancestors began flailing with sticks and throwing rocks, hands were cruder. Still, they at least had the property of grasping, which meant that tool use was possible. And as the tools grew in number—as culture evolved—biological evolution moved in lockstep, sculpting our hands into fine instruments.

A number of animals can grasp in at least a rudimentary way. Kangaroos, and for that matter kangaroo rats, are pretty dexterous. Squirrels sit up on their hind legs and fiddle with nuts. To be sure, none of these mammals has the much-ballyhooed opposable thumb. But the raccoon, a quite bright animal, comes close, with its strikingly humanlike paws (a comparison it encourages by washing food before meals). Bears, too, feed themselves by "hand." And, of course, various primates have hands deft enough for elementary tool use.

Chimps are especially handy. Among the things they've been seen doing: throwing rocks at leopards; attacking a fake leopard with sticks; taking twigs, stripping them of leaves, poking them down into a termite nest, then pulling them out and eating the termites; pounding nuts open with sticks and stones. Chimps even use sticks to brush each other's teeth.

Some chimps crumple up leaves, turning them into sponges with which to extract precious water from the hollows of trees. Chimps also use leaves to wipe the last bits of delicious brain from the skulls of freshly killed baboons. Leaves are also chimpanzee Handi-wipes, used to cleanse the body of feces and other adulteration. Jane Goodall reports that one young female chimp, dangling above a visiting scientist, put her foot on his hair and then "wiped her foot vigorously with leaves."

This sort of tool use doesn't seem to be narrowly programmed by the genes. There is innovation and, by observation, cultural transmission. Chimps at Goodall's Gombe Stream Park used sticks as levers to pry open wooden boxes full of bananas. In the Yerkes primate laboratory, one chimp learned how to use the water fountain, and pretty soon all the chimps were doing it.

Of course, there's a difference between using a water fountain and inventing one. Still, once you're in the tool-use business big-time, natural selection could move you toward inventiveness—especially if you have some other things going for you. Which brings us to the next item in our inventory of biological prerequisites for rapid gene-meme coevolution.

HEAVY SYMBOLISM

Learning by observing, teaching by threatening, and using sticks and stones can do a lot for a culture, but if your species hopes to get to the point of attending operas and anthropology lectures, the biological infrastructure for language is a must.[†] How many species have language? Well, only one if by language you mean, for example, Spanish. But if by language you mean something more generic—a symbolic code by which information is transmitted from one organism to another—then it's all over the place. Bees convey the location of flowers with their famous waggle dance. Ground squirrels emit a warning call on sighting a predator, as do many birds. Ants send out chemicals that mean everything from "Invaders!" to "Food!" Some ants even squeak "Help!" (not in so many words) when trapped, and the sound brings comrades to their rescue. East African vervet monkeys have several warning calls, depending on the predator; one means "snake," one means "eagle," one means "leopard," and each elicits an apt response (looking down, looking up, or running into the bush). Mastery of this language takes cultural fine-tuning. Young vervets may look up, see a pigeon, and give the "eagle" call. Adults then look up and, by failing to join in the call, induce an enlightening chagrin.

Why is there so much communication in the animal kingdom? Because there are so many non-zero-sum relationships. All of the above are examples of communication among kin—among organisms that have the common Darwinian interest of getting their shared genetic information into the next generation. As we've seen, the logic of non-zero-sumness is literally the reason that communication exists. And the profuseness of non-zero-sumness is the reason that communication so pervades life.

Obviously, none of the above species is close to embarking on the sort of cultural evolution that got us from the Stone Age to the sophisticated technology of the information age; none of these animals could possibly formulate a message as complex as, "Have you tried just turning it off and then turning it on again and seeing if that solves the problem?" But the point is that, once a species has the biological infrastructure for any system of communication at all, natural selection could enrich the infrastructure as necessary. In fact, it tends to. If you count all kinds of signaling—visual, auditory, chemical, etc.—vertebrates, ranging from fish to primates, usually have a repertoire of between ten and forty distinct messages. That's more than they started with.

Why have these "vocabularies" tended to grow? Perhaps because a communicating animal can find itself caught up in an intra-species arms race. If communicating helps you outreproduce your neighbor, then natural selection can favor advances in the biological hardware for language. And the better the average hardware, the better an individual's hardware has to be in order to outreproduce rivals—and so on.

But why exactly would subtlety of language help your Darwinian prospects in the first place? To some extent the answer is obvious: "Snake" is more valuable information to share with your kin than "predator." Still, in the case of our species, the subtlety of language goes well beyond a mere expansion in the number of nouns. Why? That brings us to another biological asset that can move a species toward the co-evolutionary escalator: a rich social life, which favors both linguistic skill and intellectual skill more broadly.

SOCIAL CLIMBING

Primates that live in large social groups tend to have a large neocortex—the locus, in our species, of things like speech and abstract thought. Over the past few decades, students of primate psychology (including human psychology) have started to fathom why exactly this might be so—why social skills, and thus socially fluent brains, can pay off in Darwinian terms.

Two dimensions of social life seem especially conducive to the evolution of intelligence. One is hierarchy. In many species high social status helps an animal get genes into the next generation, usually by easing access to food or mates or both. So natural selection can favor things that help animals get high status.

How does intelligence help an animal gain status? Sometimes it doesn't. The term "pecking order" comes from chickens, a not notably cerebral species; their ticket to widespread admiration is pecking other chickens into submission. And even among brainy primates, physical combat often plays a big role in social sorting. Still, with some primates, other assets come into play, such as savvy. This is especially true when social life features its second major intelligence-boosting dimension: reciprocal altruism.

Reciprocal altruism, as we saw in chapter 2, is an instinct for non-zero-sum gain, a genetically based tendency to form friendships and swap favors with friends. Hence the vampire bat's donation of blood to a needy friend, who will return the favor when fortunes are reversed. Reciprocal altruism even in this bare form calls for brains: remembering who has helped you and who hasn't—and treating them accordingly. (Vampire bats have much bigger forebrains than other bats.) But reciprocal altruism can demand whole new kinds of braininess in a context of social hierarchy, when friendships become alliances, and the favors swapped include social support.

Such is the case with two of the biggest-brained primates, baboons and chimpanzees. Many an alpha male chimp has gotten where he is not by single-handedly vanquishing rivals, but by intimidating them with the help of a faithful lieutenant or two. In return, the lieutenants get the alpha's valued support in their own squabbles (and maybe even—such is the alpha's magnanimity—easy access to ovulating females). The primatologist Frans de Waal was the first to report in depth on this dynamic. The title of his classic book, *Chimpanzee Politics,* strikes some people as recklessly anthropomorphic, but those are mostly people who haven't read it. Chimp societies demonstrate how a complex and competitive social landscape would favor various intellectual strengths—not just remembering who has helped or hurt you, but cataloguing personality quirks of allies or enemies and monitoring social dynamics, sensing shifts in allegiance. "Machiavellian intelligence" is the term of art. (A "Machiavellian" intelligence needn't include deception, but it can. The evolutionary psychologist Steven Pinker describes a chimp who was shown several boxes containing food and one containing a snake; he led other chimps over to the snake and, "after they fled screaming, feasted in peace.")

We tend to overlook the deeply social orientation of human intelligence, precisely because it is so deep. But the mental tricks that constitute it become vivid when suddenly they're missing. Autistic children have trouble putting themselves in other people's shoes. Normal four-year-olds know that people who haven't looked inside a box don't know what's in it. Autistic children lack this instinct for ascribing mental perspective to people; they are "mind blind." They may be very smart—capable of superhuman mathematical feats—but they lack key parts of our evolved *social* intelligence.[†]

So far as getting on the co-evolutionary escalator goes, one key implication of coalitional contention is the emphasis it places on communication. If your team wants to subvert the dominant male, advanced planning is advisable. For that matter, if your team just wants to go hunting, and bring back huge hunks of meat that fill the dominant coalition with envy and females with sudden affection, communication is nice. Chimpanzees have been known to hunt collaboratively, with some of them chasing monkeys while others lie in ambush. Imagine how often their plans could benefit from fine-tuning—if only they could communicate more subtly.

Again, an intra-species arms race: the better that individuals communicate, the more cohesive and subtle their coalitions can be; the more cohesive and subtle the average coalition, the better a communicator you have to be for your coalition to prevail.

Chimpanzees, lacking complex language, do the best they can. De Waal recounts the case of a female chimp named Puist, who had helped a male named Luit chase off a rival, Nikkie. Later, when Nikkie threatened Puist, she turned to Luit and held out her hand, asking for support. When Luit demurred, Puist, furious, turned on him, "chased him across the enclosure and even hit him."

Luit probably got the message—that he was being punished for "cheating," for failing to uphold his end of a non-zero-sum deal. And maybe that message would prompt Luit to make amends, restoring an alliance that could benefit both him and Puist in the long run. Still, anything that removed ambiguity from the message—such as the genetic evolution of the means for articulating grievances—would have made such restoration even more likely, and thus would have been good for both chimps. As usual, clearer communication abets non-zero-sum gain—and, as we've already seen, natural selection pays attention to non-zero-sum gain.

THE PANDA'S THUMB

So there you have it: the basic equipment needed for a species to hop on the co-evolutionary escalator: learning, learning by imitation, teaching, some use of tools, along with elementary grasping abilities, a mildly robust means of symbolic communication, and a rich social existence featuring, in particular, hierarchy and reciprocal altruism (a combination that, in turn, brings Darwinian logic that can turn a mildly robust means of communication into a full-fledged language).

As things worked out, our ancestors were the first to put this package together. If they had died out, would they have been the last? These kinds of "What if" games are a pretty mushy form of analysis. But they are a favorite of people who say that the evolution of intelligence was unlikely, so let's play them for a few paragraphs.

Stephen Jay Gould notes that tens of thousands of years ago, even as *Homo sapiens* thrived, our near relatives, the Neanderthals, died out. He asks: What if we, too, had suffered their fate? Wouldn't that have been the end of intelligent primates?

Gould fails to mention the very good chance that our ancestors caused the Neanderthals' demise—perhaps by crowding them out of their niche, or perhaps less subtly, by knocking them over the head and eating them. In other words, the passing of the Neanderthals doesn't show how easy it is for even quite brainy primates, who think the world is their oyster, to be struck down in the prime of life by some fluke; the passing of the Neanderthals shows how easy it is for quite brainy primates to be struck down by roughly-as-brainy primates. If our own ancestors *had* died out around that time, it probably would have been at the hands of the Neanderthals, who could have then continued on their co-evolutionary ascent, unmolested by the likes of us.

But anyway, so what if all members of the genus *Homo* indeed *had* been wiped out? I'd put my money on chimps. In fact, I suspect that they are already feeling some co-evolutionary push; if they're not quite on the escalator, they're in the vicinity. For that matter, their near relatives (and ours), the bonobos, have much the same ingredients for co-evolution that the chimps have, if perhaps in milder form. Some zoologists suspect that chimps and bonobos have long been "held back" by the presence of humans—kept from moving out of the jungle onto grasslands and, more generally, from filling the human niche. Obviously, if humans went extinct, that would no longer be a problem.

And what if all the apes had been wiped out—chimps, bonobos, even gorillas and orangutans? Well, monkeys, though more distant human relatives than any apes, can be pretty impressive. Baboons are cleverly coalitional, and macaques are quite creative. Japanese researchers put groups of macaques on separate islands and watched cultural evolution work. Especially noteworthy was the prodigious Imo. At age two she invented and popularized sweet-potato washing. She later discovered a way to separate wheat from sand: throw the mixture in the water and skim the wheat off the top. The idea caught on.

What if somehow the entire primate branch had been nipped in the bud? Or suppose that the whole mammalian lineage had never truly flourished? For example, if the dinosaurs hadn't met their untimely death, mightn't all mammals still be rat-sized pests scurrying around underfoot? Actually, I doubt it, but as long as we're playing "What if," let's suppose the answer is yes. So what? Toward the end of the age of dinosaurs—just before they ran into their epoch-ending piece of bad luck—a number of advanced species had appeared, with brain-to-body ratios as high as those of some modern mammals. It now looks as if some of the smarter dinosaurs could stand up and use grasping forepaws. And some may have been warm-blooded and nurtured their young. Who knows? Give them another 100 million years and their offspring might be riding on jumbo jets.

Again, I'm no big fan of "What if" games. Biological evolution, like cultural evolution, is too subtly complex to be anticipated by flow charts with an "if–then" statement at each nexus. Organic history is indeed, as Gould would say, exquisitely contingent; alter any little detail in the past, and any living species could, for all you know, be history. But, again, the same cannot be said of any living *property*. You can track down and wipe out every species with some particular valuable asset—eyesight, say, or grasping ability—but, if the asset is truly valuable, it will probably reappear sooner or later.

Consider one of Gould's favorite animals, the giant panda. At some point, apparently, its ancestors were separated from the ancestors of mainstream bears and found themselves in a land loaded with bamboo. The modern panda spends ten or twelve hours a day munching bamboo stalks. In stripping the stalks of leaves, pandas use something that mainstream bears don't have—a thumb that works strikingly like a human thumb, letting the hand grasp finely. But there's one difference: whereas a human thumb has four fingers to work with, a panda thumb has five; the panda "thumb" is the hand's *sixth* digit. Natural selection fashioned it by reshaping a small wrist bone and rerouting some muscles. Why didn't natural selection just do what it did in our lineage—turn the fifth digit into a thumb? Because, writes Gould, "the panda's true thumb is committed to another role, too specialized for a different function to become an opposable, manipulating digit." Faced with this roadblock, natural selection improvised around it.

Gould finds this example so instructive that he titled a whole book *The Panda's Thumb*. And what is the moral of the story? The absence of higher purpose in nature. "If God had designed a beautiful machine to reflect his wisdom and power, surely he would not have used a collection of parts generally fashioned for other purposes."

Maybe not. But if God were designing a machine that *designs* machines—if he were designing natural selection—he might well imbue this creative process with exactly the resourcefulness that the panda's thumb embodies. This is the resourcefulness that made the eventual advent of a second great creative process—full-blown cultural evolution—so likely. The panda's thumb illustrates that, whatever the fate of our species, one of its key properties, grasping, was all along likely to proliferate—because grasping is a useful technology. Natural selection, metaphorically speaking, seeks out and exploits technological opportunities, and it does so ingeniously, using unlikely raw materials when necessary. As Gould himself says, what is marvelous about the panda's thumb is that "it builds on such improbable foundations."

What is true of grasping ability is true of the other properties that I've listed as basic biological prerequisites for gene-meme co-evolution. Wipe out humans, even apes, even primates, and all these properties would still exist, because all have been invented independently, multiple times. Even what may be the rarest of them, reciprocal altruism, has been invented numerous times in the mammalian lineage—in primates, in dolphins, in bats, in impalas—as well as in other lineages, such as fish. Give evolution long enough, and reciprocal altruism will arise yet again—and again and again and again. (And recall, from part I, that reciprocal altruism, along with status-seeking and the coalitional impulses emanating from these two properties, does more than help animals get on the co-evolutionary escalator. Once they're at the top, and cultural evolution begins to move fast, these features help it propel the species toward great social complexity.)

As biological evolution proceeds, and more and more species possess one or another of the several key biological prerequisites for admission to the co-evolutionary escalator, it is just a matter of time before all of these properties wind up in a single species. Of course, that species, looking back, will marvel at the incredible series of lucky breaks that steered it toward the escalator. Who would have guessed that our ancestors, after spending time swinging through trees, evolving long, slender digits, would then emerge from the jungle and put deft grasping to a different and pivotal use? What mind-boggling good fortune that tree-swinging happened to be on our ancestors' résumés!

It's true. We've been very lucky. The winner of a bingo game is also very lucky. But there's always a winner.

So far, of course, all the basic prerequisites for culture have come together decisively only once; we alone have made the ascent up the co-evolutionary escalator. Then again, it's been only around 600 million years since multicelled animals made an appearance in the fossil record. And it took only 100 million years to get from crude, rodent-sized mammals to lots of brainy, socially complex mammals, several of which are very near the co-evolutionary escalator (if not on its first step). Astronomers tell us the earth has *billions* of years before the sun signs off. How likely is it that—if our species were suddenly removed from the picture—evolution could go that long without ever again happening to deposit all of these key properties in a single species?

The answer is "at least somewhat likely" if you buy Gould's view of evolution, in which a randomly fluctuating environment provides no reason for complexity to consistently grow. But if you realize that often the key evolutionary environment of organisms is other organisms, and that multicelled organisms in general tend to move toward greater complexity, turning the biosphere into a hotbed of competitive innovation and ever-ramifying diversity, the answer is "not very likely at all."

THE EVOLUTIONARY EPIC

Consider one "near miss," one species that has come tantalizingly close, but not close enough, to the co-evolutionary escalator. Dolphins are among the most distant mammalian relatives of human beings—more distant than squirrels, rabbits, and bats. Yet they have great intelligence, a complex social life featuring coalitional competition, and a still-unfathomed but probably non-trivial language. They might well be gliding up the co-evolutionary escalator even now—if only it weren't for those damn flippers! Indeed, look how robust their culture is even with that handicap. One population of dolphins in Hawaii has invented a new form of creative expression: air art.

Various dolphins, and for that matter some beluga whales, can send circles of air out of their blowholes, rising upward like smoke rings. But these Hawaiian dolphins take a whole new approach. They start by creating big underwater swirls with their fins, then turn around and blow air into the swirls. The resulting rings are big, limpid, and beautiful. Some dolphins swim through their rings. Some dolphins make two little rings and then coax them toward fusion, creating one big ring. Each artist has its own style.

And no artist was trained by humans. Though captive, these dolphins have never been rewarded for their work. They are naturally creative, culture-making animals. They are also, together, a social brain. By observing each other, and trying to improve on what they see, they've collectively turned the first crude, serendipitous creation of air art into an array of diverse memes.

Maybe the most amazing meme comes from a dolphin named Tinkerbell. She swims along on a sinuous path, releasing a string of little bubbles, and then, by brushing the bubbles with her dorsal fin, joins them together into a corkscrew pattern—or, as described by researchers, a "helix."

It isn't a *double* helix, but the symbolism is fitting enough. For memes—once a species has ascended the co-evolutionary escalator, at least—are true to the spirit of genes. Cultural evolution, like biological evolution, carries life to higher and higher levels of organization. And it does it the same way biological evolution does it: zero-sum dynamics intensify non-zero-sum dynamics; competition between entities encourages integration within them. In both evolutions, the two big barriers to non-zero-sumness—the information barrier and the trust barrier—are met and overcome with ingenious technologies.

We'll never know for sure whether another billion or so years would indeed yield another species as intensely cultural as we are. Because once gene-meme co-evolution shifts into high gear, natural selection is effectively over. Certainly it's over for the species at the top of the escalator; cultural evolution long ago supplanted genetic evolution as our key adaptive mechanism, and it has now put us on the verge of taking control of our genetic evolution, replacing natural selection with artificial, test-tube selection. And we'll increasingly be steering the evolution of other species, as well. We've long done that in a slow, ham-handed way (witness cows, pigs, and potatoes), but now we'll be doing it in a lightning-fast way (witness potatoes with built-in organic pesticides). All in all, the shape of life on this planet is now moving so fast via cultural evolution that evolution by natural selection is, for practical purposes, standing still.

The various biological adaptations that got us onto the co-evolutionary escalator—learning by imitation, language, and so on— could be said to amount to one big biological adaptation: adaptation for advanced culture. And this adaptation could be called the last adaptation—at least, the last biological adaptation that will be produced in our species by natural selection. But it is far from the end of the road, and far from the last adaptation of one sort or another that will be necessary.

Edward O. Wilson has suggested that the "evolutionary epic" serve as our binding myth in the modern scientific age—a myth not in the sense of an untruth, but in the sense of a story that explains our existence and helps us orient ourselves to the world. "Every epic needs a hero," he writes. "The mind will do." In a sense, yes; the human mind represents the triumph of our lineage against great odds; its survival, by pluck and luck, as countless others perished; its ascent toward comprehension of itself and its creator, natural selection. At the same time, there is cause to pay tribute to another mind: the "mind" that mediates natural selection. It is the relentless burgeoning of this creative biosphere, the self-accelerating growth in its innovation, that set the stage for our triumph. This giant mind all but ensured that, whether or not the human species finally reached a reflective intelligence, some species would.

We are thus generic and unique. We embody, in some essential way, the natural imperative toward intelligence (and the natural tension between conflict and integration, between zero-sum and non-zero-sum logic); yet we also bear the distinctive marks of our peculiar history.

Now humanity, having emerged from one great global mind, has finally, in the modern era, given birth to another. Our species is the link between biosphere and what Pierre Teilhard de Chardin called the "noosphere," the electronically mediated web of thought that had taken crystalline form by the end of the second millennium. This is a mind to which the whole species can contribute, and a mind whose workings will have consequences for the whole species—epic consequences of one sort or another.

PART
III

FROM

HERE

TO

ETERNITY

Chapter Twenty-one

NON-CRAZY QUESTIONS

Science, after all, is only an expression for our ignorance of our own ignorance.

—Samuel Butler

Pierre Teilhard de Chardin, writing in the mid-twentieth century, declared the world's nascent telecommunications infrastructure "a generalized nervous system" that was giving the human species an "organic unity." Increasingly, humankind constituted a "super-brain," a "brain of brains." The more tightly people were woven into this cerebral tissue, the closer they came to humanity's divinely appointed destiny, "Point Omega."

What exactly was Point Omega? Hard to say. Teilhard's philosophical writings are notable about equally for their poetry and their obscurity. As best I can make out, at Point Omega the human species would constitute a kind of giant organic brotherly-love blob.

Teilhard's superiors in the Catholic Church hewed to a more conventional theology. They forcefully encouraged Teilhard, a trained paleontologist, to confine his published pronouncements to the subject of fossils.

After Teilhard's death in 1955, his most cosmic writings were finally published. They generated buzz in some avant-garde circles, but they never gained mainstream acceptance, either in the church or the wider world. Why? In part because his notions of how evolution works were mushy and mystical, and never earned the respect of the scientific establishment. In part because Point Omega meshed so poorly with extant theology. And in part, perhaps, because comparing societies to organisms had not-so-long-ago been a pastime of European fascists, who had justified murder and repression in the name of superorganic vigor.[†]

It's amazing how fast a viewpoint can move from radical to trite. Today, with fascism seeming like an ancient relic, and the Internet looking strikingly neural, talk of a giant global brain is cheap. But there's a difference. These days, most people who talk this way are speaking loosely. Tim Berners-Lee, who invented the World Wide Web, has noted parallels between the Web and the structure of the brain, but he insists that "global brain" is a mere metaphor. Teilhard de Chardin, in contrast, seems to have been speaking literally: humankind was coming to constitute an actual brain—like the one in your head, except bigger.

Certainly there are more people today than in Teilhard's day who take the idea of a global brain literally. But they reside where Teilhard resided: on the fringe of opinion.

Are they crazy? Was Teilhard crazy? Not as crazy as you might think. And once you understand how relatively non-crazy it is to call humankind a giant brain, other aspects of Teilhard's worldview begin to look less crazy as well. Such as: the idea that there is a point to this whole exercise; the idea that life on earth exists for a purpose, and that the purpose is becoming manifest.

I'm not saying these things are true—at least, I'm not saying it confidently, the way I've been saying that organic history and human history have a direction. I'm just saying these things can't be dismissed with a wave of the hand. They don't violate the foundations of scientific thought, and they even gain a kind of support, here and there, from modern science.

ARE WE AN ORGANISM?

There are various reasons that, at first glance, you might be skeptical of this giant global brain business. One is that a real, literal brain belongs to a real, literal organism. And the human species isn't an organism; it is a bunch of organisms. But before dismissing the possibility that a bunch of organisms can themselves constitute an organism, we should at least get clear on the definition of an organism. That turns out to be harder than it sounds.

Consider the "colonial invertebrates." As Edward O. Wilson has noted, some come close to qualifying as "perfect societies"—so close, in fact, that "the colony can equally well be called an organism." The awesome, sixty-foot-long Portuguese man-of-war, for example, certainly looks like an organism—like a giant, colorful jellyfish—and indeed is usually called an organism. But it evolved through the merger of distinct multicelled organisms, which grew more specialized as they grew more interdependent: some paralyze fish, others eat the fish and then share the nutrients. Among other colonial invertebrates that blur the line between organism and society are our old friend the cellular slime mold (which vacillates between autonomous cells and unified slug) and corals (including, aptly, the "brain coral").

For that matter, even things that we all agree are organisms—such as ourselves—can have their colonial aspects. Remember the discussion of our cells and our organelles—formerly distinct creatures that merged? There is a little Portuguese man-of-war in all of us.

Indeed, as we've seen, cells and organelles have not only distinct roots, but distinct routes: different pathways that their genes take into the next generation, and hence somewhat different Darwinian interests. The organelle's DNA, relying on maternal transmission, might profit by biasing reproduction in favor of females—and in some plant species it does exactly that. So one criterion you'd think might serve as a clear-cut distinction between organism and society—complete unity of purpose among the organism's constituents—won't work.

In fact, even if we leave organelles aside, and look only at the nuclear genes—at the chromosomes constituting the genome—all is not peace, love, and understanding. The reason is that, though the genes in a genome would appear to be in the same boat, there is a brief but crucial period when they aren't. When it comes time to send a boat to the next generation—when an egg cell is created and sets sail hopefully—half of the genes must be left behind, to make room for genes that come from the sperm. Likewise, only half of a male's genes will, via sperm, make it into that egg during fertilization. As a rule, genes are assigned to eggs in an even-handed way, so that a given gene, whether from male or female, has a fifty-fifty chance of winding up in a given intergenerational boat. But if a gene could find a way to bias the assignment process, placing itself in most or all of the boats, it might proliferate by natural selection.

This has actually happened—in mice and fruit flies and, no doubt, other, less studied species. A type of gene called a "segregation dis-

torter" has only one apparent function: distorting segregation—slant-
ing the sorting process so that it can sneak onto the intergenerational
ship time and time again. It is a professional stowaway.

There is also a bigger genetic stowaway—a whole chromosome
called a B chromosome that appears in lots of organisms, including
people. Like a stowaway who steals food from the crew at night, a B
chromosome is a parasite; it can hurt the organism's chances of reproducing, delaying the onset of fertility in females. But from the point
of view of genes on the B chromosome, that's okay; if they slightly
reduce the number of ships that set sail, but manage to sneak onto all
of them, they will do better than genes that play by the rules, getting
excluded from half of the ships. It is these law-abiding genes that suffer from the shrinkage of the overall fleet.

Generally speaking, law-abiding genes do a good job of solving
such problems—preserving the rule of law, foiling would-be parasites
in various ways. The reason is that if they don't solve such problems,
and parasitism runs rampant, natural selection casts the whole lot of
them aside—parasites and law abiders alike—in favor of genes that
run a tighter ship. (In cultural evolution, analogously, societies that
don't solve the "trust" problem, that don't discourage rampant parasitism, have tended to lose out to societies that do.) This ability of
selection at the level of the organism to override selection at the level
of the gene is the reason these examples of conflict of interest within
an organism are, in the scheme of things, small potatoes. Even people
with parasitic B chromosomes—around one in fifty of the people you
see each day—have an air of organic unity about them.

Still, the fact remains that one of the things you might expect to be
a clear, bright line between society and organism—internal unity of
purpose—isn't clear or bright. As the zoologist Matt Ridley has put
it, "What is the organism? There is no such thing." Each so-called
organism, he notes, "is a collective." And not a wholly harmonious
collective—at least, not by definition.

If the line between organism and society isn't the distinction between complete and incomplete unity of purpose, then what *is* the
line? That's the problem: lacking a clear boundary, biologists are free
to differ. In 1911 the great entomologist William Morton Wheeler
published a paper called "The ant colony as an organism"—a title
that, he stressed, was not meant as mere analogy; an ant colony, in his

view, *was* a type of organism, a "superorganism." This view gained favor for a time and then fell out of fashion, but lately it has made something of a comeback. One reason may be the growing awareness of conflict within organisms—the growing sense that all organisms are in some sense societies.

You may, of course, disagree with William Morton Wheeler and his contemporary defenders. You may insist that a society consisting of distinct organisms cannot itself qualify as an organism. Still, given that the notion of the superorganism is at least seriously entertained by people whose business it is to think about such things, you can't dismiss it as out-and-out *crazy*.

Suppose we grant Wheeler's claim for the sake of argument: an ant colony *is* an organism. Then why can't we call a human society an organism? Is the key difference the extreme interdependence of ants—the fact that some castes of ants would perish were it not for the food-gathering castes?[†] That seems a dubious distinction, given the current interdependence among humans. I depend for my nourishment on the labor of many people I've never met. If you dropped me in the Rocky Mountain wilderness, without anything made by other people—no pocket knife, no clothes—I'd wind up as bear chow.

There is one other salient objection to taking the phrase "giant global brain" literally. Namely: brains have consciousness. They don't just process information; they have the subjective *experience* of processing information. They feel pleasure and pain, have epiphanies of insight, and so on. Are we really to believe that, as the Internet draws billions of human minds into deeper collaboration, a collective, planetary consciousness will emerge? (Or even fragmented planetary consciousness? Will General Motors feel spiteful toward Ford?)

Far be it from me to make this argument. My aim is more modest: to convince you that, if I *did* make this argument, it wouldn't be a sign of insanity. The question of transcendent planetary consciousness, whatever the answer, is non-crazy.

Oddly, the key to granting the question this legitimacy is to hew to a soberly scientific perspective. Indeed, the more scientific you are in pondering consciousness, the more aware you become of the limits of science; the more inclined you become to approach cosmic questions in general with a touch of humility.

COULD A GIANT GLOBAL BRAIN BECOME CONSCIOUS?

In the 1963 science-fiction story "Dial F For Frankenstein," by
Arthur C. Clarke, the world's telecommunications system comes to
life. With a global network of satellites having just interlinked the
planet's telephone switching systems, all the phones start ringing at
once. An autonomous, thinking supermind has been born.

When college students sit around late at night and wonder whether
a giant global brain could ever be conscious, they usually have a scenario like this in mind. The assumption is that when something
reaches consciousness, it starts acting like other conscious things we're
familiar with—capriciously. But the fact is that, for all we know, the
giant global brain—the intercontinental web of minds and computers
and electronic links—already *is* conscious. At least, that prospect is left
distinctly open by the view of consciousness that underlies mainstream behavioral science today.[†] For according to this view, we can
never know whether any given thing possesses consciousness.

By that I don't just mean that you can't know what it's like inside
someone else's head unless you're that someone (though that's of
course true, and it's part of what I mean). I mean that, according to
this mainstream scientific view, consciousness—subjective experience,
sentience—has zero behavioral manifestations; it doesn't *do* anything.

Sure, you may *feel* as if your feelings do things. Isn't it the sensation
of heat, after all, that causes you to withdraw your hand from the surprisingly hot stove? The answer presupposed by modern behavioral
science is: no. Corresponding to the subjective sensation of heat is an
objective, physical flow of biological information. Physical impulses
signifying heat travel up your arm and are processed by your equally
physical brain. The output is a physical signal that coerces your muscles into withdrawing your hand. Here, at the sheerly physical level, is
where the real action is. Your sensation of pain bears roughly the relation to the real action that your shadow bears to you. In technical
terms: consciousness, subjective experience, is "epiphenomenal"—it
is always an effect, never a cause.

You may disagree. You may think consciousness is some kind of
ethereal yet active *stuff*. But if that's your view, then you're probably
already tolerant of weird scenarios such as Teilhard's; you may well
have already concluded that we live in a strange universe, that science

can't illuminate all its dimensions, and that there is thus room for conjecture about higher purpose and higher consciousness.

What's interesting—and underappreciated—is that you could reach
the same conclusion if you *accept* the hard-core scientific view that
consciousness is a mere epiphenomenon, lacking real influence. After
all, if consciousness doesn't *do* anything, then its existence becomes
quite the unfathomable mystery. If subjective experience is superfluous to the day-to-day business of living and eating and getting our
genes into the next generation, then why would it have ever arisen in
the course of natural selection? Why would life acquire a major property that has no function?

People who claim to have a scientific answer usually turn out to
have misunderstood the question. For example, some people say that
consciousness arose so that people could process language. And it's
true, of course, that we're conscious of language. As we speak, we
have the subjective experience of turning our thoughts into words. It
even feels as if our inner, conscious self is *causing* the words to be
formed. But, whatever it may feel like, the (often unspoken) premise
of modern behavioral science is that when you are in conversation
with someone, all the causing happens at a physical level.[†] That someone flaps his or her tongue, generating physical sound waves that enter
your ear, triggering a sequence of physical processes in your brain that
ultimately result in the flapping of your own tongue, and so on. In
short: the *experience* of assimilating someone's words and formulating a
reply is superfluous to the assimilation and the reply, both of which
are just intricate mechanical processes.

Besides, if conscious experience arose to abet human language,
then why does it also accompany such things as getting our fingers
smashed by rocks—things that existed long before human language?
The question of consciousness—as I'm defining it here, at least—isn't
the question of why we think when we talk, and it isn't the question
of why we have *self*-awareness. The question of consciousness is the
question of subjective experience *in general,* ranging from pain to anxiety to epiphany; it is the question of sentience. To phrase the matter
in the terminology made famous by the philosopher Thomas Nagel,
the question is: Why is it *like something* to be alive?

You might think that this question would get answered in the
course of a tome called *Consciousness Explained,* the much-noted book
written by the philosopher Daniel Dennett. But when people such as

Dennett try to "explain" consciousness, they usually aren't tackling the question we're asking here. They are trying to explain how a brain could generate consciousness. Whether they succeed is debatable, but in any event our question here isn't how brains generate consciousness, but *why*—why would an aspect of life with no function be an aspect of life in the first place?[†]

The mystery of consciousness has lately been underscored by computer science. Though artificial intelligence hasn't advanced at breathtaking speed, there has been progress in automating sensory and cognitive tasks. There are robots that "feel" things and recoil from them, or "see" things and identify them; there are computers that "analyze" chess strategies. And, clearly, everything these robots do can be explained in physical terms, via electronic blips and the like. "Feeling" and "seeing" and "analyzing," these machines suggest, needn't involve sentience.[†] Yet they do—in our species at least.

Faced with the mystery of consciousness, some people—including such philosophers as David Chalmers, author of *The Conscious Mind*—have suggested that the explanation must lie in a kind of metaphysical law: consciousness accompanies particular kinds of information processing. What kinds? Well, that's the question, isn't it? But a not uncommon view is that the information processing needn't be organic. Consciousness may reside in computers, networks of computers, even networks of computers and people. The philosophers who hold this view aren't fuzzy-minded New Agers or reactionary Cartesians or mystical poets like Teilhard himself; they are people who accept a basic premise of modern behavioral science—that all causality happens in the physical world—and who also appreciate the weirdness that emerges from this premise upon sustained contemplation. Basically, their answer to the question "Could the giant global brain become conscious?" is: We wouldn't know if it were, and for all we know it is.

Teilhard considered the idea of global consciousness not just conceivable, but compelling. The reason lies partly in his broad definition of evolution. Though he saw the difference between biological evolution and what we've been calling cultural evolution, he tended to think of them as a single, continuous creative act, with the upshot in both cases being "complexification." And if cultural evolution is indeed a nearly seamless outgrowth of biological evolution, then you would expect it to evince the same basic properties as biological evolution. One of these properties, Teilhard surmised, was growth in consciousness, occurring in lockstep with growth in complexity.

How did he reach this conclusion? Aside from mystical revelation, the only possible answer is guesswork. After all, since subjective experience is only accessible to the experiencer, we can never know for *sure* that anything other than ourselves is conscious—not even our next-door neighbor, much less a chimpanzee. On the other hand, most people who have watched chimpanzees as they suffer the slings and arrows of outrageous fortune suspect that they can experience things like pain and hunger and excitement. Dogs, too, seem capable of pain and hunger and excitement (not to mention shame). Cats, too (except for shame). For that matter, lizards and snakes recoil from heat. And can we really rule out the possibility that bacteria feel some tiny, crude dose of pain? They do recoil from electric shock, after all. And they do have a brain. At least, they have a "brain"—their DNA, the onboard computer that controls their behavior.

Whether or not you think bacteria are sentient, Teilhard de Chardin did. And, in his view, when individual, mildly sentient cells merged into multicelled organisms, and then acquired a collective brain, consciousness took a great leap forward. And when brainy multicelled organisms—us—merge into large, thinking webs, constituting another collective brain, a comparable leap could presumably take place. After all, the fact that the connective tissue is now made of electronic stuff, rather than gooey organic stuff, doesn't matter so much if you consider biological and technological evolution part of the same creative process. Writing in 1947, Teilhard marveled at computers and said that radio and TV "already link us all in a sort of 'etherized' universal consciousness." But this was nothing compared to the future, when the links among human minds would grow denser and become truly global, as the "noosphere" matured. He asked: "What sort of current will be generated, what unknown territory will be opened up, when the circuit is suddenly completed?"

DOES EVOLUTION HAVE A PURPOSE?

Teilhard's conviction that biological and cultural evolution are a single creative process no doubt drew strength from his a priori faith. If you begin with the assumption that the history of this planet (indeed, of the universe) has an unfolding purpose, then you naturally see continuity in the unfolding. But suppose we proceed in the reverse direction. Suppose we try to examine the mechanics of evolution—biological, or cultural, or both—with no a priori assumptions and ask: Do they show signs of purpose? Philosophers call this the question of teleology—the question of whether a system seems built to pursue a telos, an end, a goal.

Of course, we've already established (or, at least, I've argued) that biological and cultural evolution move in a direction—toward broader and deeper complexity. But that doesn't mean they are moving toward a goal, an end. Only if evolution was *designed* to move in a particular direction does that direction qualify as a telos.

What may be the most famously wrong teleological analysis in history was performed by the British cleric William Paley. Writing just before Charles Darwin was born, he used the evident functionality of plants and animals to argue for the existence of a grand designer, a God. If you come across a rock, he wrote, you have no reason to conclude that it was made for a purpose. But if you come across a pocket watch, you know that it was made by a watchmaker, for the purpose of keeping time. Plainly, living things are more like watches and other artifacts than like rocks; animals are evidently designed to eat, to breathe, to do other things. "There is precisely the same proof that the eye was made for vision, as there is that the telescope was made for assisting it," Paley wrote.

Of course, after Darwin, Paley's stock fell. But let's be clear on why. Paley wasn't wrong to say that life is evidently functional. And he wasn't wrong to say that this functionality strongly suggested a designer. He was just wrong to assume that the designer was a being rather than a process. The eye *was* made for vision—it just wasn't handcrafted by God.

Sure, you could quibble about whether the eye was made *for* vision, since natural selection didn't set out with vision in mind and then build the eye to perform this preconceived function. Still, eyes, like telescopes, exist *by virtue of* their contribution to vision; if they didn't facilitate vision, they wouldn't be here. Compare that to a rock that is used to build a wall. Sure, the rock winds up with a function, a purpose, but it didn't come to exist by virtue of its contribution to that function. Rocks aren't evidently designed to do something; eyes, like telescopes, are. In fact, evolutionary biologists often use the words "designed" and "purpose" in talking about organisms and their organs. As in: designed by natural selection for the purpose of processing visual information about the environment.

One thing evolutionary biologists *don't* generally do is talk about natural selection itself having a "purpose" or being a product of "design." That sort of philosophical speculation isn't part of their job description. But there is no reason in principle that we can't inspect natural selection for signs of purpose, of telos.

Before doing so, though, we should recognize that we're operating under a large handicap. The reason biologists can so confidently say that animals (or plants) have a purpose is because they know what process created animals and plants. These biologists don't just inspect a few animals and say, "Hmm, they all seem bent on reproduction, so maybe transmitting genetic material is their overarching purpose; and maybe clear vision via eyesight is a subordinate goal, a purpose that serves the overarching purpose." The biologists *know* that genetic transmission is the goal an organism is designed to pursue, because they know about natural selection.

In contrast, when we inspect evolution itself for signs of purpose, we don't have the advantage of knowing what thing or process created it; this is the mystery, after all, that inspired our inspection in the first place. In other words, we're in the position of William Paley: examining something for signs of purpose without knowing what, if anything, designed it. Indeed, we're in even worse shape than Paley. He had lots of organisms to inspect: Exhibits A, B, C, ad infinitum. When we inspect natural selection, we have only one exhibit.

Now, duly warned about the overwhelming difficulty of our task, and the necessary sketchiness of our conclusions, we can proceed to inspect evolution for signs of overarching purpose, of design.

How do we proceed? For starters, let's try to get clearer on what would qualify as evidence of purpose in the first place. And to do that, let's return to the William Paley problem, the problem of examining a single living thing—a cell, an animal, a plant—for signs of design. Suppose you are an extraterrestrial version of William Paley, sent to Earth to see if it has any objects that seem teleological, any things that seem to pursue a goal. Suppose you happen to land in a greenhouse. After a few weeks you notice that the green objects surrounding you tend to get taller. Can you conclude that these green things indeed have at least one goal—to grow?

No. Directionality in the narrowest sense isn't evidence of telos. Water flows from high to low, but we don't think of water as being imbued with purpose.

Plants, though, do more than move directionally. Suppose that, while in the greenhouse, you use your awesome extraterrestrial powers to relocate the sun. You notice that within a few days the plants have reoriented themselves so that their leaves face it. And another obervation: if you cut off the end of a branch, a new twig sprouts. These plants seem hell-bent on growing upward—and, in particular, orienting themselves toward light. All told, they seem to meet what the philosopher Richard Braithwaite proposed as a rough-and-ready criterion of teleological behavior: "persistence towards the [hypothesized] goal under varying conditions."

On the other hand, by that criterion, rivers might qualify as teleological; erect a hill in their path, and they meander around it. No, there must be some other diagnostic sign of teleology—something plants and animals have that rivers don't. What is it? Philosophers have nominated various properties for this honor. What I consider the best candidate emerged in the mid-twentieth century, amid excitement over cybernetics, the study of feedback systems. Here it is: The plant adjusts to varying conditions *by processing information*. It has sensors that absorb information reflecting the state of the environment—where the light is coming from, for example—and this information guides the plant's growth accordingly. And so too with every other form of life that pursues goals under varying conditions (which is to say, every other form of life). All plants and animals sense environmental change by processing information, and this information leads to appropriate change in the plant or animal. Even the *E. coli* bacteria—the ones, discussed in chapter 17, that relocate when nutrition gets scarce—are in this sense teleological.

Does evolution itself pass this test—persistence toward the hypothesized goal under varying conditions by processing information? To answer this question, you must now become a different extraterrestrial visitor. You are extremely long-lived. To you, a billion earth-years is a week. You arrive on Earth in the year 1 billion B.C., and your mission is to see whether organic evolution seems to be a goal-seeking process. Right away, you come up with a good candidate for the goal—the creation of organic complexity, in several senses: broadening the diversity of species, raising the average complexity of species, expanding the outer limit of complexity, and expanding the outer limit of behavioral flexibility—that is, of intelligence.

What's more, as you observe this movement toward complexity, you notice the same sort of stubbornness that the other extraterrestrial scientist had noticed about plants: if you "prune" the tree of life, evolution regenerates branches. The branches may not look exactly like the ones pruned. (Even real branches don't, viewed up close.) But they're certainly reminiscent. If you prune the tree by wiping out all life on an island, the island gradually gets repopulated and, with time, the previously filled niches get filled, even if not by the exact same species that filled them before. If you prune the tree by wiping out all life that can fly, flight gets reinvented, again and again, until the air is once again full of flying objects. If you prune the tree by wiping out the most complex life, or the smartest life, replacements are forthcoming.

And, rather as a plant reorients itself when the source of light moves, evolution channels life toward the most benign environments. If lush lands, dense with flora and fauna, dry up, while dry lands become wet, the balance of the biomass shifts accordingly.

Of course, some life stays behind in the newly dry lands, adapting, managing to thrive in barren terrain. Indeed, in general what is striking is the varying conditions under which evolution creates complex life.

There is no doubt, all told, that evolution by natural selection fulfills the rough-and-ready criterion of goal-directed behavior: "persistence toward the goal under varying conditions." But what about the further condition, the one that separates living things from rivers? Does natural selection adjust to varying conditions by processing information?

Yes. It sends packets of information—genes—into the world. If they proliferate, this positive feedback signifies an environment with which they are adaptively compatible. (In fact, their proliferation *constitutes* adaptation to this environment; through this positive feedback life "senses" and reacts adaptively to environmental change.) Of course, sometimes new genes don't proliferate. This negative feedback signifies a lack of adaptive compatibility with the environment. Trial and error is a system of information processing—even if, as here, the trials are randomly generated.

Teilhard de Chardin used to speak of evolution as groping its way along a changing landscape. God forbid that we use Teilhard as a guide to evolution's ways; his mysticism often clouded his scientific vision.[†] But on this one point—the validity of the "groping" metaphor—he was defended by no less an authority than the noted geneticist Theodosius Dobzhansky. Natural selection indeed "gropes"—blindly sends out feelers. It assimilates the feedback into amendments of design that allow it to keep generating complexity amid varying conditions.

In chapter 17, I suggested that a bacterium's cyclic AMP, in signifying an environment low in carbon, and motivating behavior appropriate to a carbon shortage, constituted a meaningful symbol—at least, if we define meaning as the Peircian school of pragmatic philosophy does. You might say that genes which spread by natural selection are meaningful in the same sense. The genes for the "sweet tooth"—the genes we all seem to have, inclining us to like sweets—reflect the fact that, in the environment of human evolution, sweet fruits had nutrition. And, in addition to reflecting this fact about the environment, the genes motivate behavior appropriate to it (or, at least, behavior that was appropriate before candy bars showed up). As Dobzhansksy once wrote, "natural selection is a process conveying 'information' about the state of the environment to the genotypes of its inhabitants."

To say that natural selection "senses" the local environment isn't to say it has subjective experience. Then again, a radar-guided gun that automatically tracks and fires at its target doesn't (so far as we know) have subjective experience—yet we commonly say that such a gun "senses" movement. Indeed, this sort of "servomechanism" is a textbook example of a "goal-seeking" or "purposive" device that some philosophers have used when trying to find criteria separating the teleological from the aimless. They put servomechanisms, next to plants and animals, on one side of the divide and rocks and rivers on the other. I contend that evolution by natural selection belongs on the first side.

Let me reiterate the meagerness of my aspirations. I'm not saying there is proof that biological evolution has a purpose and is the product of design. I'm just saying that it's not *crazy* to believe this. Biological evolution has a set of properties that is found in such purposive things as animals and robots and is not found in such evidently purposeless things as rocks and rivers. This isn't proof of teleology, but it's evidence of it.

Or, to put the point another way: It may indeed be that evolution is not teleological. But if that's the case, then evolution is the only thing I can think of that exhibits flexible directionality via information processing and *isn't* teleological.

IS NATURAL SELECTION BORN OF NATURAL SELECTION?

Even if there *were* proof that evolution is teleological—a product of design, a process with a purpose—we would still be a long way from Teilhard de Chardin's worldview, complete with a God and a happy ending. The moral of the William Paley story, after all, is that something designed and possessing purpose needn't have been designed by a being, much less a divine one. Indeed, it's possible—not likely but not quite impossible—that organic evolution, like individual organisms, was imbued with purpose by a Darwinian process; that, just as the blossoming of a flower is the product of natural selection, the blossoming of planet Earth's global organic web was the product of a kind of higher-order natural selection.

Sound crazy? Well, consider the hypothesis of "directed panspermia," proposed by the presumably sane Francis Crick, co-discoverer of the structure of DNA. Crick's conjecture is that the seeds of evolution—bacteria, say—were sent to our solar system from a civilization far, far away that was itself the product of biological and then cultural evolution. Let's grant Crick this flight of fancy (a privilege we grant Nobel laureates) and combine it with a second flight of fancy: Meanwhile, in some other faraway place there had been a process of evolution that proved less fertile. It failed to generate animals clever enough to build spaceships with which to fling forth primitive life.

If you put these two flights of fancy together, you have a kind of meta–natural selection. Seeds of evolution, such as bacteria, are planted on planets and either flourish fully—yielding clever creatures with high technology and a penchant for inseminating other solar systems via rocketfuls of one-celled life—or don't. The seeds that do create fertile civilizations are the "fittest." They propagate. So, 3-billion-year generation after 3-billion-year generation, fitter seeds—seeds conducive to the evolution of clever, adventurous species—evolve, or, rather, meta-evolve. And presumably our civilization is on the verge of a new generation—poised to load up a rocket with bacteria any decade now. (Why not put people instead of bacteria on the rocket? Crick says a planet's rockets would presumably be able to travel farther with primitive life onboard than with complex life onboard.)

I don't recommend thinking about the meta-natural-selection scenario for very long, because complications quickly set in. And, though some of these kinks can be ironed out[†] on sustained reflection, some can't. For example: Assuming evolution typically takes as long as it took on this planet, there could have been only a few meta–life cycles in the 12-billion-year history of the universe. That wouldn't seem to allow for very sophisticated design.

Still, even if meta–natural selection itself isn't a scenario I'd bet money on, it has its virtues as something to ponder. In particular, it implies an intriguing reconceptualization of life on earth. In this view, the entire 3-billion-year evolution of plants and animals is a process of epigenesis, the unfolding of a single organism. And that single organism isn't really the human species, but rather the whole biosphere, encompassing all species. The human species—not to belittle the job—is just the biosphere's maturing brain. (Teilhard coined "noosphere" with "biosphere" in mind.) Just as an organism's brain, upon maturing, has stewardship of the body, the human species has now been given—for good or ill—stewardship of the biosphere.

This view may seem bizarre at first, but it would probably seem less strange if you adopted the vantage point of that hypothetical extraterrestrial—watched earthly evolution in time lapse, from a sufficiently great distance. The biosphere's persistent growth, crowned by the human species' sudden transfiguration of the landscape, might seem as relentlessly directional as a flower's life. And, if you buy the arguments for directionality in parts I and II of this book, you have to concede that the unfolding of this "superorganism"—biosphere plus noosphere—did indeed have a powerful impetus; like the growth and flowering of a poppy, it was all along highly likely, assuming a certain fertility of local circumstance, and assuming disaster wasn't inflicted by some outside force. In this narrow, probabilistic sense, we can say that, just as a poppy seed's destiny was to become a poppy, the primordial DNA's destiny was to become (loosely speaking) a global "superorganism," complete with a species sufficiently intelligent, and sufficiently high tech, to qualify as a superbrain.

If you go further—buy not just the arguments in earlier chapters, but the argument in *this* chapter—we can start talking about destiny in a *less* narrow sense. We can say that evolution has not just highly probable directionality, but evidence of purpose; that maybe life on earth writ large, like the poppy seed, heads not just in a direction, but in a direction it was designed to head in.

Big deal. Remember: the direction a poppy was "designed" to head in—toward flowering—is subordinate to a "higher" purpose that isn't terribly inspiring: transmitting genes. To be sure, the "higher" purpose of the global superorganism is unlikely to be analogous; the meta-natural-selection scenario is too wobbly to convince me that our ultimate purpose is to spread genetic information intergalactically. Still, the larger point stands: "higher" purpose needn't be very high. If evolution indeed has a purpose, that purpose may, for all we know, be imbued not by a divinity, but by some amoral creative process.

In short, this chapter, even if you buy its logic, leaves us in William Paley's shoes: having validly found evidence of design, yet having failed to find genuine evidence of the kind of designer that offers even a granule of spiritual reassurance. But there's still a slender thread of hope: there's one chapter to go.

Chapter Twenty-two

YOU CALL THIS A GOD?

We don't know what the hell it is, except that it's very large
and it has a purpose.
—*Dr. Heywood Floyd in the movie* 2010

The original meaning of the word "evolution" was "unfolding" or
"unrolling"—as in, the unrolling of an ancient scroll to get to the
end of the story. There is something to be said for this long-lost sense
of the word. Though neither biological nor cultural evolution is
scripted, inexorable in the way that a written narrative is inexorable,
both have direction—even, I've argued, a direction suggestive of purpose, of telos. The unfolding of life on this planet may be a story with
a point.

Of course, points can be good or bad. And direction plus purpose
doesn't necessarily equal goodness. Pol Pot had direction and a strong
sense of purpose. Is there any reason to believe that in the case of biological and cultural evolution, the direction is toward the good, the
purpose benign? Or, to put the question in common language: Is
there evidence not just of design, but of *divine* design? Any signs of
something worthy of the label "God"?

Historically, aficionados of directionality have tended to answer yes
in one sense or another. Hegel said his dialectic of history amounted
to the manifestation of God. Bergson said the élan vital could be
viewed as divine, and that evolution was God's "undertaking to create
creators, that He may have, besides Himself, beings worthy of His
love." (It's lonely at the top.) To Teilhard de Chardin, Point Omega
was the climactic incarnation of God's love.

I can't claim this much confidence in specifying where exactly

God fits into the picture—or even in asserting that some sort of divine being *does* fit into the picture. Still, it does seem to me that an appraisal of the state of things from a scientific standpoint yields more evidence of divinity than you might expect. Which is to say: nonzero.

What do I mean by "scientific"? I don't mean using satellite-based sensors to detect divine radio waves. I just mean examining theological scenarios by using evidence that is there for all to see—rather than invoking claims of special revelation, or mystical insights reached through meditation or through medication, or whatever. (This empirically based endeavor is sometimes called "natural theology.") Let's accept, if only for the sake of argument, the previous chapter's contention that biological and cultural evolution have some hallmarks of design. Does the design seem to embody the values that people associate with God?

In one sense, the answer has to be no. The kind of God that is hardest to find evidence of is the kind most people seem to believe in: a God that is infinitely powerful and infinitely good. After all, presumably that kind of God wouldn't have let Pol Pot happen—and wouldn't allow the various other forms of cruelty and suffering in the world (including those inherent in organic evolution, and thus in our creation). This is not, of course, some new insight that emerges from this book's vantage point. It is a very old insight—"the problem of evil"—that emerges from the most casual inspection of the everyday world: Why would a benign, almighty God let bad things happen to good people—or to people in general?

Some thinkers have solved the problem of evil straightforwardly, by denying its premise. Ancient Zoroastrians said God is not omnipotent, but rather is in pitched battle with an evil spirit, and is doing His best. More often, theologians have finessed the issue: God is good and omnipotent, so all the seemingly bad things He tolerates must have redeeming qualities that make them ultimately good. For example, maybe suffering is a prerequisite for "soul building."

This argument has often drawn the obvious rejoinder: If God is omnipotent, why doesn't He rewire the universe so that suffering *isn't* necessary for "soul building"? What would be wrong with prefab souls?

Personally, I prefer the Zoroastrian scenario. Or, perhaps, a scenario in which a good God, though not confronting an active, satanic force, is in some other sense of limited power. Maybe in creating the universe, He (She, It) faced metaphysically imposed design constraints.

Anyway, the aim of this chapter is not to describe God or explain God's ways, a task that is above my pay grade. I'm using "God" as convenient shorthand for something vaguer than what the word generally connotes. The point here is just to ask: Are there signs of *any* divinely imparted meaning in the evidence immediately before us: the history of life on earth? Granted directionality in the sense of growing complexity, is there any directionality along what you might call a spiritual or moral dimension? For that matter, is there anything you might *call* a spiritual or moral dimension?

A SOURCE OF MEANING

One odd result of material progress has been to increase the tendency of people to find life devoid of meaning. Back in the early Middle Ages, when life expectancy was around thirty and going to bed with a full stomach was a rare treat, people were sure life had meaning. In the late-modern era, as longevity became a virtual birthright in some societies, people began opining that existence is pointless. What's more, adherents of this view tend to think that they're on solid scientific ground—that modern science, by solving mysteries of life that in ages past were given divine explanation, underscores the absence of higher purpose.

What these people need is a good stiff thought experiment! Imagine a planet on which life evolves. Little bits of self-replicating material (call them genes) encase themselves (by a process we'll call natural selection) in protective armor that exhibits behavioral flexibility. One species in particular—a brainy, two-legged organism—exhibits lots of behavioral flexibility. These organisms are capable of great feats: communicating with subtlety, creating art, watching TV.

Sound familiar? Not so fast. These organisms have one other feature: the absence of consciousness—no trace of sentience; it isn't *like anything* to be them. Yes, fire burns their hands, so, yes, they're designed to withdraw their hands from fire, but, no, they don't feel pain. Or happiness, or anything. These organisms look and act just like human beings; young lovers kiss passionately, and new parents beam with pride—except without the passion and the pride. These are just robots with unusually good skin.

Obviously, such a world would lack the kinds of things many people cite as key sources of life's meaning: such feelings as undying love, devout allegiance, unmitigated triumph, and so on. But there is something else, too. Such a world would lack *moral* meaning. After all, these so-called organisms are just machines, as devoid of feeling as a computer (or at least, as devoid of feeling as we presume a computer to be). Is there anything immoral about unplugging a computer for good? And if not, then how could there be anything immoral about killing one of these insensate organisms on this emotionally barren planet, where there was never any potential for fulfillment in the first place? This is what a world truly without meaning would look like: it would offer no context in which words such as "right" and "wrong" made sense.[†]

Maybe the strangest thing about life on this imaginary, zombie-inhabited planet is this: it is precisely the kind of life that you would expect to evolve on *this* planet. Indeed, the fact that life on earth isn't inhabited by these zombies is a source of great and perhaps eternal perplexity. For, as noted in the previous chapter, subjective experience, according to the premises of modern behavioral science, lacks a function; it is redundant, superfluous.

The seeming superfluousness of consciousness has prompted the philosopher David Chalmers to remark, "It seems God could have created the world physically exactly like this one, atom for atom, but with no consciousness at all. And it would have worked just as well. But our universe isn't like that. Our universe has consciousness." For reasons unknown, God decided "to do more work" in order "to put consciousness in." The key bit of effort, so far as Chalmers can tell, was to draft a law assigning consciousness to some, and perhaps all, types of information processing.

By "God" Chalmers doesn't mean a guy with a white beard. Most philosophers use the term at least as vaguely as I'm using it: it refers to whoever, whatever—if any being, any process—specified the laws of the universe. Still, the fact that the one feature of human existence that is of mysterious, even inexplicable, origin is also the central source of life's meaning doesn't exactly discourage speculation about divine beings and higher purpose. And it renders odd the tendency of people convinced of life's meaninglessness to cite, as support, science's having "explained away" the mysteries of life. After all, it isn't just that science hasn't managed to *solve* the mystery of consciousness.

In a sense, science *created* the mystery of consciousness; the mystery emerges from a hard-nosed, scientific view of behavior and causality.

THE GROWTH OF MEANING

Of course, a law assigning sentience to complex data processing doesn't do any good unless there's some complex data processing going on. Conveniently enough—as we saw in part II of this book— organic evolution ensures as much. Over time, we see more and more complex animals that process information more and more elaborately.

It isn't just that natural selection favors *behavioral* complexity, and thus deft data processing. Complexity of biological structure itself, from the very beginning, entailed information processing. Forget about your brain and its ability to plan vacations, wondrous though this is. Just think about your lungs or kidneys, about breathing or uri- nating. These things, too, are data-rich—not just via involvement with the nervous system, but via hormonal control, via all kinds of minor bits of cellular cross-talk. For that matter, a single cell—any one of yours or any one bacterium—has at its heart an information processor of no meager sophistication, DNA.

Granted, when it comes to our most sublime, most meaningful moments—feeling love or empathy, joy or epiphany, even abject but profound remorse—kidneys and bacteria just won't get the job done. Brains are where the action is. So it's fortunate that large multicellular animals with great behavioral complexity seem to have been in the cards. My point is just that these brains are a continuous outgrowth of something at life's very essence: a primordial imperative to process information. Given the apparent connection among information pro- cessing, sentience, and meaning, it seems fair to say that evolution by natural selection was from the beginning a veritable machine for mak- ing meaning.

As we've seen, the logic by which complexity, hence data process- ing, hence meaning, grows is the logic of non-zero-sumness. The genes along a strand of DNA have a non-zero-sum relationship with one another, as do the organelles within a cell, the cells within a body. In all of these cases, the cause of the non-zero-sumness is shared Dar- winian interest—being in the same boat in one sense or another—and the result is transmitted information. For, as noted in chapter 8, the successful playing of non-zero-sum games—cooperative coordina- tion—generally involves communication.

Within any organism, it is these well-played non-zero-sum games, and the flexible coherence they bring, that let life persist in the face of mounting universal entropy; that let life defy the spirit, though not the letter, of the second law of thermodynamics. That games which call for information processing should be on the leading edge of this war against entropy makes perfect sense. As Jacques Monod observed (in chapter 17), to arrange matter into an orderly form and keep it there, in the face of the second law's tendency to mix things up, requires a "discriminative," even "cognitive" capacity. Ever since life's initial defiance of the spirit of the second law, information processing has risen to higher and higher levels, following the logic of organic coherence, which is to say, the logic of non-zero-sumness.

That biological evolution has an arrow—the invention of more structurally and informationally complex forms of life—and that this arrow points toward meaning, isn't, of course, proof of the existence of God. But it's more suggestive of divinity than an alternative world would be: a world in which evolution had no direction, or a world with directional evolution but no consciousness. If more scientists appreciated the weirdness of consciousness—understood that a world without sentience, hence without meaning, is exactly the world that a modern behavioral scientist should expect to exist—then reality might inspire more awe than it does.[†]

THE ORIGIN OF GOODNESS

The meaning imparted by consciousness, one might argue, isn't an *inherently* good thing. After all, sentience brings equally the capacity for joy and for suffering, for good and for bad. It is the existence of sentience, of meaning, that allowed Pol Pot to be a person of consequence. On that imaginary planet of zombies, devoid of meaning, the Pol Pots and Hitlers and Stalins of the world would be incapable of evil; however destructive, they could inflict no suffering, prevent no happiness, affront no dignity.

In short, the existence of meaning is morally neutral; it creates the potential for good, but doesn't, by itself, tip the scales in that direction. In this light we might hope for more from a divine architect than mere meaning, the mere *capacity* for good things. We might hope for the *realization* of good things—every now and then, at least, and the more often, the better.

On the other hand, isn't goodness a slightly naïve thing to ask of an architect whose plans included natural selection? At its core, natural selection is cutthroat. It is a zero-sum struggle for finite resources, and there are no rules. How much good could come of that?

More than you might think. As we've seen, this dynamic had the paradoxical effect of weaving ever-larger non-zero-sum webs, from a single strand of DNA all the way up to a society of multicellular animals. The point isn't just the attendant growth in data processing, hence sentience, hence meaning. Eventually, this dynamic brought some semblance of actual good. For, when the impetus of non-zero-sumness finally reached the level of animal societies, it entailed—as we saw in chapter 18—the invention and proliferation of love.

Actually, "entailed" is a bit strong. What this impetus clearly *entailed* wasn't love per se, but the evolution of altruistic behavior among close kin. This altruistic behavior in turn *seems* to have entailed—for reasons concealed in the more general mystery of consciousness—the subjective experience of love. (At least, in *our* species, love is what parents feel when they nurture and protect their offspring, and there's no reason to think chimps or dogs are any different.) Anyway, the main point is that with the advent of altruism, animals were doing something other than eat each other; they were helping each other, and feeling good about it to boot.

Intra-family altruism has evolved multiple times, and naturally so. With closely related organisms tending to start out life near each other, commonality of Darwinian interest is thick, just waiting to be harnessed by the logic of kin selection. Though maternal devotion was presumably the original form of kin-directed love (even many insects display maternal altruism), other forms followed: sibling love and—in our species and some others—paternal love.

Altruism, having established a beachhead within the family, eventually branched out beyond close relatives. As we've seen, in a number of species, including ours, natural selection invented reciprocal altruism, which, notwithstanding its underlying cold calculation, involves heartfelt obligation, even affection. This tendency of human beings to form bonds beyond the family would become crucial as cultural evolution began the long geographic expansion of non-zero-sumness. Biological evolution, having created goodness by inventing altruism, would now surrender center stage to the second great evolutionary force, with which any hopes for expanding goodness would now lie.

But before we get too rhapsodic about all this bonding, a word is in order about affection's oft-underplayed downside. Ever hear of the "Texas cheerleader mom"? She was convicted of plotting to murder her daughter's rival for a high-school cheerleading slot. The good news is that this woman is manifestly not typical of mothers in Texas—or anywhere. The bad news is that she nonetheless illustrates, if in grotesque proportion, a ubiquitous point: love is, by design, an invidious emotion. The problem isn't just that love gets extended selectively, often coming to a screeching halt at the bounds of family. The problem is that love is often deployed to the active detriment of people beyond the family. It's a jungle out there, after all, and we want our loved ones to triumph.

This seamy underside of affinity isn't confined to intra-family affinity. One common purpose of reciprocal altruism in primates is to cement coalitions which then compete with other coalitions, sometimes violently. In general, as the biologist Richard Alexander has observed, the flip side of "within-group amity" is "between-group enmity."

This dour equation seems almost to have been a constant of human history. Lengthened and strengthened bonds have tended to involve deepened fissures. Consider those nostalgic reveries about wartime. Soldiers talk about the indelible devotion to their comrades in arms, and civilians recall the sense of brotherhood that suffused a whole nation. Sounds great. But as amity thus reached national scope, the petty enmities of daily life weren't so much erased as displaced—piled up, sky-high, along the nation's border: a mass of hatred between peoples. It almost seems as if one of the basic laws of the universe, right next to "conservation of mass" and "conservation of energy," is "conservation of antipathy."

Here again we encounter the problem of evil: you wouldn't expect a benign and omnipotent God to embed such a law in the universe. Yet the law—or, at least, the "law"—does seem fundamental. It is grounded in the basic paradox of creation: non-zero-sumness, wondrous though it is, was created by, and for, zero-sumness, and is thus naturally prone to malicious use. Kant's "unsocial sociability" lies in the very logic of natural selection.

THE GROWTH OF GOODNESS

This conservation-of-antipathy business threatens to put a damper on some of the celebrating we did earlier in this book. Remember the brief rhapsody about the expanding circle of moral consideration? How ancient Greeks had come to concede the humanity of Greeks who lived in distant cities? How the moral compass kept growing in tandem with the scope of non-zero-sumness, so that people increasingly recognized that inhabitants of foreign lands, speakers of foreign languages, adherents of foreign faiths, are human beings nonetheless? Obviously, we would have to feel less festive about this progress if it turned out that every iota of advance entailed regression. Is "conservation of antipathy" truly a law of nature—or, at least, a rough tendency so deeply embedded in human affairs as to doom moral progress?

No, for two reasons. First, martial fervor is not the only source of social bonding. If I water my neighbor's plants while he's away, and he returns the favor, our mutual amity grows just a bit—without any necessary growth in my dislike of anyone else. And much of the growth of non-zero-sumness over the past few millennia has been of this sort—people being "pulled" together for common gain, not "pushed" together by a common enemy.

One of the main pulling forces, of course, has been economic. Granted, commerce can be a cool affair, and often fails to expand the web of affection, but it does expand the web of *tolerance*. You don't have to love your grocer, but you shouldn't assault him. You don't have to love the people who built your Toyota, but it's unwise to bomb them—just as it's unwise to bomb the people overseas who are buying the things *you* made.

The second reason that the alleged "conservation of antipathy" doesn't doom moral progress has to do not with these "pulling" forces of economics but with the "pushing" forces, under which people unite to thwart a common threat. Though war is the time-honored example of such a force, we saw in chapter 15 how things can change as social organization approaches the global level. At that point, barring extraterrestrial invasion, conquest isn't the peril that brings people together. Rather, they cooperate to evade such things as terrorism, international crime, environmental calamity, and economic collapse. More than before, non-zero-sumness can thrive without zero-sumness as its ultimate source. To whatever rough extent "conservation of antipathy" *has* held as a general law of history, it seems to be in the process of being repealed.

Of course, zero-sumness hasn't vanished. Corporations compete with corporations, politicians compete with politicians, soccer teams compete with soccer teams. What's more, some of the new sources of common peril—terrorists and international criminals—are playing emphatically zero-sum games: they have interests quite opposed to those of society at large, and will accordingly inspire their share of antipathy. Still, as non-zero-sumness has grown, finally reaching global extent, a particular *kind* of zero-sum dynamic has begun to weaken. And it is the most pernicious kind: bitter struggle between geographically separate groups, featuring blind hatred of whole peoples. The *spatial* dimension of zero-sumness, historically its most abhorrent dimension, has begun to fade. In this sense—a limited but far from trivial sense—we can say that non-zero-sumness is on the verge of having "won" in the end.

Obviously, cultural evolution's movement toward this moral threshold isn't proof of a benign universal architect, any more than biological evolution's expansion of meaning or its invention of goodness was. But, like those biological developments, this cultural development is closer to being evidence of divinity than its opposite would be. Once you've accepted that evil is, for whatever reason, built into the fabric of human—indeed, organic—experience, the basic trend lines don't look all that bad.

THE FUTURE OF GOODNESS

The prospect of peaceful, even respectful coexistence among the world's peoples might seem enough to satisfy anyone. But Teilhard de Chardin hoped for more. After all, if peace and tolerance grow *only* out of non-zero-sum calculation—*only* out of rational self-interest—then there is something cool and mechanical about it all. And Teilhard preferred warm and fuzzy things. In his musings about humanity's approach, via globalization, to Point Omega, he wrote: "Humanity, as I have said, is building its composite brain beneath our eyes. May it not be that tomorrow, through the logical and biological deepening of the movement drawing it together, it will find its *heart,* without which the ultimate wholeness of its powers of unification can never be fully achieved?"

As is often the case with Teilhard, his exact meaning is not clear. But at the very least he had in mind the expansion of brotherly love, of Christian charity, to planetary breadth. Is there any hope for such a thing? That would certainly be good news, on a number of grounds. In particular, as noted in chapter 16, any such pervasive fellow-feeling could help bring security to the world without massive intrusions on liberty.

The new technologies of interdependence do sometimes bring flashes of something richer than mere tolerance. Occasionally, in the e-mails that flit around the globe, true empathy transpires. Occasionally, watching TV and seeing the suffering of foreigners in a superficially alien culture, a viewer is struck by the realization that, fundamentally, all human beings are alike. Certainly charity in the material sense—donation to the needy—has reached unprecedented geographic scope this century.

Of course, it may forever remain true that nothing brings people together, heart to heart, quite like a war. And that sort of bonding, thankfully, is unavailable on a planetary scale. But other common challenges—environmental distress, for example—are not devoid of bonding power.

Indeed, the classic experiment on inter-group solidarity suggests that inanimate threats can be quite unifying. Several decades ago, the psychologist Muzafer Sherif used boys in a summer camp (unbeknownst to them) to study human nature. He divided them into two groups and put them in a series of zero-sum games, with cherished perks going to the winning team. Jingoism blossomed; there was one full-fledged riot, and even after all zero-sum games had ended, contact between the groups brought slurs and fistfights. Then Sherif put the groups in a series of non-zero-sum situations, where all boys faced some mutual threat. Sure enough, antagonism was so dampened that some erstwhile enemies became lasting friends. And the mutual threats that did this congealing weren't invasions from a neighboring camp, but rather such things as the breakdown of the truck all campers depended on, or of the pipeline that brought water into camp.

This doesn't mean that combatting global warming will lead to a transnational lovefest. But it is evidence that, as global interdependence thickens, long-distance amity can in principle grow even in the absence of external enmity. And it's something to build on. There is no telling what it could mean as technology keeps advancing; as the World Wide Web goes broad bandwidth, so that any two people anywhere can meet and chat virtually, visually (perhaps someday assisted, where necessary, by accurate automated translation). One can well imagine, as the Internet nurtures more and more communities of interest, true friendships more and more crossing the most dangerous fault lines—boundaries of religion, of nationality, of ethnicity, of culture.[†]

The common interests that support these friendships needn't be high in gravitas. They can range from stopping ozone depletion to preserving Gaelic folklore to stamp collecting to playing online chess. The main thing is that they be far-flung and cross-cutting. Maybe this is the most ambitious realistic hope for the future expansion of amity—a world in which just about everyone holds allegiance to enough different groups, with enough different kinds of people, so that plain old-fashioned bigotry would entail discomfiting cognitive dissonance. It isn't that everyone will love everyone, but rather that everyone will like enough different kinds of people to make hating any given type problematic. Maybe Teilhard's mistake was to always use "noosphere" in the singular, never in the plural. Maybe the world of tomorrow will be a collage of noospheres with enough overlap to vastly complicate the geography of hatred. It wouldn't be Point Omega, but it would be progress.

THE FUTURE OF GOD

One might hope that moral progress will get a boost from the further evolution of the world's spiritual traditions. Certainly, in the past, religious doctrine has offered an expanding spiritual basis for the technologically driven expansion of non-zero-sumness. The growth of Islam more than a millennium ago not only created a network of Muslim traders who could trust one another; by preaching tolerance of Christians and Jews—"people of the Book"—Islamic scripture smoothed the path of commerce all the more.

Judeo-Christian doctrine was similarly pragmatic in its evolution. It is easy to miss the continuity here—to see an abrupt shift between the Old Testament, with its wrathful, tribalistic God, and the New Testament, with its universally loving God. Indeed, in early Christian times, some thinkers—such as Marcion, in the second century—used this dichotomy to solve the problem of evil: the God of the Old Testament, the creator of life, was an evil god, and the God of the New Testament, the God of love, had now come to help us make the best of a bad situation. But the truth is that even in the Old Testament,

God can be seen turning into a nicer guy. In the millennium before Christ arrived, as commerce drew distant lands into deeper contact (thanks partly to the debut of coins), the scope of God's sympathy grew. In two books composed in the second half of the millennium—Jonah and Ruth—God's love reaches beyond tribal bounds, to gentiles.

The Book of Jonah is especially instructive. God tells Jonah, a Jewish prophet, to go warn the city of Nineveh that its sinful ways will bring devastating divine wrath. Jonah resists. Nineveh, after all, is a despised gentile city. Why give it a heads-up? But, after attempting to flee this charge by ship and finding himself in the belly of a giant fish, Jonah reconsiders. Upon emerging, he grudgingly follows God's orders. The people of Nineveh, now warned, mend their evil ways, and God spares them, filling Jonah with disappointment. In the closing verse, God asks a sulking Jonah: "Should I not pity Nineveh, that great city, in which there are more than a hundred and twenty thousand persons?"

Earlier, when God reproves Jonah for being angry at the gentiles' escape from death, there is an interesting ambiguity of translation. In the King James translation, God asks: "Doest thou well to be angry?" In the New Revised Standard Version God asks: "Is it right for you to be angry?" I have no idea which is truer to the original, but the juxtaposition is profound. For the key to the expanding compassion that the Book of Jonah marks is indeed the link between doing well and doing right. As the realm of non-zero-sumness has grown, material prosperity has entailed acknowledging the humanity of an ever-larger fraction of the human species. Jonah is on the cusp of the trend. In an increasingly interconnected land, he would do well to do right—to mute his anger toward whole groups of people he's never met.

In effect, as cultural evolution has progressed, the size of society writ large—the number of people with whom one's fortunes are intertwined, with whom one does well to do right—has grown. And one role of religious doctrine has always been to congeal societies. (The word "religion" comes from the Latin *ligare,* "to bind.") Émile Durkheim went so far as to say that "the idea of society is the soul of religion." Depending on how it is put, the point can sound a little dispiriting—as when Durkheim wrote: "In the last analysis men have never worshipped anything other than their own society." Still, if worshiping your own society finally, in a global age, involves not denigrating other peoples but, rather, recognizing the moral worth of human beings everywhere, then there is something to be said for worshiping your own society. The equation between doing well and doing right may sound crass, but if, over tens of millennia of cultural evolution, it brings moral enlightenment, then my hat is off to it.

TODAY'S SERMONETTE

Is it really possible that new, improved religions might help congeal the world? If so, what might their core doctrines be? Oh, the usual—universal brotherhood—except this time with feeling. Also, the admonitions against greed that are scriptural boilerplate could stand to get dusted off and read with an eye to (among other things) slowing the rate at which the planet becomes a giant cauldron of garbage dumps, melting ice caps, and Mercedes-Benz utility vehicles.

Of course, one difficulty with pinning any hopes on religion is its much-discussed ongoing erosion at the hands of science, an erosion that is one alleged source of modern and postmodern nihilism and ennui. But one point of this book has been to challenge the conventional belief that science really has dispelled deep mystery and all evidence of purpose.

As for the scientific assault on mystery: a truly scientific perspective shows consciousness—the fact that it is like something to be alive—to be a profound and possibly eternal mystery, and a suggestive one to say the least. And divinity isn't the only thing it suggests; by its nature, the open question of consciousness underscores the continued openness of various other questions, such as free will.

As for the scientific assault on purpose: A strictly empirical analysis of both organic and cultural evolution, I've argued, reveals a world with direction—a direction suggestive of purpose, even (faintly) suggestive of benign purpose. Life on earth was, from the beginning, a machine for generating meaning and then deepening it, a machine that created the potential for good and began to fulfill it. And, though the machine also created the potential for bad—and did plenty of fulfilling on that front—it now finally shows signs of raising the ratio of good to bad; or, at the very least, of giving the human species that option, along with powerful incentives to exercise it.

This recent uptick in the moral stock market, coming several billion years after the creation of life, may strike some people as underwhelming. If you really sit and ponder all the suffering that has been caused by—in fact, was built into—biological and cultural evolution, you may find it hard to muster a lot of gratitude toward the universal architect.

Or you can take the opposite tack. Maybe that's what I've done—spent so much time pondering the horror intrinsic in the past that I'm grateful for small things. I gave up so long ago on an omnipotent and benign deity that I'll take a few wisps of good karma and hope they signify something larger.

But, whether or not I'm straining to find divinity, I don't think I'm straining to find meaning. The point isn't just that, for reasons that are exceedingly hard to fathom, we have consciousness, and thus are playing for real moral stakes. The point is that we are playing for the highest stakes in history. More souls are crammed onto this planet than ever, and there is the real prospect of commensurately great peril. At the same time, there is the prospect of building the infrastructure for a planetary first: enduring global concord.

And if we did that—if we laid a foundation for peace and fulfillment around the world—that would counterbalance a lot of past evils, given the number of people now around to enjoy the benefits. It may literally be within the power of our species to swing nature's moral scales—which for so long tended to equilibrate near dead even, at best—decisively in the direction of good; maybe it is up to us, having inherited only the most ambiguous evidence of divinity, to *construct* clearer evidence in the future. Maybe history is, as various thinkers have suggested, not so much the product of divinity as the realization of divinity—assuming our species is up to the challenge, that is. (One theologian has paraphrased Teilhard as believing that "God must become for us less Alpha than Omega.")

My belief that some workable infrastructure for concord will very likely emerge does nothing to drain the drama from the present, for one plausible route to long-run success is near-term catastrophe. However close to inevitable stable world governance may be in the long run, here and now we are playing for the highest stakes that have ever been played for, and winning will depend in no small part on continued moral growth. Which is to say: winning will depend on not wanting other peoples to lose.

There are many ways to react to all this, but nihilism and ennui don't seem to me among the more logical.

IN THE BEGINNING . . .

In the New Testament, the Gospel of John begins, "In the beginning was the Word, and the Word was with God, and the Word was God. . . . In him was life." More than one science writer of a cosmic bent has juxtaposed this verse suggestively with the modern scientific view of organic evolution: in the beginning was, if not a word, at least a sequence of encoded information of *some* sort.

Fair enough. But if cosmically suggestive juxtaposition is what you're after, you needn't stop here, for the biblical word "Word" is richer than it sounds. It is a translation of the Greek *logos,* which can indeed mean "word" but can also mean many other things, including "reason." And you might say that, once self-replicating genetic information existed, a line of reasoning, a chain of logic, had been set in motion. A several-billion-year exercise in game theory had commenced.

Logos also means "argument," and it is tempting to view biological and cultural evolution somewhat as Hegel viewed human history—as a very long argument. Competing ideas about how to organize organic entities clashed. And non-zero-sumness won in the end.

One scholar has rendered *logos* as the "point," the "purpose"— the end that one has in mind. And, indeed, the religiously inclined might speculate that the spiritual corollary of the triumph of non-zero-sumness—the expansion of humanity's moral compass—was the purpose of history's game-theoretical argument all along. In the beginning, you might say, was the end, and the end was a basic truth— the equal moral status of all human beings.

The idea that a kind of *logos* might be the force guiding a directional history is far from new. In fact, this was the theory of Philo of Alexandria, member of an ancient philosophical school that some scholars believe was the conduit through which *logos* entered Christian scripture. Permeating human history, Philo said, was a "divine *Logos,*" a rational principle that was immanent in the world but, at the same time, was part of God's transcendent mind. And in what direction was *Logos* moving history, in Philo's view? "The whole world," he wrote, "may become, as it were, one city and enjoy the best of polities, a democracy." Not bad, as two-thousand-year-old predictions go.

Of course, Philo didn't have access to game theory, so he couldn't talk about non-zero-sumness. Then again, game theorists weren't the first people to recognize the logic of interdependence, and Philo certainly grasped it. Mutual need, he believed, was what wove God's diverse creatures—people, plants, animals—into a whole.

God "has made none of these particular things complete in itself, so that it should have no need at all of other things," Philo wrote. "Thus through the desire to obtain what it needs, it must perforce approach that which can supply its needs, and this approach must be mutual and reciprocal. Thus through reciprocity and combination, even as a lyre is formed of unlike notes, God meant that they should come to fellowship and concord and form a single harmony, and that a universal give-and-take should govern them, and lead up to the consummation of the whole world."

Amen to that.

In real life, of course, the story has been more complex than Philo's story. In a sense, it has been a better story—not better in moral terms, but better in literary terms, in dramatic terms. It has featured, ever since the first bacterium, growing knowledge—and, with the arrival of human beings, growing *self*-knowledge. It has also featured amity and strife, good and evil—the two forces vying with each other, yet inextricably bound together. And now, in the past century, as knowledge has grown exponentially, so have the stakes of this contest. More than ever, there is the real chance of either good or evil actually prevailing on a global scale. War and other forms of mass slaughter, other manifestations of massive hatred, could be ended—or, on the other hand, they could set new records for death and destruction; they could even, conceivably, end us. And the outcome may hinge on the further spread of knowledge—not just empirical knowledge, but moral knowledge.

Talk about a page turner! Maybe, in the end, this is the best argument for higher purpose: that the history of life on earth is too good a story not to have been written. But, whether or not you believe the story indeed has a cosmic author, one thing seems clear: it is our story. As its lead characters, we can't escape its implications.

APPENDIX 1

APPENDIX 2

ACKNOWLEDGMENTS

NOTES

BIBLIOGRAPHY

INDEX

APPENDIX 1

On Non-zero-sumness

I seriously considered titling this book "Non-zero-sumness." But a number of trusted advisers opined that non-zero-sumness is a very ugly word. Some people seemed to find the sound of it literally painful. "There's something about the 'umness' part," one of them said. So I backed down.

But I did liberally pepper the book with the term—and with such allied terms as "non-zero-sum" and "negative sum," and so on. What's more, I did title the book "Nonzero."

Why am I so attached to the terminology of game theory? Does it really add anything to more familiar words? Can't we just say, for example, that zero-sum games are competitive and non-zero-sum games are cooperative? There are several reasons that I think the answer is no—that there's no substitute for game theory as a way of looking at the history of our species.

For starters, there are a number of cases in which people comply with non-zero-sum logic, and yet "cooperate" is a misleading word. I have a non-zero-sum relationship with the people in Japan who built my Honda minivan, but neither I nor they ever chose to cooperate with each other.

Still less do individual genes on my chromosomes think of themselves as cooperating with one another, though they behave in accordance with non-zero-sum logic. Part of the argument of this book is that the logic of biological integration and of social integration can be subsumed in a single analytical framework. And it seemed to me that, if I wanted a vocabulary that would apply not just to people, but to genes, then it was better to minimize the use of fuzzy terms like "cooperate" and stick mainly with cold, precise terms such as "non-zero-sum." (A particular appeal of using game theory in biology is that in Darwinian theory the "payoff" is clearly and quantitatively defined—as genetic proliferation; so actually adding up the nonzero sums is in theory doable without making artificial assumptions.)

The terminology of game theory helps unify not just human history and organic history. *Within* each of these realms, the terminology can be unifying. If you ask what is common to reciprocal altruism and kin selection (two basic biological routes to social integration), the answer is non-zero-sum logic. And more than once I have found, in the literature on cultural evolution, lists of factors conducing to social integration that seemed diverse but, on close inspection, all embodied non-zero-sum dynamics. (Such as: the need to spread risk, the benefits of irrigation, the efficiency of division of labor, the need for military defense, the need to avoid overexploiting the fish population, etc.)

Another virtue of using the analytical framework of game theory is that it has well-established principles applicable to the subject of this book. The most important of these is that, if self-interested entities are to realize mutual profit in a non-zero-sum situation, two problems typically must be solved: communication and trust. I found this principle powerful when applied to human history (see especially chapters 2, 8, 13, and 15) and, surprisingly, valuable even when applied to organic history (see chapter 18).

But perhaps the ultimate rationale for using the word "non-zero-sum" is that, no matter how far you stretch the word "cooperation"—even if you're willing to apply it to genes and to the relationship between me and Honda's employees—the two words still are not comparable. A non-zero-sum relationship is not a relationship in which cooperation is necessarily taking place. It is (usually) a relationship in which, if cooperation *did* take place, it would benefit both parties. Whether the cooperation *does* take place—whether the parties realize positive sums—is another matter. And, aside from "non-zero-sum," I can't think of a word that captures this relationship. To call it a relationship in which "self-interest rationally pursued could lead to cooperation" just isn't very economical. Even a relationship "conducive to cooperation" isn't very economical, and is vague, even misleading, to boot.

All of which brings us back to "non-zero-sumness."

Non-zero-sumness is a kind of potential. Like what physicists call "potential energy," it can be tapped or not tapped, depending on how people behave. But there's a difference. When you tap potential energy—when you, say, nudge a bowling ball off a cliff—you've reduced the amount of potential energy in the world. Non-zero-sumness, in contrast, is self-regenerating. To realize non-zero-sumness—to turn the potential into positive sums—often creates even more potential, more non-zero-sumness. That is the reason that the world once boasted only a handful of bacteria and today features IBM, Coca-Cola, and the United Nations.

If alternatives for "non-zero-sum" are awkward at best, alternatives for "non-zero-sumness" aren't much better. Probably the most attractive is "potential synergy" (which has the nice property of rhyming with "potential energy"). Yet "potential synergy" doesn't suggest, as strongly as "non-zero-sumness" does, that we're talking about something that is a kind of *stuff*. I don't mean physical stuff, but I do mean something real enough and important enough to be a one-word noun. (It's the "umness" part that I like.)

There's another problem with "potential synergy." It sounds as if the choice is between realizing zero sums and realizing positive sums. And often, to be sure, this *is* the choice (which is why I used "potential synergy" several times in the text); thus, we read in the newspaper of corporations merging to realize synergy. But sometimes in non-zero-sum situations, the object of the game isn't to reap positive sums but simply to avoid negative sums. And here "synergy" would sound strange; we don't read in the newspapers about two nuclear powers signing an arms control treaty in order to find synergy, even though the treaty is the successful outcome of a non-zero-sum game.

In the long run—maybe decades down the road—I expect that the strongest validation for using game-theoretical terminology to describe biological evolution and cultural evolution will turn out to be, in a sense, technological. Biologists and social scientists have lately started to use the computer to simulate evolution. I don't doubt that someday, as their computer models grow more and more elaborate, we will see on the computer screen, for example, full-fledged chiefdoms evolve out of a network of villages. And, to judge by existing evolutionary computer simulations, at the heart of these programs will be game theory. Particular payoffs will be assigned for particular kinds of interactions, and these payoffs will shape the unfolding of evolution.

The seminal exercise in computer-simulated evolution is described in Robert Axelrod's classic book *The Evolution of Cooperation*. It involved that most famous of non-zero-sum games, the prisoner's dilemma.

Actually, the fame is in some ways unfortunate, because the prisoner's dilemma has a couple of quirks that impede intuitive comprehension. For one thing, in this game the object is to get the *lowest* score, since the score represents how many years each player has to spend in prison. The second counterintuitive feature is that in this game to "cheat" is to tell the truth and to "cooperate" is to not tell the truth. But, for better or worse, the prisoner's dilemma is the textbook non-zero-sum game, so we'll here explore it by way of showing how, with the help of game theory, evolution can be simulated on a computer.

In the prisoner's dilemma, two partners in crime are being interrogated separately. The state lacks the evidence to convict them of the crime they committed but does have enough evidence to convict both on a lesser charge—bringing, say, a one-year prison term for each. The prosecutor wants conviction on the more serious charge, and pressures each man individually to confess and implicate the other. She says: "If you confess but your partner doesn't, I'll let you off free and use your testimony to lock him up for ten years. And if you *don't* confess, yet your partner does, *you* go to prison for ten years. If you confess and your partner does too, I'll put you both away, but only for three years." The question is: Will the two prisoners cooperate with each other, both refusing to confess? Or will one or both of them "defect" ("cheat")?

As noted in chapter 8, the prisoner's dilemma has some key features that have shaped the unfolding of non-zero-sumness during human history.

First, there's the importance of communication. If the two prisoners can't communicate with each other, and both behave logically, they will almost certainly both suffer as a result. To see this, just pretend you're one of the prisoners and run through your options one by one. Suppose, first of all, that your partner cheats on you by copping a plea. Then you're better off cheating on him and confessing: you get three years in jail, as opposed to the ten you'd get by staying silent. (In the payoff matrix shown below, this fact is signified by comparing the first digit in the upper-left and lower-left quadrants.) Now suppose your partner doesn't cheat—doesn't confess. You're still better off cheating, because then you get out of jail, whereas if you stayed mum like your partner, you'd each get a one-year jail term. So the logic seems irresistible: don't cooperate with your partner; cheat on him.

		HIS STRATEGY	
		Confess (cheat)	Stay mum (cooperate)
YOUR	Confess (cheat)	You: 3 years Him: 3 years	You: 0 years Him: 10 years
STRATEGY	Stay mum (cooperate)	You: 10 years Him: 0 years	You: 1 year Him: 1 year

But if both of you follow this logic, and both cheat, then you'll both get three years in jail. (That is, you both wind up in the upper-left-hand quadrant.) And if both of you *hadn't* cheated—if both had stayed mum—you would have just gotten one year in jail. So mutual mumness is, relatively speaking, the win-win outcome. But it makes no sense for either of you to stay mum unless you've both been assured by the other that he will stay mum, too. That is why communication is vital.

A second key feature of the prisoner's dilemma is the importance of trust. It is crucial that when your partner and you assure each other you will stay mum, you believe each other. After all, if you suspect that your partner may renege on the deal, then you're better off repaying his cheating with cheating, copping a plea yourself: a three-year prison sentence as opposed to a ten-year sentence. What's more, your suspicions are hardly irrational, since your partner does have a temptation to cheat: If he confesses while you honor your deal and stay mum, he gets to walk. (What are you going to do? Sue him?)

Axelrod organized a tournament that amounted to a simulation of biological evolution. Several dozen people submitted computer programs that embodied particular strategies for playing the prisoner's dilemma. The programs were then allowed to interact with each other—as if they constituted a kind of society. Upon each interaction, the two programs involved would "decide"—on the basis of their algorithms—whether to cheat or cooperate. (Often, in making this decision, they would draw on their memory of how the other program had behaved in past encounters.) Depending on what each had decided, both would receive a score representing the outcome of that encounter.

Then each would move on to the next encounter, with another program. In each round, there would be enough encounters so that every program interacted with every other program 200 times. At the end of each round, the scores for each program, each "player," were added up. Programs were then allowed to "replicate" in proportion to their score. So the better your program did in one round—one "generation"—the more copies of it there would be in the next generation.

The winning program was called "Tit for Tat," submitted by Anatol Rapaport (whose 1960 book *Fights, Games, and Debates* is a very nice introduction to game theory). Tit for Tat's strategy was very simple. On its first encounter with any given program, it would cooperate. On subsequent encounters, it would do whatever that program had done on the previous occasion. In short, Tit for Tat would reward past cooperation with present cooperation and would punish past cheating with present cheating. Generation by generation, Tit for Tat came to dominate the population, so that, more and more, Tit for Tats spent their time interacting with other Tit for Tats. Such interactions invariably blossomed into stable, cooperative relationships. As the game wore on, the "society" of players in Axelrod's computer exhibited more and more amity and order.

One striking thing about this evolution of cooperation is that it transpired *without* the players being allowed to communicate with each other—even though communication, in the generic non-zero-sum situation, is considered a prerequisite for a reliably positive outcome. The reason this could happen is that players would encounter the same players again and again (making this an "iterated" prisoner's dilemma). Thus, by observing what a given player had done on the last occasion, another player could, in effect, gather information about the player's likely future behavior. (This is, in a sense, a kind of de facto communication—and certainly a form of information transmission.) What's more, players could punish each other for past cheating and reward each other for past cooperation.

By showing how cooperation could evolve without formal communication, Axelrod had shown how reciprocal altruism could evolve in animals that don't do much talking—including chimpanzees and vampire bats. (See Wright, 1994, chapter 9, for elaboration.) He had also shown how stable, cooperative relationships could form in a very small society of humans without much explicit discussion; so long as the same players encounter each other day after day—as in a small hunter-gatherer society—trust could develop even with little explicit commitment.

Of course, through cultural evolution, the settings for non-zero-sum games have gotten much less intimate than a hunter-gatherer society. Chances are you've never met the person who made your shoes. In fact, chances are that any one person who had a hand in making your shoes has never met all the other people who had a hand in it. A key feature of cultural evolution has been to make it possible for such non-zero-sum games to get played over great distances, among a large number of players. And in these kinds of situations, typically, there *does* need to be explicit communication (however circuitous), and there *do* need to be explicit means of sustaining trust. Hence the importance of evolving information technology in expanding the scope and complexity of social organization. Hence, too, the importance of evolving "technologies of trust" (often, though not always, in the form of laws enforced by a government) in helping to realize the non-zero-sum potential that new information technologies (and other technologies) create.

Axelrod later used the computer and game theory to simulate the cultural evolution of norms. (See Axelrod, 1987.) I suspect that, with computer power now dirt cheap, and graduate students always hungry for a fresh angle on old subjects, we will begin to see a lot more computer simulations of cultural evolution (not involving the prisoner's dilemma necessarily, but involving zero-sum and non-zero-sum games). In fact, it would surprise me if such efforts aren't already under way. If these efforts prove fruitful, that will be a kind of vindication of the use of game-theory terminology in arguments about the dynamics of cultural evolution.

APPENDIX 2

What Is Social Complexity?

Leslie White began a 1943 essay in the journal *American Anthropologist* with this sentence: "Everything in the universe may be described in terms of energy." The essay was called "Energy and the Evolution of Culture." It launched White's crusade to restore the notion of cultural evolution to the stature it had enjoyed in the nineteenth century, before being shrouded in ill-repute by Franz Boas and his students.

White's thesis lived up to the essay's opening grandeur. Cultural evolution, he announced, could be measured by a single gauge, calibrated in terms of energy. "Culture develops when the amount of energy harnessed by man per capita per year is increased"; or when "the efficiency of the technological means of putting this energy to work is increased"; or both. In short, "culture evolves as the productivity of human labor increases."

Few if any cultural evolutionists would deny that energy technology is important stuff. But White went beyond saying that energy technology matters in cultural evolution. He virtually *equated* cultural evolution with the evolution of energy technology. The efficiency with which energy is captured and employed wasn't just *a* cause, or *a* gauge, but *the* cause, and *the* gauge, of cultural evolution.

You can see the appeal of this idea. Cultural evolutionists usually speak of human societies evolving toward greater "complexity" or higher degrees of "organization." And precisely defining complexity or organization is such a notoriously frustrating task that many people give up and fall back on an intuitive definition, like Supreme Court Justice Potter Stewart's famous definition of pornography: "I know it when I see it." Energy and its consumption, in contrast, can in principle be measured with precision.

Still, to let energy's quantifiability alone earn it a place at the core of our analysis would be to repeat the folly of the man in the old joke about the lost car keys. He is looking for the keys under a streetlamp when his wife asks why he's searching so far from the place where he dropped them. "The light's better over here."

One scholar who took on the stiff challenge of looking for the keys far from the streetlamp was Robert Carneiro, White's student. In the early 1960s, Carneiro actually tried to quantify social complexity and then document its growth. He started by taking 100 diverse societies—hunter-gatherer and agricultural, literate and pre-literate—that were well described in the anthropological or historical literature. He listed some of their conspicuous traits. Did they have special religious practitioners? Temples? Craft specialization? Craft specialization for trade?

It wasn't enough to just invent such traits out of thin air. Their significance had to be validated by showing, through a technique called Guttman scale analysis, that they were logically interrelated. In particular: some traits had to imply the existence of others with high probability. Sometimes these implications were almost obvious. Thus all societies with temples had religious specialists. Sometimes the implications were less obvious. All societies with formal codes of law had towns with over 2,000 people. The reverse, of course, was often not true: not nearly all societies with religious specialists had temples, and not all societies with towns of over 2,000 had legal codes.

Carneiro ranked the traits in order. The most common—such as trade between communities—were at the bottom, and the less common—calendars, cities of over 100,000—were at the top. In general, societies with any given trait would tend also to have the traits below it. I stress the word *tend*. The "higher" traits don't imply the lower with 100 percent confidence. And that's especially true of traits at nearly the same level. Thus, communities with clearly marked social status (trait number 7) don't *always* have formal political leadership (trait number 3), but societies with markets (trait number 26) do.

All told, when you stack up Carneiro's fifty traits in order, you find considerable predictive power. If a society has a given trait, then the chances are over 90 percent that it will have any randomly selected "lower" trait.

Of course, it would be nice if the figure were 100 percent. And, in general, it would be nice if laws in the social sciences were airtight, like laws in the "hard" sciences, rather than just statistically predictive. But, alas, human society is the most complex phenomenon in the known universe, so identifying the key to its dynamics is a messy task. If you want to find measurements and theories in the social sciences that are as neat as in chemistry, go look under a streetlamp.

What did these results mean? Carneiro postulated that this list of traits was a kind of ladder, a general evolutionary sequence. If you could have watched these different societies evolve, rather than just look at them statically—if you had movies instead of snapshots— you'd see them all acquire these traits in *roughly* the same order, from the bottom of the list to the top. Cultures in general would tend to have trade between communities and special religious practitioners and craft specialization long before they had temples and laws and big cities.

Years later, to test his model, Carneiro looked at the evolution of Anglo-Saxon England from the fifth through the eleventh centuries. In historical sources he found 300 cultural traits whose order of appearance he could document, and 33 of those traits had been among the 100 on his ladder. He looked at every pair of traits among those 33—all 528 combinations—to see, in each case, if the two had indeed appeared in the order predicted by his ladder. The answer was yes in 86.5 percent of the cases. Moreover, almost all of the exceptions—the 13.5 percent—came when comparing traits that had been very near one another on the ladder to begin with, such as "towns of 2,000 or more" (trait number 27) and "state or church employs artisans (28)"; or "taxation in kind" (24) and "military conscription" (25). In other words: they were pairs of traits whose relative order you would have guessed would vary from one case of cultural evolution to the next.

All of this gave Carneiro confidence that he had indeed found a meaningful list of *cumulative* cultural traits: traits that tend to appear in a particular order and, having appeared, tend to persist. Hence, he suggested, a rough gauge of cultural complexity: add up the number of these traits a society possesses, and thereby assign it a rank. As it happened, another scholar, Raoul Naroll, had earlier developed a quite different measure of social complexity, involving, for example, the number of craft specialties and the number of "team types." And these two measures of complexity ranked societies in almost exactly the same order. This finding, Carneiro wrote in 1969, "strengthened my conviction that cultural complexity is something real, objective, and measurable."

Notwithstanding Carneiro's confidence, there is today no consensus on how to measure social complexity. Some analysts stress the number of "levels of hierarchical control" in social organization.[†] And many, like Naroll, consider the degree of division of labor to be relevant (rather as some biologists use the number of "cell types" to gauge an organism's complexity). But the Potter Stewart litmus test—"I know it when I see it"—retains a certain appeal.

In defense of cultural evolutionism, it should be noted that the "hard" sciences aren't doing much, if any, better when it comes to defining complexity. Physicists and chemists have rigorous definitions for "order." (A pure substance with its many identical molecules neatly arrayed is the ultimate in order.) And they have rigorous definitions for "entropy." (The ultimate in entropy is a heterogeneous substance with its many different kinds of molecules randomly distributed.) And they know that complexity is something between the two—something that has pockets of order yet isn't *pure* order. But there is no consensus on what exactly is the essence of complexity, or on how to quantify complexity.

ACKNOWLEDGMENTS

A number of scholars kindly agreed to critique chapters of this book in manuscript form. Grouped by the parts of the book they focused on, they are: Robert Carneiro, Timothy Earle, Allen Johnson, William Peace, and Pete Richerson (pre-state societies); C. C. Lamberg-Karlovsky, Jeremy Sabloff, and Norman Yoffee (ancient states); G. W. Bowersock, Norman Cantor, and Gregory Clark (Europe since the dissolution of the western Roman Empire); Timur Kuran and Peter C. Perdue (Asian and Islamic history); John Tyler Bonner, James Gould, David Queller, and E. O. Wilson (evolutionary biology); Paul Romer (economics); and John Haught (theology). I'm very grateful to all of these people. Obviously, none of them is responsible for any flaws that remain in this book.

Gary Krist—in what has become his customary unenviable role—read the book in raw, first-draft form and, as usual, combined sharp big-picture criticism with vital encouragement. Mickey Kaus waited for a more refined version but, even so, found countless opportunities for further refinement. His relentlessly trenchant critique, at both the micro and macro levels, was a godsend (if occasionally an annoying godsend). Henry Finder, John Horgan, and Ann Hulbert also served on my focus group, providing valuable feedback on much or all of the book, even though all three of them have better things to do with their time.

Various scholars generously submitted to interrogation, formal or informal: Richard E. W. Adams, Robert Allen, Lloyd Anderson, Jean Arnold, Robert Bagley, William Boltz, G. W. Bowersock, Robert Braidwood, Robert Carneiro, Brad De Long, Douglas Deur, Timothy Earle, Barry Eichengreen, Kent Flannery, Steven A. Frank,

Jonathan Haas, Marvin Harris, Brian Hayden, Harry Jerison, Allen Johnson, Gary Johnson, David Keightley, Timur Kuran, Werner Loewenstein, Bruce Mazlish, Deirdre McCloskey, William McNeill, Rick Michod, David Papineau, Peter C. Perdue, Ken Pomeranz, David Queller, Paul Romer, Jeremy Sabloff, William Sanders, Thomas Schelling, Bruce Smith, Gil Stein, Warren Treadgold, Patty Jo Watson, Rich Wilk, David Sloan Wilson, and Norman Yoffee.

In 1996, I happened to meet John Lewis Gaddis, and he invited me to present my then-inchoate thesis to his graduate students in history. The result was a bracing seminar that sharpened my thinking. Given the aversion of most historians to bigthink theories, it was heartening for someone of Gaddis's stature to take my project seriously. I also profited from presenting my thesis at a 1997 conference on cultural evolution held at Windover, in the Adirondacks, and organized by Oliver Goodenough—and there I benefited, too, from conversations with my cabin-mate, Pete Richerson. Also valuable was a 1996 Star Island conference on the Epic of Evolution sponsored by the Institute on Religion in an Age of Science.

Lots of people courteously responded to e-mail queries, or gave other bits of aid along the way—so many, in fact, that I'm sure I'll leave many out if I start listing them. Nonetheless, here goes: James Beniger, Elizabeth Cashdan, John W. Deming, Lee Dugatkin, Ursula Goodenough, Jeff Herf, Gary Johnson, Floyd Lounsbury, Robert Perrin, Matt Ridley, Loyal Rue, Martin Seligman, Richard Steckel, Peter Swire, Christopher Wills. Assistance in matters ranging from title consultation to moral support came from Connie Barlow, Sarah Boxer, Mike Kinsley, John McPhee, Barry Munger, Cullen Murphy, and Rich Parker.

My research assistant, Michael Price, possesses the generically ideal qualities for the job—smart, efficient, diligent—but he also brought special skills to it. A graduate student in anthropology, he has an interest in both biological and cultural evolution, and is conversant in the literature. So he read and critiqued almost the entire manuscript, in addition to the usual duties of tracking down journal articles and checking obscure facts, services he performed tirelessly to the very end. At the outset I had no plans to use a research assistant; I stumbled onto Michael by happenstance. But it's now hard to imagine having finished the project without him—at least, having finished it before the next millennium.

My agent, Rafe Sagalyn, gave me excellent early advice and has been generous with his time since. My editor at Pantheon, Dan Frank, is known as a writer's editor, and with this book, as with my last book, he showed why; his editorial advice and his unflagging support have meant a lot to me. Once the book was done, Jennifer Weh, Altie Karper, Marge Anderson, Fearn Cutler de Vicq, Jeanne Morton, and others at Pantheon helped steer it to actual publication (a process complicated by the fact that I consider "done" a more elastic concept than Altie does).

My wife Lisa once again served as in-house editor, plowing through the entire manuscript. But this time her performance was more heroic than last time, because now we have two children, and her devotion to them consumes a lot of time. So my thanks are twofold: for remaining a uniquely trusted source of editorial (and other) guidance notwithstanding pervasive distraction, and for doing such a good job with the distraction itself.

It is common in acknowledgments for authors to apologize to their children for time stolen by the writing of the book, and I suppose that's in order here. God knows I've spent a lot of time up in my study incommunicado. Yet my memories of writing this book will be inextricable from memories of my family. The sound of my daughters playing outside, or one floor below me, was a kind of music to work by (though their occasional bursts of sibling rivalry were a kind of music not to work by). And their brash intrusions on the sanctity of my study—though formally discouraged—gave me, I can now confess, some of my deepest joy over the last few years. It's hard to imagine having been as maniacally focused on this book as I've been, and still stayed more or less sane, without having them to love. So, naturally, this book is dedicated to them.

And finally, a note to Mark Graven, my colleague at the *Red Bank Register* nearly two decades ago: You can have your Teilhard de Chardin books back now.

NOTES

INTRODUCTION

3   epigraph: Teilhard de Chardin (1969), p. 85.
3   "pointless" universe: Weinberg (1979).
3   "outside Time and Space": quoted in Wright (1988), p. 271.
4   "secret of life": Watson (1969).
6   †New technologies permitting new non-zero-sum games: This general expectation is consistent with a solid body of economic theory and data relating to the "Pareto optimum." By definition, the Pareto optimum has been reached when it is no longer possible to make anyone in the society better off without making someone else worse off. If there is a mutually profitable transaction that you and your next-door neighbor could conduct, but haven't yet conducted, then the Pareto optimum hasn't been reached. Thus, to say that the Pareto optimum has been reached is to say that there are no remaining unplayed win-win games—or, as economists would typically put it, that all the gains from exchange have been exhausted. (Technically speaking, this isn't the same as saying that all possible non-zero-sum games have been played, since not all non-zero-sum games are simple win-win games. Still, for present purposes we can think of the Pareto optimum as being, loosely speaking, the point at which all "non-zero-sumness"—as defined later in the introduction—has been realized.) It is now well established that technological innovation often raises the Pareto optimum. That is, technological evolution creates new win-win games (even if social organization may have to change in some minor or major way in order for the games to be productively played). A landmark in this line of thought was a paper by Robert M. Solow (1957) showing that growth in economic output couldn't be accounted for merely by capital and labor inputs—and that, by implication, technological change was playing a role. One important current extension of Solow's line of thought is "new growth theory." See Romer (1990).

1. THE LADDER OF CULTURAL EVOLUTION

13   epigraph: Mackay (1981), p. 102.
13   "It can now be asserted": Morgan (1877), p. vi.

13 "three distinct conditions": Morgan (1877), p. 3. Morgan was not the first to take an evolutionary view of culture. The British anthropologist Edward Burnett Tylor, in his 1871 book *Primitive Culture,* had championed a version of the savagery-barbarism-civilization typology. Some have argued that Tylor and Morgan were thinking in evolutionary terms before Darwin's *Origin* was published in 1859. (See Nisbet, 1969.) Certainly Herbert Spencer had an evolutionary view of social change before 1859. Then again, there were versions of *biological* evolutionism that predated Darwin's (including Spencer's). So notions of biological evolution may have influenced even pre-Darwinian versions of cultural evolutionism.

14 Sitting Bull: Hays (1958), p. 45.

14 "Democracy in government": Engels (1884), p. 759. See Service (1975), p. 33, on the relationship among Marx's, Engels's, and Morgan's ideas. See also Farb (1969), pp. 131–32.

14 John Stuart Mill: Mill (1872).

14 "We have stood out": quoted in White (1987), p. 108

15 "the most inane, sterile": quoted in White (1987), p. 100.

15 Berlin: Berlin (1954), p. 24.

15 Popper: Popper (1957).

15 " 'metahistorian' . . . deviations": Dray (1967), p. 253. A notable exception to the historians' aversion to "metahistory" is William McNeill, whose much-lauded 1963 book *The Rise of the West* addressed the full sweep of human history from a new theoretical perspective. Also, Immanuel Wallerstein has inspired a school of history, sometimes called "world-system theory," which sees globalization as a very old and somewhat exploitative process, featuring the dominance by capitalist states of less affluent polities.

16 Leslie White: Other thinkers who helped drive the revival of cultural evolutionism at mid-century include the anthropologist Julian Steward and the archaeologist V. Gordon Childe. A bit later, within sociology, other variants of evolutionism came to life, including "modernization theory." But by 1993 a historian could report that for the past decade the term "modernization theory" had been "most often brought up only to be denounced." (Grew, 1993, p. 229.) See also Fukuyama (1993), p. 69.

16 most archaeologists don't espouse: See Stein (1998) on recent skepticism about cultural evolutionism within archaeology.

2. THE WAY WE WERE

18 epigraph: Morgan (1877), p. 562. Morgan did believe in *some* innate mental differences among races, as did virtually everyone back then.

18 "wretchedest type": Farb (1969), p. 36.

18 "I feel quite a disgust": Wright (1994), p. 181.

18 "The tendency to value": Hays (1958), pp. 266–67.

19 "psychic unity of humankind": Morris (1987), p. 99. On the modern conception of a universal human nature, see Wright (1994), Pinker (1997), Tooby and Cosmides (1990).

20 "The Irreducible Minimum": Farb (1969), p. 43. The term "Shoshone" encompasses a variety of societies, some more complex than others. (See Johnson and Earle, 1987, p. 31.) I'm describing the least complexly organized Shoshone society, documented by Julian Steward (1938). Service (1975, p. 68) attributed the social fragmentation of the least socially complex Shoshone partly to depredation at the hands of Shoshone who had acquired horses, but Steward, who studied the Shoshone firsthand, attributed it to a barren habitat.

20 "entire political organization": Farb (1969), p. 44.

20 Shoshone . . . !Kung: For elaboration on this comparison, see chapter 2 of Johnson and Earle (1987). Often a !Kung camp will have at its core a single group of relatives, such as a brother and sister. But they commonly have spouses, whose siblings may also live in the camp, as may these siblings' spouses. So even when a camp is loosely based around a single group of close kin, the camp can comprise multiple families and include many individuals who are not biological kin. (See Lee, 1979, pp. 54–76.) Still, Johnson and Earle classify the !Kung as family-level foragers.

20 giraffe meat spoilage: See Lee (1979). The !Kung do have a crude meat preservation technology—the making of a kind of jerky. But the process is laborious, and the jerky is ruined by the first rainfall.

20 "someone else's stomach": Farb (1969), p. 66.

21 hunting a chancier endeavor: See Ridley (1996), p. 102.

21 social complexity of big-game hunters: Johnson and Earle (1987), p. 59.

21 "rabbit boss": Johnson and Earle (1987), pp. 34–35.

22 "gambling, dancing": Thomas (1972), p. 145.

22 functions of the fandango: Johnson and Earle (1987), p. 36.

23 reciprocal altruism: See Trivers (1971) and Wright (1994), chapter 9.

23 "generalized reciprocity": See Lee (1979), p. 437. Sahlins (1972), pp. 193–94, takes a slightly more nuanced view.

24 "generalized familistic way": Lee (1979), p. 460.

24 "accusations of improper meat distribution": Lee (1979), p. 372.

24 [†]anti-cheating technologies via both cultural and biological evolution: See Ridley (1996) for an excellent discussion.

25 zero-sum dimensions within non-zero-sum games: See Rapaport (1960), chapter 11, for an exercise in extricating the zero-sum from the non-zero-sum dynamic and then addressing the two problems in turn as a way to solve the overall game.

26 the pursuit of social status: See Wright (1994), chapter 12, and Williams (1966).

27 "Through the desire for honor": Kant (1784), p. 44.

28 "an Arcadian, pastoral existence": Kant (1784), p. 45.

28 "hidden plan of nature": Kant (1784), pp. 50, 52.

3. ADD TECHNOLOGY AND BAKE FOR FIVE MILLENNIA

29 epigraphs: Smith (1937), p. 13; Marx (1846), p. 136.

29 Eskimo boats: There is a chance that the Eskimo came across the Bering Sea by boat after the deluge. (A study of dental morphology suggested that the Eskimo-Aleut population entered America around the time of the deluge. See Fagan, 1995, p. 181.) In any event, the fact that the indigenous languages of Siberia and Alaska belong to wholly different language families (see Ruhlen, 1987) suggests essentially distinct cultural evolution in recent millennia. The main point, for purposes of this chapter's analysis of the Nunamiut and the Tareumiut, is that the distinctive Eskimo culture—kayaks, harpoons, the hunting of large sea mammals, and so on—evolved in the New World within the last two millennia. See Fagan (1995), p. 201. (There was brief North American contact with a Viking expedition around the end of the first millennium A.D., but apparently of no great moment.)

30 †independent examples of onging change: There has lately been publicity about the possibility of earlier and possibly momentous contact between the New World and non-Asian lands. The occasion for much of this publicity was the discovery in North America of the remains of "Kennewick Man," who seemed to have been a Caucasian, presumably of European origin. But later analysis showed Kennewick Man to be more like "a cross between the Ainu and the Polynesians" (*Newsweek,* April 26, 1999). A relationship to the Ainu—the indigenous people of Japan—would hardly overturn the notion that all of America's indigenous inhabitants resulted from migration over the Bering Strait. Nor would a Polynesian relationship, since the Polynesians themselves were ultimately of Asian ancestry. And certainly Kennewick Man couldn't have come from Polynesia by boat across the Pacific Ocean, since the Polynesian migration didn't even make it as far as Hawaii until A.D. 400, many millennia after his death. For the same reason, any speculation about early waterborne Polynesian migrants to America are, for our purposes, irrelevant; by the earliest time that a Polynesian boat could have quasi-plausibly managed to make it to the Americas, indigenous American culture had already reached the state level of cultural evolution.

In general, the several recent findings in the Americas of very old human remains that seem anatomically unlike native Americans turn out, like the discovery of Kennewick Man, not to upset conventional notions about the essential independence of American cultural evolution. These skeletons resemble Polynesians, Ainu, or other East Asians, and are thus consistent with the notion of an Asian migration across a Bering land mass. (See aforementioned issue of *Newsweek;* its cover's promise of a radical revision of current thinking is undermined by the story's actual content.)

30 †unprecedented array of cultures: Of course, none of these cultures was literally pristine, as the act of observing invariably has an influence. What's more, often European influences had in one sense or another diffused to a given native American culture before European witnesses encountered it. These imperfections in the ethnographic record have to be kept in mind, but the record is still uniquely valuable.

31 "we don't let people starve": Johnson and Earle (1987), pp. 137, 198.

31 !Kung camps a big, extended family: See Lee (1979), pp. 54–76.

32 "walk across their backs": See Farb (1969), p. 169.

32 Nootka technology: Coon (1971), pp. 35–37, 66–67, 126–29, 134–35, 143–46, 271.

32 Chilkat robes: Coon (1971), pp. 271–73.

32 crafts handed down through families: Farb (1969), pp. 180–81.

33 fish traps: Coon (1971), pp. 143–46.

33 †public goods: Johnson and Earle (1987), pp. 166–68. A "public good" is a good whose benefits are inherently available to anyone. The classic example is a lighthouse—anyone who sails by can use it for guidance, whether or not they helped pay to build it. Public goods are considered to naturally fall within the public sector, among benefits that the government should provide—precisely because of the difficulties an entrepreneur would face in financing them by charging the particular beneficiaries. But technology can change the definition of a public good.

Modern electronic information technology, for example, makes it at least *somewhat* more feasible for an entrepreneur to build a lighthouse and charge a toll to ships that use it. Even the invention of money—which the Northwest Coast Indians of course didn't have—makes it more feasible to provide certain goods and services via the private sector.

34 "fishing warden": Douglas Deur, personal communication. These "fishing wardens" were found among the Kwakwaka'wakw and some other Northwest Coast peoples. Even if the "fishing warden" and the Big Man were not one and the same, Deur notes, the Big Man almost certainly bolstered the warden's authority. So did religious beliefs about the need to "appease" the fish by exploiting them in a manner that met with their approval. In general, societies often come up with a means of limiting the exploitation of such resources as fisheries. See Acheson (1989), p. 358.

34 blunting misfortune: Johnson and Earle (1987), p. 170. See Murdock (1934), pp. 239–40 on forms of stored wealth.

34 " 'tribal' and linguistic boundaries": Johnson and Earle (1987), p. 198.

34 "social security . . . savings account": Johnson and Earle (1987), p. 170.

35 a single, central authority: The bounds of authority could be complex. Thus, a village might consist of two separate clans (a clan being a group of families who claimed a common ancestor). And each clan might have a different Big Man.

35 "vigorous and overbearing": Benedict (1959), p. 173.

35 tokens for future blubber: Johnson and Earle (1987), pp. 170, 198.

36 corruption of the Potlatch: Farb (1969), p. 188, Coon (1971), p. 282.

36 the pristine Potlatch: See Johnson and Earle (1987), pp. 322–23; Murdock (1934), p. 244.

36 Potlatch as debt recorder: Boas (1899), p. 681.

36 "largely based on credit": Boas (1899), p. 681.

36 Morgan's tightly technological stages: Morgan (1877), pp. 10–11.

36 Service's stages: Service (1962), pp. 142–47.

36 fuzzy stages: Some evolutionists, mindful of the ways that societies of comparable overall complexity can differ structurally, speak of a "multilinear" evolution, as distinguished from the more ladderlike "unilinear" scheme used by Morgan in the nineteenth century. The term comes from Julian Steward (1955). But even Steward depicted diverse cultures as passing through a somewhat ladderlike sequence of stages. See, e.g., table 5 on p. 190. On the general fuzziness of evolutionary "stages," see Rothman (1994) and—for the view of a skeptic of evolutionism—Yoffee (1993).

37 dissenting from Service's labeling: See Carneiro (1991), p. 168.

37 "I have won": Johnson (1989), p. 65. See also, on the Big Man competitive feast as an institution, Hayden (1993), pp. 244–47.

37 social status and its rewards: See Wright (1994), chapter 12.

37 "We shall eat Soni's renown": Harris (1991), p. 106.

38 "is usually a good speaker": Johnson (1989), p. 65.

38 "Brothers Under the Skin": Hays (1958). See Richerson and Boyd (1999) for various other ways that cultural evolution mutes natural human tensions that would otherwise impede the advance to higher levels of social complexity.

39 social complexity in Florida and California: For example, the Chumash of California had multicommunity governance—see Arnold (1996), pp. 105–6—as did the Calusa of Florida, to be discussed in chapter 7.

40 archaeological evidence of complex hunter-gatherers: See Fagan (1995), pp. 167–69, and Hayden (1993), pp. 200–206. The term "prestige technology" is Hayden's (1998). See also Brown and Price (1985), pp. 437–38, Arnold (1996), pp. 103–4, 114, and Price and Brown (1985), p. 16.

40   "global trend toward great complexity": Fagan (1995), p. 167.

41   Upper Paleolithic and Mesolithic technology: See Hayden (1993), chapter 6, and Fagan (1995), chapters 4 and 5. In the Upper Paleolithic, some scholars have surmised, there may also have been crude snares and traps. Made of wood, they would not have been preserved.

41   Australian aborigine technology: Lourandos (1985), esp. pp. 392, 408. The boomerang may have been invented in Europe, but even if so the Australian invention was independent.

41   "traditional static model": Lourandos (1985), p. 412.

42   !Kung technology: Lee (1979), pp. 124–34. The arrowheads are now made of metal (still with bone shaft), but that is due to contact with modern cultures.

42   various hunter-gatherer technologies: Coon (1971), pp. 16–20.

42   "mustache lifters": Murdock (1934), p. 164.

43   "We are all materialistic": Hayden (1993), p. 269.

43   Shoshone and Nahuatl languages: See Farb (1969), p. xviii, Ruhlen (1987).

43   genetic difference among Northwest Coast Indians: Cavalli-Sforza (1994).

4. THE INVISIBLE BRAIN

45   epigraph: Braudel (1981), p. 401.

46   Kuikuru and surplus: See Johnson and Earle (1987), p. 72.

46   "rarely . . . designing cathedrals": Wenke (1984), p. 186.

46   two factors conducive to division of labor: Smith (1937). See also Ridley (1996), p. 43. Smith also cited the size of the market as a factor conducive to division of labor. But of course there may be overlap between this and the other two factors; if the market is enlarged by rising population density, then per capita costs of transport and communication have dropped.

47   †diversity of natural resources: The diversity of resources was reinforced by the geographic diversity of the Northwest. With mountains near the sea, it was practical for coastal peoples to trek up to high altitudes to exchange candlefish oil for animal pelts that were rare and prized in coastal areas. This sort of "micro-environmental variation" has been cited as a factor to explain differences in social complexity among peoples. (Allen Johnson, personal communication; see Johnson and Earle, 1987.) There is, for example, relatively little such variation in Amazonia, compared to parts of South America where cultural evolution reached the state level. Another possibly potent factor—which, like micro-environmental variation, applies to the Northwest Coast Indians—is extreme seasonal variation that makes survival precarious for part of the year, and thus raises the value of structures of governance that diffuse risk among large groups of people. Both of these factors—diversity of resources among nearby peoples, and a risk of scarcity confronting nearby peoples—make relations among peoples more non-zero-sum. As Johnson has noted, taking such factors into account helps explain why some areas with fairly high population density—such as New Guinea—feature relatively low social complexity.

48   skyrocketing manioc production: Carneiro (1970), p. 734. Carneiro documented this surplus production not among the Kuikuru themselves, but among various comparable manioc-producing Amazonian peoples.

48   group size and innovation: See Kremer (1993). On new growth theory generally, see Romer (1990). There is sometimes a blurry line between the effects of population growth and of elevated population density. Population growth is important only if communication among individuals is practical—otherwise individual innovations can't spread through the whole group. And the higher the population density, the more practical the communication. By the way, among the first theorists to note that increased population would increase the rate of cultural innovation was William Ogburn. See Ogburn (1950), pp. 110–11.

48   †Northwest Coast Indians as an invisible brain: Of course, in an economy with centralized planning—the Soviet economy of the Cold War, or the Big Man economy of the Northwest Coast—the brain is not very diffuse, and thus not so invisible; the economy has a visible headquarters. Even so, signals must be sent and received over a social "nervous system," and the fastest nervous system is a compact nervous system. Besides, there was some barter in the Northwest. In particular, trade *among* villages and peoples had no central governor, so a network of nearby villages and peoples amounted to an invisible brain of the more diffuse, Smithian sort.

48   anthropologists ignoring lubricating effects: See, e.g., Maschner (1991), p. 931. He summarizes the three types of theories about what drove the complexification of Northwest Indian societies. None of the theories is along the lines of the argument presented here. Still, some cultural evolutionists have noted the importance of the low information costs afforded by population density. Arnold (1996, p. 95) cites "sufficient population densities for certain kinds of social interaction and communication to occur" as one of the conditions under which complex hunter-gatherer organization is found.

49   "Irresistible reproductive pressures": Harris (1991), p. xii. Johnson and Earle (1987) embrace and elaborate Harris's paradigm, showing how at different levels of social organization stress and intensification could lead to new forms of social organization.

49   persistent population growth: See Kremer (1993), table 1, p. 683. The population growth rate, though low, has been positive since long before our ancestors crossed the threshold to *Homo sapiens* status several hundred thousand years ago.

50   Polynesian complexity: See Diamond (1997), p. 62. Diamond's explanation for the correlation differs from mine. See also Kirch (1989) and Sahlins (1963).

50   Middle Paleolithic population and technology: On population growth, see Kremer (1993), table 1. On technological innovation, see Lenski et al. (1995), table 5.1, and Hayden (1993), figure 6.1.

50   human species encompasses Old World: Price and Brown (1985), p. 13.

50   Upper Paleolithic technology: See Klein (1989), p. 360, and Lenski et al. (1995), table 5.1.

50   Mesolithic population and technology: See Kremer (1993), table 1, and Lenski et al. (1995), table 5.1.

50   mutual positive feedback: Johnson and Earle (1987), who are in the Harris school and thus stress the "downside" of population growth, write (pp. 324–25): "The primary motor of change in the subsistence economy is the positive feedback relationship between population growth and technology."

51   migrating pierced seashells: Ridley (1996).

51   invisible social brains interlinking: The exchange of prestige items among hunter-gatherer peoples has been documented from Australia (see Lourandos,

1985) to pre-Columbian America (see Bender, 1985, p. 56). There are periods, as Arnold (1996) notes on p. 111, when long-distance trade decreases. Such a time is seen in southwest California's archaeological record. At the same time, though, there was an increasingly massive circulation of prestige goods and foods found closer at hand. And during this period the overall complexity of the cultures in question grew. In short, the reliance on distant sources of prestige items may decline for various reasons, including the newly adept exploitation of nearby resources and the newly adept manufacture of prestige items by nearby peoples. Even so, as cultures thus flourish, and population grows symbiotically, their eventual contact with distant cultures grows only more assured. More generally, if indeed, as some have argued, there is a decline in long-distance trade during the Mesolithic/Archaic periods compared to the Upper Paleolithic (see Price and Brown, 1985, p. 6), there is at the same time growing cultural complexity within local regions, and growing diversity of cultural tradition among regions—a diversity that would later stimulate long-distance economic intercourse.

51 New World lagged behind Eurasia in population by several millennia: See Kremer (1993), tables 1 and 7. In 1500, when Old World population was around 400 million, the New World's population was 14 million. In 3000 B.C., the whole world's population had been 14 million—and almost all of that, no doubt, was in the Old World.

51 New World post-agricultural social evolution: Carneiro, personal communication.

52 Tasmania and Australia: Kremer (1993), p. 709. And see Lourandos (1985), pp. 410–11, re the low social complexity of Tasmania as compared to Australia. Kremer doesn't use these data in reference to population *density* and its reduction of information costs. He uses them to bolster the new growth theorists' emphasis on the overall *size* of a continuous population and its correlation to the rate of innovation. Still, the data apply to both arguments, which are wholly compatible. Kremer also makes the interesting point that the *rate* of population growth has risen as a function of total population—a fact that could be explained by the tendency of absolute population size to speed technological innovation, in conjunction with the tendency of the ensuing technological evolution to facilitate higher rates of population growth.

53 modern Darwinian theory of status: See Wright (1994), chapter 12, and Williams (1966).

5. WAR: WHAT IS IT GOOD FOR?

54 epigraph: quoted in Carneiro, ed. (1967), p. xlv.

54 "don him as a trophy poncho": Keeley (1996), pp. 7, 99.

54 "pose as Tasadays": *New York Times,* May 8, 1997.

55 !Kung homicide rates: Keeley (1996), p. 29.

55 Eskimo homicides: Keeley (1996), p. 29.

55 Ainu warfare: Murdock (1934), p. 176.

55 Yanomamo warfare: See Chagnon (1968). The Yanomamo are horticulturalists, not pure hunter-gatherers, but when first studied by anthropologists they were less altered by direct contact with modern societies than many hunter-gatherer peoples on the anthropological record and thus may offer an unusually valuable glimpse into the nature of prehistoric war.

55 "like eggs in a basket": See Keeley (1996), p. 38.

55 violent death in complex hunter-gatherers: Price and Brown (1985), p. 12.

55 marriageable Naga warrior: Davie (1968), p. 44.

55 decorative brass tray: Davie (1968), p. 42.

55 "Damaging beyond hope": Carneiro (1990), p. 197.

55 "unpleasantly bellicose": Keeley (1996), p. 183.

56 †non-zero-sumness among comrades: Comrades don't have a purely non-zero-sum relationship. Though they often perceive themselves as having a non-zero-sum relationship, and this perception is often valid, a sensitivity to zero-sum dynamics can lurk beneath the surface, and can lead to such things as desertion and mutiny.

56 "Only by imperative need": Service (1975), p. 40.

57 Northwest military technology: Murdock (1934), pp. 241–42; Johnson and Earle (1987), p. 164.

57 Haida as Vikings: Murdock (1934), p. 241.

57 "requiring prompt combination": See Carneiro, ed. (1967), p. xlix.

57 "the tamest are the strongest": Service (1975), p. 41.

57 †"the prisoner's dilemma": See appendix 1 for details.

57 "We fight against the Rengyan": Evans-Pritchard (1969), p. 143.

57 "Each segment is itself segmented": Evans-Pritchard (1969), p. 142. Note that a "segmentary" society is considered a *type* of society that illustrates this dynamic with particular clarity (the Nuer being an example). Still, at some level of abstraction, this dynamic characterizes all societies.

58 Delian League: Starr (1991), p. 292.

58 Iroquois: Farb (1969), pp. 129–30.

58 "a very great slaughter": 1 Samuel 4:10, Revised Standard Version.

58 "We will have a king": 1 Samuel 8:19–20, Revised Standard Version. Similarly, the introduction of the bow and arrow is said by some to have helped coalesce societies on Alaska's northwest coast. See Arnold (1996), p. 104.

59 †pushing and pulling as fuzzy: The first complication with characterizing war-induced organization as resulting from "push" is that the logic of an attacking society is different from the logic of the attacked. Attackers are drawn together by the prospect of mutual gain through cooperation. Defenders are drawn together by the prospect of avoiding, through cooperation, mutual loss. That is, the former game is positive-sum, the latter is negative-sum. (See discussion of this distinction later in this chapter.) Thus we might say war "pushes" people together when the people in question have been attacked, but that it "pulls" people together when the people in question launch an attack.

The second complication with attributing war-induced organization to pushing forces and economic organization to pulling forces is suggested by the previous chapter. Evolutionists who emphasize the "intensification" or "stress" scenarios are in effect viewing economic organization—like the organization that governs the salmon trap—as negative-sum or "push"-driven; they see its purpose as being to avoid mutual calamity.

It is tempting, in light of some of these complexities, to opt for an alternative bifurcation: between those non-zero-sum dynamics that are "zero-sum-dependent" and those that aren't. In this view, all non-zero-sumness created by war—whether among aggressors or defenders—is zero-sum-dependent. This distinction—zero-sum-dependent versus not zero-sum-dependent—may well be the most analytically productive tool to use, but it is not the most literarily convenient. For purposes of this book I'll stick with "push" versus "pull."

59   "Given the universal disinclination": Carneiro (1990), p. 191.

59   "benefits of participation in the total network": Service (1962), p. 15. Service did not, alas, provide examples. He may have had in mind cases where small clusters of Northwest Coast Indians, unhappy with their leadership, chose to shift allegiance to a new Big Man. See Johnson and Earle (1987). (Service, unlike some evolutionists, considered the Northwest Coast Indians to have attained chiefdom-level social organization.)

60   "Force, and not enlightened self-interest": Carneiro (1970), p. 734.

60   Service and Carneiro conditions: Insko et al. (1983), pp. 977–99.

60   †consolidating to fend off aggression: One much-cited case is the formation of the Iroquois League out of five previously warring tribes in sixteenth-century North America. The Iroquois ascribed the league's genesis to a prophet born of a virgin mother. He dreamed of an immense evergreen, stretching up to the "land of the Master Life," and resting on five roots representing the Iroquois tribes. Inspired by the prophet, Hiawatha, a Mohawk, embarked on a frenzy of diplomacy, convincing leaders of the other four tribes that it was time to usher in "the Great Peace."

Historians presume that Hiawatha's argument was strengthened by the growing presence of menacing white men, some of whom inhabited alarmingly large boats. But even with this stimulus of external hostility, the Iroquois League remained essentially that: a league, a confederacy. Its governing council could act only on unanimous vote, and the actions were strictly military; tribes were autonomous in their internal governance. So the Iroquois, though an example of voluntary consolidation, are not a clear-cut example of the voluntary surrender of sovereignty. See Farb (1969), pp. 129–30.

60   little evidence of warfare in the rubble: Drennan (1991); Jonathan Haas, personal communication. In reply, Carneiro argues that warfare is not always reflected in archaeological remains.

60   †negative- and positive-sum games: Some game theorists, such as Thomas Schelling, frown on the "negative-sum/positive-sum" distinction. The reason is that they frown on the "zero-sum/non-zero-sum" terminology to begin with. What is commonly called a "zero-sum" game they would rather call a "fixed-sum" game. After all, if the winner of a tennis match gets $100 and the loser $50, that is not, strictly speaking, "zero-sum," but it has the essential property we mean to imply with the term "zero-sum"; the two players' interests are wholly at odds, and good news for one equals bad news for the other. And the source of this property is not that the sum of the benefits they get from playing the game is zero—which it isn't—but that the sum of the benefits is "fixed," in this case at $150. By the same token, what is commonly called a "non-zero-sum" game Schelling would rather call a "variable-sum" game (to describe, say, the dynamic between two *teammates* in a doubles match that pays $100 to the winning team and $50 to the losing team). Schelling is surely right that a zero-sum game is merely a special case of a fixed-sum game—the case in which the sum is fixed at zero. So the "fixed-sum/variable-sum" terminology is, strictly speaking, more generally applicable to life. But "zero-sum/non-zero-sum" is the terminology used by the founders of game theory, John von Neumann and Oskar Morgenstern, and it remains in common use. And its implication that there are negative-sum and positive-sum games—though slippery on close inspection—is useful for present purposes.

61   Sun Tzu's advice: Cotterell (1995), p. 243.

61   †heedlessly launching war: The reasons for this are complex and variable, but one is that within a given society, perspectives differ. War may be negative-sum for opposing infantries, yet zero-sum for opposing leaders, since one of them will survive to inherit the spoils. (War may even be *positive*-sum for the leaders, if the external threat congeals support for the leader.) And leaders, being leaders, often make their perspective prevail. Still, there is often much popular support for war, at least going into it. At the outset of World War I, the German masses were at least as gung-ho as Kaiser Wilhelm II. Whatever the impending reality, Germans were by and large conceiving of war as zero-sum. (But German soldiers eventually came to grasp non-zero-sumness. Amid trench warfare, opposing infantrymen staged lengthy de facto truces. See Axelrod, 1984, chapter 4.)

61   " 'waging' peace in ever-widening social spheres": Service (1975), p. 61.

61   "War is bad": See Keeley (1996), p. 145.

61   "can always find refuge": Murdock (1934), p. 240.

62   "an inherently dangerous world": Johnson (1989), p. 71. Lourandos (1985), p. 407, describes a similar dynamic among Australian aborigines.

62   war among crowded Yanomamo: See Chagnon (1973) and Carneiro (1970), p. 737. Population density may be conducive to political complexity simply by virtue of posing a kind of crowd-control problem. Various studies suggest that non-complex societies have trouble maintaining a large aggregation without interpersonal tensions becoming serious. See Cohen (1985), p. 105. Though this problem has often been solved by mere migration—the "fissioning" of hunter-gatherer and horticultural villages once they get unmanageably large—sometimes this solution wouldn't be feasible, as when surrounding lands are barren, or are inhabited by enemies. In this case, political mechanisms for ensuring harmony may evolve. This dynamic figures in Carneiro's "circumscription theory" of the origin of the state. See Carneiro (1970).

62   larger villages and firmer leadership: Carneiro (1970), p. 737.

62   "to reduce the possibility of warfare": Chagnon (1973), p. 136.

62   phony trade: Chagnon (1968).

62   †"Where warfare is intense": Chagnon (1973), p. 136. The difference between the Carneiro and Service interpretations of the expanded social organization to which Chagnon attests could be described this way: Carneiro sees the social organization as founded in recognition of non-zero-sumness that emanates from an *external* zero-sum dynamic—from an enemy external to the two villages that have now combined into one unit. Service sees the social organization as founded on recognition of non-zero-sumness that originates *internally*—in the negative-sum potential of war between the two villages. Either way, the expansion of social organization accords with non-zero-sum logic.

63   urinating in the morning: Keeley (1996), pp. 145–46.

63   "ground to bits": quoted in Johnson and Earle (1987), p. 164.

64   make love or fight: I've heard the remark attributed to Irvin DeVore.

6. THE INEVITABILITY OF AGRICULTURE

65   hotter climate, extinctions: Harris (1991), pp. 29–31.

65   "no need to complicate the story": Braidwood (1960), p. 134.

66   "not yet satisfactorily explained": Wenke (1981), p. 114. Fourteen years later, an archaeology textbook reiterated the challenge: "The problem for anthropologists is not only to account for why people took up agriculture but also to explain why so many populations adopted this new, and initially risky, economic transition in such a short time." (Fagan, 1995, p. 233.)

66 "the original affluent society": Sahlins (1972), pp. 1, 37. See Fagan (1995), p. 235, for studies concluding that farming is more labor-intensive than hunting-gathering.

67 shorter lives and more rotten teeth: Diamond (1997), p. 105; Bruce Smith (personal communication).

67 Kumeyaay: See Smith (1995), p. 17.

67 †juicier grapes: Some anthropologists have argued that, with some plant species, at least, the sheer act of replanting can lead to a kind of autodomestication, because for various reasons the seeds that get replanted are of higher quality than the seeds that fall by the wayside. See, e.g., Smith (1995).

67 aborigines and yams: Smith (1995), p. 17.

67 Shoshone: Keeley (1995), pp. 259–62.

68 Hunter-gatherers knew enough to farm: Cohen (1977), p. 9, writes: "The principles of domestication are universally understood. Man did not need education as much as he needed motivation." Cohen, unlike many who have said such things, takes issue with the notion of "equilibrium" (see below), though not for the same reasons that I take issue with it.

68 "change in any patterned fashion": Hayden (1993), p. 148. Maybe it's a sign of how deeply and subtly the assumption of equilibrium influences anthropological thought that this quote comes from Brian Hayden, whose "competitive feasting" theory (mentioned later in this chapter) seems at odds with it. In its heyday several decades ago, the notion of natural equilibrium was grounded in Lewis Binford's "population-density equilibrium" model. Binford's model was in turn grounded in a biological theory which has since been discredited: that animal populations naturally keep themselves in check lest overpopulation should threaten the food supply.

68 "What keeps hunter-collectors": Harris (1991), p. 26.

68 meeting *collective* needs: For a welcome corrective to the tendency to see societies as unified, see Brumfiel (1992). Though focusing on state-level societies, her analysis is relevant to the origin of agriculture; she notes that exclusive focus on the society as a whole, and its adaptive relationship to the environment, "leads us to overestimate the external as opposed to the internal causes of change" (p. 553).

69 work of Douglas Deur: Deur (1999), and Deur and Turner, eds. (in press).

70 †further along in the evolution toward agriculture: That's not to say that the Northwest Coast Indians, if left alone for another millennium, would wind up looking just like known horticultural societies. Cultural evolution isn't that "unilinear." In fact, known horticultural societies often have less in the way of capital projects than some complex proto-agricultural hunter-gatherer societies have had. (Johnson, 1989, p. 75, and personal communication.) And this may indicate that the two kinds of societies are taking a different "route" to complexity. Consistent with this conjecture is the fact that the two types of societies are often found in different types of environments. (Observed horticultural societies have often been in jungly equatorial regions. They may be atypical of past horticultural societies, rather as observed hunter-gatherer societies are often atypical of past hunter-gatherer societies by virtue of being on relatively undesirable land.) On the other hand, it's conceivable that as a hunter-gatherer society slowly evolved toward horticulture, it would jettison such capital projects as storehouses because families were now more securely self-sufficient, less exposed to periodic scarcity. At the same time, though, other non-zero-sum rationales for social organization could grow— notably the importance of military organization in an environment with high population density and land that is worth conquering. (See below, explanatory note labeled "conflict . . . intensified by farming.") As Johnson and Earle (1987, pp. 158–59) note, among agriculturalists, "the primary cause of organizational elaboration appears to be defensive needs. . . . In hunting and fishing economies and among pastoralists, economic causes are more prominent in promoting group formation and regional networking."

70 Yanomamo gardens: Chagnon (1968), p. 37.

70 †Big Men and Head Men: The term "Big Man" is sometimes used loosely, and that is the way I'm using it here. Not all societies featuring Big Men in this loose sense are "Big Man" societies in the sense defined by Johnson and Earle (1987). Johnson and Earle call the Yanomamo and some of the so-called New Guinean "Big Man" societies "acephalous local groups." Still, though these societies may not have quite as much hierarchy and structure as the Northwest Coast Indians, their Big Man is functionally comparable to the Northwest Coast Big Man.

70 †a thirst for status: High status for women was, during evolution, also conducive to the spreading of genes, but less so than for men, and as a result women on average seem to pursue status less intensely than men. See Wright (1994), chapter 12. This helps explain why human cultural history has witnessed many "Big Men" and few if any "Big Women."

71 five wives: See Oliver (1955), p. 223.

71 "brideprice": Deur, personal communication.

71 "many wives produce more food": Hayden (1993), p. 242. See also p. 247.

71 "the envy of rivals": Johnson (1989), p. 73, paraphrasing Bronislaw Malinowski.

72 "the *lifting* of the constraints": Cashdan (1980), pp. 119–20. See also Wright (1994) and Chagnon (1979).

72 conspicuous disparities in status: See Keeley (1995). In Keeley's data table, 46 percent of ninety-six hunter-gatherer societies are classified as showing some class distinctions, based on wealth and/or parentage. But among those thirty hunter-gatherer societies that plant, sow, or cultivate wild species, the figure is 73 percent. Oddly, Keeley doesn't seem to have performed this calculation. His analysis is devoted to attacking Brian Hayden's "competitive feasting" hypothesis, described below. In pursuing that cause, Keeley aggregates various indices of social and cultural complexity, sometimes in a fairly arcane way. Meanwhile, he ignores this simple, straightforward calculation that supports Hayden's hypothesis.

72 "empire builders": Hayden (1993), p. 241, and personal communication.

73 recalculating hunter-gatherer workday: Hill et al. (1985), p. 44. Hawkes and O'Connell (1981).

73 "crucial secondary staples": Keeley (1995), p. 260. The Northwest Indians seem to have had similar motivation to plant. They recited tales of disasters that had afflicted past generations (Deur, personal communication).

73 "secure sources of food energy": Johnson (1989), p. 51.

74 war and manpower: See Diamond (1992), pp. 189–90.

74 Jericho: Smith (1995), pp. 2–3, 80.

74 Jericho's wall: Wenke (1984), p. 175; Fagan (1995), pp. 252–53; and Smith (1995), pp. 2–3. The wall was not built for centuries after the first farming in Jericho, but that doesn't mean war wasn't a problem from the outset. Anything as (then) novel and forbiddingly labor-intensive as the construction of a huge wall would have been unlikely to happen as soon as it was needed. Ofer Bar-Yosef argues that the wall was for flood control, not defense, but the theory is controversial.

74   †conflict . . . intensified by farming: During the Mesolithic, as hunter-gatherer societies grew more complex and affluent, the stakes of war had grown. Stockpiles of food, elaborate tools, and exotic tokens of wealth raised the bounty of conquest, and thus the incentive to attack. Further raising the incentive was the growing value of real estate. As population expands, land gets precious, especially land so rich that previously itinerant hunter-gatherers begin to adopt it as a year-round abode.

74   Mesolithic war: See Hayden (1993), pp. 207–8.

74   conspicuous status competition: See Hayden (1993), pp. 200–6.

74   Pomo: Sahlins (1972), p. 219. See also Bender (1985). Note that, even if a horticultural and a hunter-gatherer society are equally efficient in meeting their basic food needs, when it comes time to create a *surplus* of food (for trade, etc.), the horticultural society will often be more efficient. The reason is that hunter-gatherers exploit the nearest food sources first, so they are walking farther to obtain each additional increment of food. For horticulturalists, creating surplus is often just a matter of clearing more nearby land of wild vegetation—and clearing land is no more time-consuming for the last crop planted than it was for the first crop.

74   Jericho as trade center: Hayden (1993), p. 373.

75   exhausting nature's cornucopia: This dynamic is stressed especially in Cohen (1977) and Cohen (1985). Note that this sort of overpopulation could bring extinctions of favored prey. Indeed, some believe overhunting was a factor in the demise of various prey species, such as the mammoth, around 13,000 B.C. (See Harris, 1991, p. 21.) Thus it may be right to list such extinctions as one "cause" of agriculture, yet wrong to see in this scenario the play of contingency; any human-induced extinctions were simply examples of the inexorable strain on food resources that was bound to catch up with the human species sooner or later.

75   Upper Paleolithic and Mesolithic technological evolution: Hayden (1993), chapter 6, and Fagan (1995), chapters 4 and 5.

75   157 plant species: See Diamond (1997), pp. 144–45. A few of the plants were used for such things as dyes or medicine, not food.

75   sedentism and domestication: Mesoamerica has long been thought an exception to this rule, with domestication preceding sedentism. Recently, though, evidence to the contrary has emerged. See Price and Gebauer (1995), p. 9.

75   "more costly in terms of procurement": Price and Brown (1985), p. 11.

75   "no great conceptual break": Cohen (1977), pp. 15, 285. Cohen's model emphasizes only one of the three factors I cite—the struggle against scarcity.

76   capital projects and trade: Cohen (1985), pp. 100, 104–5. In *some* respects, to be sure, agriculture can reduce non-zero-sumness. (See above, explanatory note labeled "further along in the evolution toward agriculture.") Ultimately, as we'll see in the next few chapters, agriculture would considerably raise non-zero-sumness. And much of this potential would be realized via new forms of social organization.

76   farming arises five to seven times: See Diamond (1997), p. 100, and Smith (1995), chapter 1. Smith lists the seven primary centers of agricultural development as the Near East, South China, North China, Sub-Saharan Africa, South Central Andes, Central Mexico, and Eastern North America.

76   readily domesticable species: See Diamond (1997), Steward (1955), p. 192.

7. THE AGE OF CHIEFDOMS

78   epigraph: quoted in Aron (1967), vol. 2, p. 45–46.

78   Natchez: Farb (1969), pp. 193–95.

79   institutionalized political leadership: Sahlins (1963) notes that the Polynesian chief had powers of office, whereas the generic Melanesian Big Man had to rely almost entirely on his powers of persuasion.

79   chiefdoms precede states: Carneiro (personal communication). For a dissenting view on the universality of the chiefdom-state transition, see Yoffee (1993).

79   special access to divine force: See Service (1975), p. 16.

79   tattoo-like engraving: Hayden (1993), p. 301.

79   force-fed wives: Hayden (1993), p. 289.

79   Tahiti: Fagan (1995), p. 308; Kirch (1989), p. 167. Ziggurats: Johnson and Earle (1987), p. 207.

79   Easter Island: Fagan (1995), p. 308. According to Kirch (1989), each of these statues represented constituent clans—presumably meaning minor chiefs, not the paramount chief. Even so, the vast number (hundreds) and uniformity of them suggests a large, highly integrated society of fairly affluent clans, which usually means a chiefdom. In any event, we know from ethnographic observation that in the Polynesian islands the typical indigenous social organization is the chiefdom.

79   †other hallmarks of chiefdoms: In some areas monumental architecture is less characteristic of chiefdoms (and of states) than in others. But there are other clues that, taken together and weighed judiciously, can also serve as an archaeological litmus test: a large village (home of the paramount chief) among smaller ones; large, central food storehouses; diverse technology; sheer population size, along with signs of sharp status differences. Especially suggestive—since status is typically hereditary in chiefdoms—is the lavish burial of an infant: graves with alabaster statues and copper ornaments, as in the Middle East, or, in Mesoamerica, basalt-column tombs loaded with jades. On infant burial, see Flannery (1972), p. 403.

79   Calusa: Carneiro (1992), p. 27.

80   megaliths: Hayden (1993), pp. 313–14.

80   Caesar in Gaul: Widmer (1994), p. 125.

80   immense Mesoamerican heads: Fagan (1995), pp. 481–83; Hayden (1993), p. 266.

80   "the most common form of society": Widmer (1994), p. 126. Widmer may be using a looser definition of a chiefdom than some scholars now favor. Most indigenous societies of Melanesia were at a level of complexity just under the chiefdom level as most strictly defined. Still, some Melanesian societies, as on Fiji and the Trobriand Islands, are commonly considered chiefdoms.

80   division of labor: See Sahlins (1963), p. 296, on how Polynesian chiefs, by subsidizing crafts, achieved a technical division of labor not matched in the Big Man societies of nearby Melanesia.

80   the occasional dam: Hayden (1993), p. 304.

81   "tremble with feare": Carneiro (1992), p. 33. Re demigod status: Carneiro (1991).

81   "Chiefs co-opt": Hayden (1993), p. 292. On the prevalence of this sort of perspective within archaeology, see Stein (1998), pp. 8–10.

81   Polynesian chiefdoms blossom: Kirch (1989), pp. 13–15.

81   *mana* and *tapu*: Kirch (1989), pp. 37–40, 68.

81   talking chiefs: Service (1978), p. 271.

81   "stands to the people as a god": quoted in Kirch (1989), p. 166.

82   communal paste pool: Kirch (1989), pp. 134–35.

82   financing irrigation: Johnson and Earle (1987), p. 236.

82   four hundred fish ponds: Kirch (1989), p. 180.

82   canoes: Kirch (1989), p. 181.

82   [†]"redistributive" ritual: Service (1978), p. 269. The "redistributive" function of chiefs isn't emphasized these days as much as it once was (see Earle, 1997, pp. 68–70). Still, this function can indeed be performed by chiefs. And, as Johnson and Earle (1987, p. 216) note, even if the redistribution takes place via intra-chiefdom trade, so that the chief doesn't directly orchestrate it, he facilitates it by keeping internal peace.

82   patronage for subordinates and relatives: Kirch (1989), pp. 167, 260. Intense nepotism is characteristic of chiefdoms.

82   making chiefly tombs: Kirch (1989), p. 181.

82   "part of their nature": Service (1962), p. 152.

82   "chronic state of war": Murdock (1934), p. 63.

82   "incessant warfare . . . constant fighting": quoted in Kirch (1989), p. 195.

83   "universal, acute": Carneiro (1990), p. 193.

83   the nature of status-seeking: See Wright (1994), chapter 12, and Williams (1966).

83   "eat the power . . . too much": Sahlins (1963), pp. 297–98. Fear of public disfavor helps explain why Polynesian chiefs sometimes used the "talking chief" to gather public opinion about contemplated policies (de Sola Pool, 1983, p. 12).

84   [†]games people won't play: Note that what these people see as unacceptable falls short of pure parasitism; the "chief" in the experiment isn't trying to keep *all* the surplus that emerges from synergistic cooperation. Rather, these people are just being offered what they see as a bad bargain. Why are people so averse to bad bargains that they'll turn one down even when—as here—turning it down leads to an even worse outcome? The answer favored by most evolutionary psychologists is that, during human evolution, getting a reputation as someone who would tolerate exploitation could lead to repeated exploitation, even to the point of diminishing your prospects for survival and procreation. Thus genes for "pride"—the sort of pride that keeps you from accepting a mere $20 out of $100—could, in the long run, do better than alternative genes that permitted exploitation.

84   cross-cultural findings: See Camerer (1997), p. 169, and Camerer and Thaler (1995), pp. 210–11. (Since the hardcover edition of this book came out, I've learned of ongoing experiments that suggest greater cross-cultural variance, with people in one society accepting quite low payoffs. But, as the paperback edition was being finalized, these data were still unpublished.)

84   [†]Nash and bargaining imbalances: See Davis (1989), Nasar (1998), and Nash (1950). Note that, in the game-theory experiment described above, a comparable imbalance of bargaining power is evident: the first player has the power to set the terms, and the second player has only the power to accept or reject them. This imbalance of power leads to outcomes that one could argue are unjust. It seems unfair for the two players to split the money by, say, 70–30, since neither did any actual work for it, yet most players consigned to the less powerful position do accept a mere 30 percent of the loot if it is offered.

84   "vote with their feet": Johnson and Earle (1987).

84   no easy way out: Allen Johnson, personal communication.

84   Natchez: Widmer (1994), p. 136.

85   monopoly on legitimate force: Like so many of the criteria used to distinguish among different levels of social organization, this one isn't absolute. There have been a few societies—such as Anglo-Saxon kingdoms—that are considered state-level but permit retaliation by victims of certain types of crimes. (Robert Carneiro, personal communication.)

85   impaled skulls of past enemies: See Carneiro (1991), p. 173.

85   "spectre of rebellion": Kirch (1989), p. 167. Sahlins (1963), p. 297, writes that "even the greatest" Polynesian chiefs knew that "generosity was morally incumbent on them."

85   "rat will not desert": See Service (1975), p. 95.

85   "put to death": Kirch (1989), p. 261.

86   "restriction of access": Hayden (1993), p. 311.

86   "control of water": Hayden (1993), p. 300.

86   †Hawaiian dam: I certainly don't mean to suggest that the chief's handling of dam construction was the optimally efficient solution to this particular non-zero-sum problem. As Earle (1997), pp. 75–82, notes, Hawaiian irrigation systems were usually small—often comprising no more than five farm families, and almost never extending beyond a single community; so a more local level of governance, even ad hoc cooperation, could have gotten the job done. No doubt Earle is right to suggest that the chief horned in on these local non-zero-sum problems as a means of increasing his power. Still, the fact is that to get the power he had to solve the problems. How much inefficiency and unfairness his intrusion introduced—that is, how much more the irrigation system "cost" the farmers than it would have cost if they'd been left to build it themselves—is a difficult empirical question. (And the fact that the chief, having handed out irrigated plots in return for labor, continued to "tax" the plot owners long after their labors had ended, isn't itself determinative, since the chief performed other public functions, and at least some of the taxes supported them.) My point is simply that, though the inefficiency and unfairness were no doubt of non-trivial proportions, they were not the whole story. It has become fashionable to write—as in the textbook quoted above—as if a chief's self-aggrandizement entailed no public service, but this is manifestly not the case.

86   "apex of the socio-political pyramid": Kirch (1989), p. 39.

86   "viewed within the living context": Harris (1991), pp. 112–13.

87   "stigma of low *mana*": Kirch (1989), pp. 68, 196–97. In the simple chiefdoms of the Trobriand Islands, another feedback mechanism punished chiefly incompetence: the chief's status rose or fell according to his performance in public ceremonies. (Johnson and Earle, 1987, pp. 222–23.) Since this performance depended on material well-being—on the recent productivity of his people—it was in some ways a good gauge of the quality of his governance.

87   "invent supernatural sanctions": Hayden (1993), p. 292.

87   kudos for Marx: "Infrastructure" is a term used by late-twentieth-century Marxists to encompass what Marx called the "material forces of production" and the "relations of production."

88   memes as non-genetic information: The tidy definition of cultural information as any information transmitted from person to person non-genetically comes from Bonner (1980).

89   people repelling memes: See Pinker (1997), p. 210. To be sure, as cultural evolution has created an environment more and more removed from the environment in which human evolution took place, it has become easier for memes to "dupe" brains. See Blackmore (1999). Again, heroin addiction is an example; had this threat existed during human evolution, people might be better at resisting it.

89   "religious memes . . . parasitically exploiting": Dennett (1997), p. 41.

89   God as a virus: Dawkins (1976).

90   memes displace other memes: For a mathematical treatment of cultural transmission via imitation, see Boyd and Richerson (1985).

91   Indo-Europeans: See Wright (1991).

91   maritime words: Kirch (1989), p. 53.

91   *mana* and *tapu:* Kirch (1989), p. 63.

91   Tiv and Iroquois: Emile Durkheim saw a *mana*-like concept in the Iroquois. See Aron (1967), vol. 2, p. 50. Elman Service (1975), p. 92, found one in the Tiv. The concept of *tapu,* in the elemental sense of preserving an awe-inspiring distance between ruler or priest and the larger populace, is no doubt widespread.

91   natural fissioning: See, e.g., Cohen (1985), p. 105; Farb (1969), p. 191; Widmer (1994), p. 140.

92   allocating water: Kirch (1989), p. 175; Johnson and Earle (1987), p. 232.

92   Hawaiian word for "law": Kirch (1989), p. 175.

92   from chiefdom to state: Hayden (1993), p. 367. Carneiro (1970), p. 736, asserts a gap of "two or three" millennia between the transcending of village autonomy and the "rise of great empires."

8. THE SECOND INFORMATION REVOLUTION

93   epigraph: *New York Times,* February 20, 1998.

93   temple receipt: *New York Times,* November 11, 1997.

95   Atlantis: Coe (1992).

95   Chinese writing: Keightley (1989a).

95   *rongorongo:* Kirch (1989), p. 273.

95   Sumerian and Mesoamerican script: See, e.g., Lamberg-Karlovsky and Sabloff (1995), p. 360, and Adams (1997), pp. 57–58. On China, see Adams (1997), p. 22.

95   "It should be Fu Hao": Keightley (1983a), p. 548.

95   "no disaster": Keightley (1983a), p. 527.

95   perishable Chinese writing materials: Ho (1975). See also Keightley (1989a). Ho has found pottery shards, with symbols he interpreted as numbers, that precede the oracle bones, and Keightley also sees this as evidence of early, rudimentary writing.

96   theory of Sumerian writing: Schmandt-Besserat (1989).

97   Sumerian division of labor, public works: See, e.g., Saggs (1989), pp. 62–68.

97   buttressing ruler's authority: On Easter Island, the one part of Polynesia where pictographic writing had begun to evolve, it seems to have been used to record royal genealogies and sacred chants. See Kirch (1989), p. 273.

99   "pay the Sun-God's interest": White (1959), p. 344; Norman Yoffee, personal communication.

99   a shekel to harvest a field: Saggs (1989), p. 163.

99   biting a nose: Saggs (1989), pp. 164–65.

100   Mesopotamian seals: Lamberg-Karlovsky and Sabloff (1995), p. 358.

100   digging canals: Oates (1979), p. 44.

100   "Hammurabi is the Prosperity": Norman Yoffee, personal communication.

101   "He who shall not observe": Goody (1986), p. 98.

101   sealing alliance via intermarriage: Flannery (1972), p. 421.

101   "You are my brother": Lamberg-Karlovsky and Sabloff (1995), p. 166.

101   "Between Kings there is brotherhood": Lamberg-Karlovsky and Sabloff (1995), p. 166.

102   "Piles of corpses": Lamberg-Karlovsky and Sabloff (1995), p. 165.

102   "donkey express": Lamberg-Karlovsky and Sabloff (1995), p. 168. Note that here, too, the information is being used for non-zero-sum purposes. A king's relationship with a prince governing a distant city is non-zero-sum; each benefits from the continued dominance of the ruling regime.

102   Aztec relay runners: Bray (1991), p. 88.

102   Dahomey information technology: Herskovits (1938), vol. 2, pp. 70–81.

102   *quipu* recording nonquantitative data: Lloyd Anderson, personal communication. And see Larsen (1998), p. 185. It may be that rather than fully representing nonquantitative data, the *quipu* conveyed mnemonic cues helpful to "memory experts"—a vocation found in various complex societies that lack a fully developed script. In fact, the word for the partly evolved script found in Easter Island—*rongorongo*—seems to be derived from the word for memory expert. See Kirch (1989), p. 273.

102   Incan roads: See Adams (1997), p. 123.

103   Incan error suppression: Adams (1997), p. 124.

103   rate of Incan data travel: *Encyclopaedia Britannica* (1989), vol. 6, p. 277.

103   "talking drums": Service (1978), pp. 356–57.

103   Bronze Age: Bronze was also in use among pre-urban peoples in Thailand and the Balkans. In general, archaeologists put less stock in the "Bronze Age" and "Stone Age" designations than they used to.

104   Wool and trade: Goody (1986), pp. 69–71.

104   thirty-five jars of beer: Nissen, Damerow, and Englund (1993), p. 46.

104   Mesopotamian literacy rate: Stein (1994a), p. 12. For evidence of the hierarchical nature of Mesopotamian economy and society around 3100 B.C., at the beginning of the age of writing, see Nissen (1985), especially pp. 358–60.

104   goddess of fertility: Nissen, Damerow, and Englund (1993), p. 106.

105   rich man bribes teacher: Nissen, Damerow, and Englund (1993), p. 109. In general, on scribes and social status, see Lenski, Nolan, and Lenski (1995), pp. 181–83.

105   driven like cattle: Lenski, Nolan, and Lenski (1995), p. 185. See also Goody (1986), p. 117.

105   †what lower classes lacked: For Mesopotamian wool spinners, having a scribe could have been the first step toward cutting out the monopolistic middle man, and marketing their goods directly to the foreigners who craved them. Or, at least, a more widespread literacy might have brought more people into the middle-man business, giving spinners bargaining leverage.

105   slave revolts: Herskovits (1938), vol. 2, p. 319.

105   "You arrested me": Baines (1988), p. 203.

105   writing and concentration of power: Baines (1988). Baines explicitly says that the *concentration* of writing abilities is the culprit.

106   oxen as currency: Smith (1937), p. 23. Actually, it's not as bad as it sounds. Oxen were a standard unit of accounting—of calculating payment due—but the payment itself could be made in the form of other goods.

106   political and economic pluralism: See Yoffee (1995), especially p. 295; Stein (1994a), especially p. 15; and Brumfiel (1992).

9. CIVILIZATION AND SO ON

107   epigraph: McNeill (1980), p. 37.

108   "pristine" and secondary states: See, e.g., Fried (1983), p. 470.

108   West Africa: See, e.g., Diamond (1997), pp. 282–83.

108   diffusion linking the Americas: Lamberg-Karlovsky and Sabloff (1995), p. 267.

109   "generalizing from a single case": Dray (1967), p. 255.

110   Mesopotamia's pre-state evolution: Lamberg-Karlovsky and Sabloff (1995), pp. 101–11, 114; Stein (1994b); Fagan (1995), pp. 367–68. As Stein notes, some archaeologists are reluctant to apply the label "chiefdom" to the Ubaid culture, since some of the classic hallmarks of chiefdoms, including sharp status differentiation and political instability, appear clearly only around the end of the fifth millennium, just before the emergence of cities. Still, they do appear. Moreover, Stein finds evidence that, earlier, "Ubaid Mesopotamia consisted of a series of small, localized chiefdoms based on staple finance" (p. 41).

110   information technology in 3500 B.C.: Lamberg-Karlovsky and Sabloff (1995), p. 358.

110   urbanization of Mesopotamia: See Lamberg-Karlovsky and Sabloff (1995), pp. 108, 139. Uruk in 2800 B.C.: Fagan (1995), p. 368.

111   king in neck-stock: Saggs (1989), p. 41.

111   Sargon's daughter as priestess: Flannery (1972), p. 420.

111   "King of the Four Quarters": Lamberg-Karlovsky and Sabloff (1995), p. 162.

111   Egypt's slow cities: It was long thought that Egypt's development was so different from Mesopotamia's that to speak of early Egyptian "cities" was erroneous. Rather, what anchored Egyptian polities were towns of an administrative or priestly nature. But recent evidence suggests that Egypt's residential structure, by the early third millennium, the time of the First Dynasty, was more like Mesopotamia's than was previously believed. See Lamberg-Karlovsky and Sabloff (1995), pp. 128–29.

111   hieroglyphics by 3100 B.C.: Lamberg-Karlovsky and Sabloff (1995), p. 134.

111   Egyptian bureaucracy: Saggs (1989), p. 27. On the power and efficacy of the Egyptian bureaucracy, see Berlev (1997), pp. 87–89. The bureaucracy was especially important in preserving an adequate food supply notwithstanding year-to-year fluctuations in the level of the Nile.

112   east Asian chiefdoms: Fagan (1995), pp. 432–33; Chang (1983), pp. 512–14. (Chang's "level 2" corresponds to the chiefdom level.) See also Service (1975), pp. 263–64.

112   13,000 men: Bagley (1999).

112   10,000 workers and eighteen years: Fagan (1995), p. 435, and chapter 18 generally.

112   royal Chinese graves: Keightley (1989b), Fagan (1995), p. 436.

112   Shang China as fragmented: Robert Bagley, personal communication.

112   Shang chiefdom or state?: See Keightley (1983).

112 Ch'in law: Hsu (1989), p. 74. See also Fairbank (1992), pp. 54–56.

112 Chinese infrastructure: Gernet (1996), p. 106; Fairbank (1992), p. 56.

112 Shih Huang-ti book-burning: Gernet (1996), p. 109.

113 terra-cotta warriors: Fairbank (1992), p. 56.

113 Shih Huang-ti's caprice, demise: Starr (1990), p. 637; Gernet (1996), p. 108.

113 "king of this great earth": Starr (1990), p. 277.

113 "general governor and reconciler": Hammond (1989), pp. 270–73.

113 "savior of all mankind": Starr (1990), p. 557.

113 Monte Alban and Uruk: Lamberg-Karlovsky and Sabloff (1995), pp. 347–48, 351–52.

113 pre-state evolution in Uruk, Monte Alban: Lamberg-Karlovsky and Sabloff (1995), p. 348; Fagan (1995), p. 486; Stein (1994b); Stein, personal communication.

114 Monte Alban population: Fagan (1995), p. 487; Adams (1997), pp. 50–51.

114 Teotihuacán population: Adams (1997), pp. 43, 49.

114 Tenochtitlán: Area in 1519 was 80,000 square miles. Population estimates range from 150,000 to 300,000. See Farb (1995), p. 211; Bray (1991), p. 98; Lamberg-Karlovsky and Sabloff (1995), p. 324.

114 "the most beautiful city": Farb (1995), p. 211.

114 canoes: Adams (1997), p. 84.

114 Cortez on Tenochtitlán: Bray (1991), p. 111.

114 136,000 skulls: Bray (1991), pp. 171–73. Estimated annual human sacrifices are from 10,000 to 50,000, "mostly war captives, but also slaves and children who were obtained by purchase."

114 exotic imports in the provinces: Smith (1997), pp. 76–83.

114 market inspectors: Bray (1991), pp. 111–12.

114 "Indians appear in a better light": Bray (1991), p. 84.

115 Tenochtitlán's hydraulics: Bray (1991), p. 98.

115 roads: Johnson and Earle (1987), p. 269; Adams (1997), p. 103.

116 lima beans: Adams (1997), p. 106.

116 throat-slittings, goblets: Adams (1997), pp. 106–7; Fagan (1995), p. 529.

116 Chimú and Huari roads: Adams (1997), pp. 112–17.

116 Incan conquest: Johnson and Earle (1987), pp. 263–64.

116 "Four Quarters of the World": Adams (1997), p. 118.

116 Pax Romana–like productivity: Johnson and Earle (1987), p. 269.

116 domesticable animals: Diamond (1997); Steward (1955).

116 rolling animal toys: Bray (1991), p. 87.

116 llama: Adams (1997), p. 103.

117 smallpox and Pizarro: Adams (1997), p. 119.

117 "pecuniary and market motives": McNeill (1980), p. 34.

118 "normal and expected": McNeill (1990), p. 13.

118 wheat, meat, fish: Lenski, Nolan, and Lenski (1995), p. 194.

118 benefits slowly trickle down: Still, for a long time, the benefits of *trade* would continue to accrue mainly to the upper classes. See Drummond (1995), p. 111.

118 †onward and upward: In assessing economic progress due to trade and technological change, it is easy to confuse two issues. Some people, minimizing progress, argue that there's little evidence that the standard of living—per capita economic output—rose before the modern era. True, but that doesn't mean that productivity—economic output per *worker*—wasn't rising. When population is growing rapidly, as it has for the past several millennia, the productivity of individ-ual workers can grow significantly without the standard of living rising, since the average worker has a growing number of mouths to feed. This, in fact, is the pre-vailing interpretation of the millennia before the industrial revolution—long-run productivity growth obscured by long-run population growth. (Bradford De Long, personal communication.) Also, population growth can force people to farm on less desirable land, so that merely keeping productivity from falling could signify improved technology.

118 Egypt and China: On contrasts among ancient states generally, see Trigger (1993). He stresses a distinction between "territorial states" (Egypt, Shang China, the Inca) and the more urban states (Mesopotamia, the Aztecs).

119 church and state: See Lenski, Nolan, and Lenski (1995), p. 212.

119 Pythagorean theorem: Goody (1986), p. 80. Actually, in Greece the Pythagorean theorem may have been formulated not by Pythagoras but by later followers of the Pythagorean philosophy.

120 adding vowels: Craig et al. (1990), p. 81.

120 minting coins: Hooker (1995), p. 30.

120 Greeks and contracts: Martin (1994), p. 76.

120 vibrance and democracy of Athens: Hooker (1995), pp. 4–9. Sparta, which forcefully discouraged commerce on the part of its citizens, was less democratic.

120 Mesopotamian assemblies: Yoffee (1995), p. 302; Goody (1986), p. 120.

121 deliberative assemblies, "city hall": Yoffee (1995), p. 302; Yoffee, personal communication; Gil Stein, personal communication.

121 Dalmatian coast: Starr (1991), p. 482.

122 commerce ventures beyond politics: See McNeill (1990), p. 12.

10. OUR FRIENDS THE BARBARIANS

124 epigraph: Toynbee (1947), vol. 1, p. 420.

124 "What is safe?" Friedrich (1986), p. 27.

124 "whole world perished": Jones (1986), vol. 2, p. 1025.

124 Gupta fall: McNeill (1963), p. 384.

124 Persia and Huns: McNeill (1963), p. 386; Starr (1991), p. 705.

125 *oikoumene:* Starr (1991), p. 581. The word had come from the Greeks.

125 Hannibal's brother: Starr (1991), p. 486.

125 Goth children: Friedrich (1986), p. 36.

126 Christian lanterns: A. N. Wilson (1997), p. 9.

126 Titus's slaughter: Friedrich (1986), p. 32.

126 "fierce barbarians": Augustine (1958), p. 44.

126 barbarian plow: Singer et al. (1956), pp. 87–89.

126 barbarian stirrup: McNeill (1963), p. 384; Gies and Gies (1995), p. 55.

126 barbarian horse archers: McNeill (1963), p. 238.

126 soap: Gies and Gies (1995), p. 31.

127 dissuading Attila: McNeill (1963), p. 392.

127 Hun shopping list: Thompson (1948), pp. 170–73.

127 Germanic literacy: McNeill (1963), p. 392.

127 "nomadic, boastful": Hayden, p. 354.

127 salt, metals, wine: Starr (1991), p. 535.

127 art heads north: Fagan (1995), p. 474.

127 horseshoes, locks: Hayden (1993), p. 354.

127 "short sword": Fagan (1995), p. 475.

127 money and alphabet: Starr (1991), pp. 535–36.

128 "stable and enduring": Goffart (1980), p. 28.

128 "more civilized" and terrifying: Starr (1991), p. 700.

128 "To be ignorant": Elvin (1973), p. 41.

129 sacking of Chin capital: Elvin (1973), p. 41. These barbarians were not part of the Toba, described below.

129 Hun "empire": McNeill (1963), p. 388.

129 Toba: Gernet (1982), pp. 190–92; McNeill (1963), p. 386.

129 high ratio of income to work: Goffart (1980), p. 102.

129 "under new management": Goffart (1980), p. 3.

130 praising Euclid: Riché (1976), p. 69, footnote 114.

130 Dionysius: Riché (1976), pp. 68–70.

130 adapting Roman law: Riché (1976), pp. 71–72; Robinson (1997).

130 Justinian's "reclamation": Hollister (1974), p. 38.

130 barbarian waves of second millennium B.C.: McNeill (1963), pp. 117–19.

131 Aztecs and Toltec: Lamberg-Karlovsky and Sabloff (1995), p. 345; Farb (1969), p. 208.

131 Greeks disdaining Romans: Starr (1991), p. 488.

131 "The Greeks, captive": McNeill (1963), p. 294.

131 Greeks as civilization wreckers: Tainter (1990), p. 63; Starr (1991), p. 193; Brogan (1989), p. 190.

131 "sacker of cities": Hayden (1993), p. 347.

132 "superior to their own": McNeill (1963), p. 391.

132 military strategy and weaponry: See Elvin (1973), p. 18.

132 ironmaking and Mongols: Elvin (1973), p. 18.

132 "The precocious developing society": Service (1975), p. 321.

133 "their own evil natures": Shelton, ed. (1988), p. 182.

133 Rome and labor-saving innovation: Drummond (1995), p. 110. See Mokyr (1992), p. 167, for an alternative view of why such technologies as harvesting machines didn't spread rapidly.

133 slavery waning: See Drummond (1995), p. 120. A dissenting view is in Mac-Mullen (1988), pp. 17–18.

133 peasants *less* free: Jones (1986), p. 795; Starr (1991), p. 676.

133 freeze on changing vocations: Jones (1986), p. 861.

133 "closer to a caste system": Starr (1991), p. 676.

134 "They make a desert": Friedrich (1986), p. 28.

134 corrupt, oppressive Rome: MacMullen (1988); Drummond (1995), p. 110.

134 kissing the hem: Drummond (1995), p. 106.

135 centuries of ossification: Starr (1991), pp. 123–24.

136 minting coins for government: Drummond (1995), p. 112.

136 slaves: Treadgold, personal communication; this is a consensus view, though not definitively established.

136 eastern empire less afflicted: Jones (1986), p. 861.

136 more integrated economy: Treadgold, personal communication. See Treadgold (1997).

136 "if new ideas are to have a chance": Starr (1991), p. 124.

138   epigraph: Bowersock (1988), p. 174.

138   Conan the Barbarian: Actually, Genghis Khan, a real-life barbarian, is reputed to have said something rather like this. (See Jones, 1987, p. 229.) Yet even his own Mongols evinced the general barbarian penchant for milking civilization rather than destroying it.

139   collapse in Roman vicinity: See Nef (1964), p. 7.

140   *western* civilization: Clark (1969), p. 18: "Without Charles Martel's victory over the Moors at Poitiers in 732, western civilisation might never have existed."

140   deciding whether a culture has collapsed: On the general difficulty of determining when—and, in a certain sense, even *if*—the Roman Empire actually collapsed, see Bowersock (1988).

140   Maya robustness: William Sanders, personal communication.

140   Indus valley: Kenoyer (1998).

141   "transalpine Europe": Harris (1991), p. 254.

141   feudalism and chiefdoms: Service (1975), pp. 80–83; Johnson and Earle (1987), p. 249. Many historians trace the roots of feudalism both to longstanding elements of Germanic social organization and to elements of Roman social organization that had materialized as the western empire disintegrated. See Cantor (1993), pp. 197–98.

141   "justice and equity": Roberts (1993), p. 335.

142   †schematic simplification: The very meaning of "feudalism" is much debated. Some scholars say relations between the highly dependent serfs and their lord should be categorized separately under the label "manorialism," whereas others consider these relations simply an aspect of European feudalism, just like the more voluntary relations between vassals and lords. Still others, such as Gernet (1996), p. 53, throw up their hands and say the term *feudal* has been "so often misused that it has lost all meaning. It is better to do without it. . . ." See Cantor (1993), pp. 195–97, and Service (1975), pp. 81–83, for discussions. I use a broad definition of feudalism—encompassing "manorialism"—in part because feudalism more narrowly defined shares key properties with manorialism: a relationship of mutual obligation that can function in the absence of money and when necessary can provide a degree of local self-sufficiency, whether economic, governmental, or even military. These properties collectively give the entire system its resilience. Incidentally, there are various other complexities of the system that I'm not bothering with. For example, the bounds of a village and the bounds of a lord's manor might not coincide. Also, sometimes a vassal would serve more than one lord (drawing separate compensation in land from each), and messy conflicts of interest could ensue.

142   ten levels deep: Hollister (1974), p. 119.

143   counts and kings: Strayer (1955), p. 62.

144   Vikings melt into social fabric: Hollister (1974), p. 119.

144   reconstitution after collapse: See Service (1975), p. 82.

144   "feudal" elements: One example is fourth-century China. See Elvin (1973), p. 42. Japan may offer one of the world's more precise past parallels with European feudalism. See Duus (1993).

144   Greek regrouping: Ferguson (1991). As Ferguson notes, many scholars consider Mycenae to have been only an incipiently state-level society, a "petty bureaucratic state."

144 Chinese reconstitution: Elvin (1973), pp. 42–43.

145 "Had the destruction been complete": Cahill (1995), p. 58.

145 agricultural technology: Gies and Gies (1995), pp. 44–47.

146 spinning wheel: Pacey (1990), pp. 23–24.

146 navigational technology: Roberts (1993), pp. 428–29, Gies and Gies (1995), pp. 23, 159.

147 Roman mills: Evidence for the Roman use of the sawmill is dubious. See White (1964), pp. 82–83.

147 new uses for mills: White (1964), p. 89; Cipolla (1994), pp. 140–43.

147 French fullers: Cipolla (1994), pp. 140–43.

147 windmill: Cipolla (1994), pp. 143–44; White (1964), p. 88.

147 cotton gin: Gies and Gies (1995), pp. 118–23.

148 "rings, belts, and pins": Gies and Gies (1995), p. 123.

148 prayed, fought, labored: Cipolla (1994), p. 118; Craig et al. (1990), p. 372.

148 abacus: Gies and Gies (1995), p. 159.

149 "God shall have granted": Cipolla (1994), pp. 161–62.

149 new ancillary metatechnologies: Cipolla (1994), pp. 160, 163.

149 stock exchanges: Amsterdam's stock exchange, started in the seventeenth century, is often called the world's oldest, though Antwerp had a less modern version in the sixteenth century, and more primitive stock markets are found earlier.

150 towns' self-government: Hollister (1974), pp. 148–50; Cipolla (1994), pp. 117–22.

150 founding towns: Hollister (1974), p. 149.

150 decentralized nature of feudalism: Landes (1998), pp. 36–37, writes, "Ironically, then, Europe's great good fortune lay in the fall of Rome and the weakness and division that ensued"—and in turn cites Patricia Crone, who wrote in *Pre-Industrial Societies* (1989, Basil Blackwell) that "Far from being stultified by imperial government, Europe was to be propelled forward by constant competition between its component parts. . . . Europe failed [to sustain its ancient empire]: had it succeeded, it would have *remained* a pre-industrial society." Scholars who thus emphasize western Europe's productive decentralization often focus on the early modern period, with its competing nation-states. See Kennedy (1987), p. 20; E. L. Jones (1987); Roberts (1993), p. 415. But the basic logic—the creative force of competition among polities—applies equally to the decentralized Middle Ages.

151 "of the merchants": Robert S. Lopez, quoted in Landes (1998), p. 36.

151 "honorable positions": Cipolla (1994), p. 121.

151 "good animals": Cipolla (1994), p. 122.

151 ashes in the Tiber: Cantor (1993), p. 404.

151 Frederick gives in: Cantor (1993), p. 405; Strayer (1955), pp. 110–11. The Lombard League's military success wasn't due only to urban wealth. In Italy, unlike in Germany, the feudal nobility had taken residence within city walls, so the Lombard League had great military experience to draw on. Also, the league enjoyed the support of the pope, who was leery of Frederick's ambitions in Italy.

151 "Town air": Cipolla (1994), p. 119.

151 money economy in countryside: Cantor (1993), pp. 472–73; Hollister (1974), pp. 152–53.

152   "an interesting problem": Strayer (1955), p. 224.

152   †Marx: Marx may have suffered from an underappreciation of information technology, grounded in an underappreciation of information generally. His "labor theory of value" seems to ignore the fact that various people—investors, wholesalers, retailers—process information about the demand and supply of goods and raw materials, a service that saves consumers (and laborers) time and effort and is thus worthy of recompense. The value of information processing, in short, is a legitimate component of retail price, no less than the value of the labor involved in manufacture.

153   tradition of consulting nobles: The king's council goes at least as far back as the Anglo-Saxon witenagemot. See Hollister (1974), pp. 241, 242. Europe's medieval representative institutions are sometimes credited to the influence of Roman law, but as Cantor (1993, p. 457) notes, England was barely touched by that influence—and it is in England that the evolution of representative government on a national scale was most advanced.

153   expanding representation: Hollister (1974), pp. 244–45.

154   †independence of European democracy from Greek heritage: This isn't to say that the codification of ideology doesn't matter. Words can inspire people, and can lend power to social, religious, and economic movements, thus shaping history in the time frame of decades and even centuries. Still, in the longest run, the basic trajectory of technology renders some forms of social organization unworkable and others irresistible.

154   Mayan "mercantile pragmatism": Sabloff and Rathje (1975).

12. THE INSCRUTABLE ORIENT

155   epigraph: Gernet (1996), Introduction.

155   "magical traditionalism": Weber (1961), p. 128. See also Blaut (1993), p. 103.

155   "Oriental Despotism": The theory is Karl Wittfogel's, laid out in his 1957 book *Oriental Despotism*. For discussions, see, e.g., Adams (1966), pp. 66–68; Blaut (1993), pp. 83–87; Jones (1987), p. 10. One oft-noted problem with this theory is that in various societies, including China, irrigation systems have been locally managed, and thus don't easily account for the state-wide centralization of power. But Landes (1998), pp. 27–28, defends the theory.

155   "weird": Landes (1998), p. 57.

156   "freak of fortune": Jones (1987), p. 202.

156   "stagnation at the best": Jones (1987), p. 231.

156   "supermiraculous": Jones (1987), p. 238.

156   "conquest and booty": Landes (1998), p. 393.

157   *commenda:* Udovitch (1970), pp. 38, 48.

157   †Hanafite school: In later centuries, those parts of Islamic civilization where Hanafite law dominated—such as Turkey—would be more business-friendly than those parts—such as Arabia—that had featured Islamic legal traditions less conducive to commerce. (Timur Kuran, personal communication). One eleventh-century Hanafite scholar captured the purpose of the *commenda* elegantly: Without such a contract, "the possessor of capital may not find it possible to engage in profitable trading activity, and the person who can find it possible to engage in such activity may not possess the capital." (Udovitch, 1970, p. 49.) Jones (1987), p.

187, stresses that medieval Islamic governments did a good job of protecting private property from arbitrary seizure, and of fostering commerce more generally.

157 Baghdad checks: Hollister (1974), p. 67.

157 "the need for trade": Udovitch (1970), p. 40.

157 "doing God's work": Landes (1998), p. 393.

157 the Kabah: Armstrong (1994), p. 135.

157 †common religion and language: The "common religion" wasn't universal; the Muslims didn't insist on conversion the way Christian conquerers usually did, so there was often religious diversity in Muslim-controlled lands. Still, local merchants were often Islamic, a fact that lubricated trade. Similarly, though Arabic of course didn't become a universal language across the whole Islamic empire, it spread sufficiently to lubricate trade.

157 taxation replaces booty: Waldman (1989), p. 108.

157 Muslims retain information technologies: See Roberts (1993), p. 270.

157 al-Khwārizmī: *Encyclopaedia Britannica* (1989), vol. 23, p. 605.

158 Islamic algorithms and High Middle Ages: See Abu-Lughod (1989), pp. 15–17, 177.

158 "flying money": Gernet (1996), p. 265, puts this innovation in the ninth century, and Elvin (1973), p. 155, in the eighth. These certificates of deposit seem to have begun under government sponsorship, tied into the system of collecting taxes, but merchants began issuing their own checks. In the eleventh century, the idea evolved into government-issued paper money.

158 banking in China: Elvin (1973), p. 157. An account of this practice during the Sung dynasty, written a century or so later, says that the merchants issuing the checks would "print off more such notes of exchange . . . with as much celerity as one would mint coins" and then use them to "buy up huge stocks" of commodities. These de facto bankers thus came to own "a profusion of inns, shops, mansions, gardens, fields and precious goods." This practice—printing new checks and using them for investments—may sound different from taking deposited money and spending it (as the logic of banking is typically described), but there is no functional difference. In both cases the sum total of outstanding investments plus cash on hand equals the total value of checks issued, and in both cases the premise is that not all people possessing checks will draw on their accounts at the same time.

158 sixty merchants: Elvin (1973), p. 144. The merchants built these fleets in spite of having to surrender *half* of the fleet to the Chinese government as a kind of tax—a testament to the power of profit's lure.

158 investing in trade expeditions: Elvin (1973), pp. 143–44.

158 "I give you my word": Elvin (1973), pp. 144–45.

158 "the world had ever seen": Gernet (1996), p. 321.

158 "petty entrepreneurs": Elvin (1973), p. 167.

158 merchant associations: Elvin (1973), p. 172.

159 sales tax: Gernet (1996), p. 322.

159 silk and iron factories: Elvin (1973), pp. 174–75.

159 150,000 tons of iron: Jones (1987) p. 202.

159 *Pictures and Poems; Mathematics; Remedies:* Elvin (1973), pp. 116, 181, 191.

159 *Citrus; Crabs:* Gernet (1996), p. 338.

159 "must be obeyed": Adams (1996), p. 54.

159   "constructor of steam engines": *Encyclopaedia Britannica* (1989), vol. 2, p. 884. Some would say the second law was "derived" from Carnot's work by Kelvin and Clausius, but here "derived" means logically deduced; Clausius himself said that Carnot's assertion was a basic postulate, tantamount to the second law.

159   polarity, magnetic induction: Gernet (1996), p. 460.

160   "threshold": Elvin (1973), p. 179.

160   "Heaven-shaking thunder": Elvin (1973), p. 88.

160   "several times cheaper": Elvin (1973), p. 195.

160   300 years: Jones (1987), p. 202.

160   "the two great civilizations": Gernet (1996), p. 347.

160   equal billing: In agricultural technology, many would say China and India led the world in the thirteenth century. See Elvin, 1973, p. 129.

160   "the line of advance": Elvin (1973), p. 198. Even E. L. Jones, who in his 1987 book *The European Miracle* declared an indigenous Chinese industrial revolution a virtual impossibility, acknowledged, perhaps in a moment of weakness, that "by the fourteenth century A.D., China had indeed achieved such a burst of technological and economic progress as to render suspect the frequently expressed belief that industrialization was an improbable historical process" (p. 202). In a later book, *Growth Recurring* (1998), he went beyond parenthetically noting this suspicion, and seriously entertained it.

160   "The mystery lies": Landes (1998), p. 55.

161   Mongol horsemanship: Elvin (1973), pp. 85–90.

161   value-added tax: Morgan (1990), pp. 101–2.

161   "less risk and lower protective rent": Abu-Lughod (1989), pp. 154, 158. Morgan (1990, p. 102) says the purpose of Mongol taxation in general was "quite simply the maximum conceivable degree of exploitation." Yet he also says the Mongols realized that "it was in their interests to encourage the greatest possible amount of trading activity." Obviously, the latter goal rather constrains the pursuit of the former, and resulted in a rather modest tax, at least some of which can be justified as payment for keeping trade routes safe.

162   "prosperity pandemic": Abu-Lughod (1989), p. 356.

162   silk trees: Abu-Lughod (1989), pp. 159–60.

162   "individual monster": Morgan (1990), p. 177.

162   the plague: McNeill (1977), pp. 144–50.

162   twentieth-century depression: Brad De Long, personal communication. The negative impact of the western economic downturn on Japan was clear; the impact on China is debatable.

163   truth more complicated: Abu-Lughod (1989), pp. 343–45.

163   constricting overland commerce: Gernet (1996), p. 403; Elvin (1973), p. 217.

163   perfume: Elvin (1973), pp. 217–18.

163   size of Chinese fleet: Fairbank (1992), p. 138. The Spanish Armada had 185 ships on the eve of its epic defeat.

163   "can-do group": Fairbank (1992), p. 137.

164   *Marvels; Treatise:* Gernet (1996), p. 402.

164   Ming retreat from sea: Mokyr (1992), p. 220; Fairbank (1992), pp. 138–39; Elvin (1973), pp. 220–21.

164   "China retired": Fairbank (1992), p. 139.

164   "malady of the mind": Blaut (1993), p. 215.

165   "protocapitalism": Blaut (1993), p. 167.

165   absence of empire: Chirot (1985) writes (p. 183), "Had the Ming and Ch'ing Dynasties of China failed to unite China, it is quite likely that the first capitalist, industrialized societies would have developed there."

166   patent rights: Mokyr (1992), p. 79.

166   "free cities": Toyoda (1989), pp 314–15.

166   "extraordinary cultural innovation": Reischauer (1981), p. 73. He is characterizing the entire Ashikaga period, from 1333 to 1573, but his emphasis, in context, seems to be on the latter part.

167   "compete on terms of equality": Reischauer (1981), p. 73. Even after Japan's reunification, power was sometimes diffuse by comparison with China. Under the Tokugawa shogunate, beginning in 1603, the landscape remained basically feudal—with firm centralized control at first, but with more and more local autonomy as time wore on (Reischauer, pp. 74–86). In other ways, Tokugawa Japan and China were comparable. Both nations purposefully shut themselves off from the outside world, largely severing their connection to the emerging Eurasian brain, and in both cases they fell behind the west technologically. Landes (1998), pp. 354–65, sees, and stresses, the connection between Japan's diffuseness of power and its prosperity. But, here as elsewhere, his perceiving structural, as opposed to purely cultural, factors that are sufficient to account for differences in national history doesn't lead him to de-emphasize the purely cultural factors. Thus (p. 353), he tells us that "The Japanese were learners because they had unlimited aspirations. Their mythology told of a ruler descended from the sun goddess and a land at the center of creation." But of course, virtually all peoples have had mythologies that placed them at the center of the universe. Indeed, Landes notes that the Ming emperors seem to have considered their empire to be the center of the world—only in this case he uses such ethnocentric mythology to explain a *lack* of interest in the larger world.

167   Europe's and China's political geography: Various scholars (e.g., Kennedy, 1987, p. 17) have attributed these differences in political geography to differences in physical geography—the relatively unbroken landscape of China having conduced naturally to political unity. But Peter Perdue (personal communication) argues that China's geography is in fact extensively broken up by mountains, and, perhaps as a result, features great linguistic diversity. The political unity of China may have more to do with the state's deftness, going back to ancient times, in laying an infrastructure for unity—building canals and roads, standardizing a written language intelligible to speakers of different dialects, etc.

167   †laboratories for testing memes: It may be no coincidence that China's "golden age," the Sung period, was preceded by a half-century of political fragmentation—the "Five Dynasties and Ten Kingdoms" period. What's more, the Sung dynasty itself lived in an unusually competitive east Asia. Facing well-organized and formidable—if "barbarian"—polities on its borders, the Sung, unlike the typical Chinese dynasty, operated within "a true multi-state system," the historian Morris Rossabi (1983) has noted (p. 11). Though these Asian states may not have been as "communicative" as later western European states—not so valuable a source of innovative diffusion—they were at least as competitive, and thus provided a strong argument against technological stagnation.

167   Ming and Manchu technology: Kenneth Pomeranz, personal communication. During the Ming and Manchu periods, China saw incremental but consequential technological advances along with their efficient diffusion across China.

These innovations include the use of crushed soy beans for fertilizer and the spinning of cotton in high-humidity cellars; previously the cotton had to be spun in lands with naturally high humidity. Thus it is an oversimplification to attribute economic growth during the Ming and Manchu wholly to population growth, as some scholars have.

168   European technological interdependence: Palmer and Colton (1965), p. 271; Mokyr (1992), pp. 84, 107.

168   "sooner or later": Landes (1998), p. 368—and see p. 360, where he more explicitly contrasts this verdict on Japan with his view of China.

169   "changed the appearance": Eisenstein (1993), p. 12.

169   Baghdad paper mills: Pacey (1990), pp. 42–43.

169   *Exploitation of the Works:* Mokyr (1992), pp. 222–23. The common claim that China lost its competence in algebra turns out to be wrong, according to a Ph.D. dissertation by Roger Hart (Peter Perdue, personal communication). In any event, by the time the algebra was supposedly "lost," it had already diffused into Japan, where it remained robust. See Elvin (1973), p. 194.

169   "exceptional in human history": Mokyr (1992), p. 171.

169   lebensraum: Elvin (1973), p. 85.

170   Buddhist pawn shops: Gernet (1962), p. 69.

170   "attain Buddhahood": Landes (1998), p. 363. Curiously, Landes doesn't seem aware of the tension between this observation and his general emphasis on the Judeo-Christian heritage, including its Calvinist manifestation, in explaining Europe's economic success. (See Fukuyama, 1993, p. 227, for other examples of Buddhist doctrine being conducive to economic productivity.)

170   "excellent examples": Kennedy (1987), p. 13.

170   the pope on usury: Kennedy (1987), p. 19. Even while the papal ban on usury was technically in effect, people found various ways to lend money.

170   Economic ecumenicalism: Abu-Lughod (1989), p. 354.

170   Islamic industrial revolution unlikely: For a critical review of the main theories on why Islamic civilization became less conducive to economic development over time, see Kuran (1997).

170   Ottoman decay: See Jones (1987), pp. 178, 184, 187–88.

171   Akbar's reforms: Islamic doctrine already tolerated "people of the book"— Christians and Jews.

171   "discourage flies": Jones (1987), p. 197. Some, e.g., Embree (1981), p. 631, believe that the oppressiveness of the government was less a problem than were the costs of territorial expansion.

173   "everything that we do and suffer": Berlin (1954), p. 30.

13. MODERN TIMES

174   epigraph: Eisenstein (1993), p. 152; Cameron (1999), p. 101.

174   "neither holy": *Encyclopaedia Britannica* (1989), vol. 6, p. 22.

177   Lombardy law book: Cipolla (1994), p. 148.

177   three separate editions: Eisenstein (1993), p. 152.

177   "It almost appeared": Eisenstein (1993), p. 150.

177   "Gospel is driven forward": Eisenstein (1993), p. 150.

177   "better internal lines": Anderson (1991), p. 39.

178   German peasant revolt: Craig et al. (1990), p. 467.

178 "Did not Abraham": Rice (1981b), p. 532.

178 wars of religion: See Palmer and Colton (1965), pp. 104–31, 142.

178 Netherlands Calvinists fight Philip II: See Roberts (1993), pp. 465–66. The Calvinists were for a time allied with Catholics against the king. Note, by the way, that the Hapsburgs were also stymied in their quest for European mastery by various other polities in which the printing press had long lubricated nationalist sentiment, such as Britain.

178 †nationalism: The term "nationalism" has various meanings. Some scholars use it to refer simply to coherent national sentiment—a sense of common belonging, common interest, and common destiny of the sort that was emerging in places such as England and France by the fifteenth century. Other scholars mean by nationalism a more self-conscious, overtly purposeful national sentiment, often accompanying a subjugated peoples' quest for nationhood and often inspired by a stated ideology that justifies the quest. This sort of nationalism has been a force since the late eighteenth century. Neither kind of nationalism, I would argue, could have assumed anywhere near the power it finally assumed without the printing press. On the meanings of nationalism, see the introduction to Hutchinson and Smith, eds. (1994). On the different forms that the latter type of nationalism has assumed, see Kohn (1994).

179 "the ruler supplied justice": Jones (1987), p. 128. See also Mundy (1981), p. 404.

179 twelfth-century England and France: Strayer (1955), pp. 106–26, 143–46.

180 "the walls of their castles": Roberts (1993), p. 402. See also Kennedy (1987), p. 21; Jones (1987), pp. 130–31. Though iron cannons existed in the early fourteenth century, the powerful cast-iron cannons didn't arrive until the fifteenth century.

180 "in return for peace": Palmer and Colton (1965), p. 61. To be sure, the consent to be taxed was often not as formal a matter as it is in a modern democracy. In France during the early 1500s, for example, the king did not have to convene the Estates General to approve of new taxes. But he did have to negotiate with town councils and other assemblies. See Ranum (1981c), p. 570.

180 "one dialect shaded": Watson (1992), p. 144.

181 "unified fields of exchange": Anderson (1991), p. 44.

181 "news pamphlets": *Encyclopaedia Britannica* (1989), vol. 8, p. 661. See Martin (1994), p. 295, on *occasionnels,* which performed a comparable function in Germany and France.

181 "imagined communities": Anderson (1991).

181 newspapers, journals, books: Palmer and Colton (1965), pp. 402–6, 438, 439.

182 Spanish and German nationalism: Palmer and Colton (1965), pp. 401, 402–6.

182 Ottomans aid recalcitrants: Watson (1992), p. 177.

182 postal services: Thomas (1979), pp. 331–32. Another factor: even before the printing press, paper costs plunged thanks to waterwheel paper mills, making communication cheaper. See de Sola Pool (1983), p. 12.

182 power migrates upward and downward: See, e.g., Palmer and Colton (1965), pp. 452–53; McNeill (1963), p. 578.

182 †arbitrary empires: Of course, the "arbitrariness" of these empires was itself in some sense arbitrary. That is: the boundaries between national groups—the boundaries that make the larger empire seem like an arbitrary jumble of groups—are products of historical happenstance, and besides, whatever their origin, are to some degree irrational; there is no purely economic reason, in other words, that a Greek Orthodox Christian in the Ottoman Empire should prefer trading with a Christian to trading with a Muslim. Still, as a matter of fact, such cultural divisions can impede trade, in part by erecting "trust" barriers to non-zero-sum flow. And, when those cultural barriers are linguistic barriers as well—as they typically are in the case of nation-states—the "information" barriers are also formidable.

182   dysfunctional borders: See McNeill (1963), p. 583.

183   Germany after unity: Palmer and Colton (1965), p. 661. See Kindleberger (1993), chapter 7, on how unification attracted foreign capital. The same thing happened after Italian unification (chapter 8), although for various reasons young Italy did not see the sustained boom that young Germany saw.

183   Russian and Ottoman empires: See Kohn (1973), pp. 327–28. In east and central Europe as of 1840, most people were still "untouched by nationalism . . . their interests were confined to a narrow local outlook. Communications were still slow, travel was largely unknown, the literacy rate very low."

183   Vuk Karadžić: Palmer and Colton (1965), p. 441.

183   rulers fight the press: Palmer and Colton (1965), pp. 87, 292; Roberts (1993), p. 538.

183   "enormities and abuses": de Sola Pool (1983), p. 15, spelling updated.

183   eight hundred prisoners: de Sola Pool (1983), p. 15. French censorship had been fairly light since 1750. See Palmer and Colton (1965), p. 292.

184   *Journal des débats:* Martin (1994), p. 418.

184   "appendages of the market": Anderson (1991), p. 62. He is referring to newspapers in North and South America, but much the same applies to western Europe.

185   †England and the Netherlands: See Roberts (1993), p. 538; Chirot (1985), p. 186; De Long and Shleifer (1993), p. 672. Of course, there are many reasons that England entered the industrial revolution before such other powerful European nations as France. Among those commonly cited: lower internal tariffs, weaker guilds, more social interchange between the aristocracy and the moneyed middle class, less war, and a geography in various ways conducive to commerce. See Childers (1998) and Mokyr (1992), chapter 10. Some have stressed the security of English property rights, though others (see Clark, 1996) consider this emphasis unfounded.

185   De Long and Shleifer: See De Long and Shleifer (1993). This isn't to say that bursts of absolutism had no economic benefits. During the seventeenth and early eighteenth centuries, Louis XIV of France cleared away many barriers to internal trade left over from the Middle Ages. See Palmer and Coulton (1965), pp. 156–63. But once this was accomplished, France lacked the atmosphere of free enterprise that helped Britain lead Europe into the industrial age.

185   Spain's Golden Age: Ranum (1981b), p. 729; De Long and Shleifer (1993), p. 672.

185   "If the citizen is deterred": Kant (1784), p. 50.

185   1848: See, e.g., Palmer and Colton (1965), pp. 470, 481, 492, and chapter 12 generally; and Albrecht-Carrie (1981), p. 804.

186   Alexander II: Palmer and Colton (1965), pp. 534–36.

186   Ottoman reforms: Palmer and Colton (1965), p. 629.

187   French copyists: Jones (1987), pp. 60–61.

187   guilds accept innovation: See McNeill (1963), p. 583.

188   Sung printing revolution: Gernet (1996), pp. 332–37; Hucker (1975), p. 336. Printed texts date back to at least the ninth century.

188   Sung and European Renaissance: Fairbank (1992), p. 88.

188   thousands of characters: Hucker (1975), p. 336; Martin (1994), p. 22; Peter Perdue, personal communication.

188   Religious flexibility: Peter Perdue, personal communication.

188   Economic integration: Peter Perdue, personal communication; Hucker (1975), p. 138.

189   feedback mechanism: Gernet (1996), p. 303.

189   "social leveling process": Hucker (1975), p. 336.

189   "limelight to their ministers": Gernet (1996), p. 305. Guarding against gross regional imbalances in this newly decentralized power was a quota system which ensured that even regions with sub-par test performance were represented in the bureaucracy.

189   Ming and Manchu: Gernet (1996), p. 304. If we judge meritocracy by the number of degrees won by men from poor or uneducated backgrounds, Hucker (1975, p. 339) finds only slight falloff during the Ming but a sharp drop during the Manchu.

190   first Ming emperor: Fairbank (1992), p. 130.

190   mediocre Ming emperors: Peter Perdue, personal communication; Fairbank (1992), pp. 129–30. See also Gernet (1996), p. 396: "Whereas the political system of the Sung was based on the co-existence of independent organisms which checked each other and of various different sources of information, and whereas political decisions in that empire were the subject of discussions in which contradictory opinions could be freely expressed, the Ming government was characterized, as early as the end of the fourteenth century, by a tendency to the complete centralization of all powers in the hands of the emperor, to government by means of restricted, secret councils, by the isolation of the imperial authority, and by the development of secret police forces entrusted with the task of supervising the administration at its various different levels." Note that the first Ming emperor did bring prosperity to China, but it seems to have been prosperity of the short-term sort that is often seen during the autocratic mobilization for war. The ensuing years were years of stagnation, ended only by the "renaissance" of the sixteenth century.

190   †China complies with law of history: Never again under the Mongols, the Ming, or the Manchus does the Sung's rate of technological progress seem to have been equaled. Moreover, the economic resurgence of the sixteenth century (whose high point was the half-century after 1560. See Gernet, 1996, p. 429) overlapped partly with one of the more just and enlightened periods of Ming rule, first under Lung-ch'ing (1567–1573) and then during the early years of the Wan-li era. (See Gernet, 1996, p. 430, and Spence, 1990, p. 16.) But during the early seventeenth century, thanks largely to the dreaded eunuch Wei Chung-hsien, arbitrary and repressive rule returned—see Gernet (1996), p. 433; Fairbank (1992), p. 141; Spence (1990), pp. 17–18—and at about this time China began a long economic slide. (See Spence, 1990, p. 20.) To attribute the Ming economic resurgence to just, tolerant rule, and the subsequent decline to despotism, would be too simple. (The resurgence was visible before the laudible regimes of the late sixteenth century, and there were hints of decline before Wei Chung-hsien wreaked his havoc.) Still, the economic history of modern China, viewed broadly, certainly poses no great challenge to the general expectation that liberty, political pluralism, and the consistent enforcement of just laws is more conducive to prosperity than is the opposite.

Of course, the great advantage enjoyed by European states during the modern period is the frequency with which this basic law of history has been driven home to them. As we saw in the last chapter, living in a dense neighborhood of peer states has its virtues. One of them is having nearby examples of the wealth that a little liberty can bring. Thanks to these nearby competitors, the blunders of European nations have often come back to haunt them in instructive fashion. Before the industrial revolution, when France persecuted Protestants, some of its finest craftsmen emigrated in droves, helping Britain, an equal-opportunity employer, reach dominance in such fields as clockmaking. See Mokyr (1992), p. 241.

190   "chains of inspiration": Mokyr (1992), p. 255.

190   intellectual property law: Mokyr (1992), p. 247; Jones (1987), p. 60; Clough (1981a), p. 836.

191   locomotive feedback loop: Kelly (1995).

191   independent inventions: See Ogburn and Thomas (1922).

192   trains and telegraphs: See Beniger (1986), e.g., pp. 19, 230.

192   dating globalism: See Waters (1995), pp. 23–25, and Spybey (1996), pp. 25–27, on when various scholars date the beginning of a single world system.

192   ancient exotic trade items: McNeill (1963), p. 584.

192   "splendid and trifling": Jones (1987), p. 87.

192   Roman wheat: Jones (1986), pp. 841–45. Carrying grain long distances by land, in contrast, rarely made economic sense.

193   coffee prices: Clough (1981b), p. 848.

193   †benefits of peace and stability: In addition, there was growth in the awareness that non-zero-sum, mutually beneficial arrangements were often easier than zero-sum, exploitive relationships. For millennia, a primary impetus for empire-building had been to gain access to raw materials: you conquer a place, and it becomes a reliable supplier. But slowly the notion of achieving enduring exchange without conquest settled in, and toward the end of the eighteenth century it got an especially prominent boost. After the American Revolution, people noticed something: England kept trading with America even though it was no longer a colony. (See Palmer and Colton, 1965, p. 331.) You could do regular, reliable business with a country without dominating it. In fact, given the costs of domination, maybe this was a better way to do things!

193   resident ambassadors: Roberts (1993), p. 476; Rice (1981a), p. 489.

193   "the high seas": Schwarzenberger (1989), p. 725.

193   Grotius: Palmer and Colton (1965), p. 283; Schwarzenberger (1989), p. 725.

193   Bruce Mazlish: Mazlish (1993), p. 16.

14. AND HERE WE ARE

195   epigraph: "Ode to W. H. Channing."

195   "There can be no prediction": Popper (1957), p. vii.

195   "by closing down or controlling": Popper (1957), p. 154.

196   "the 'rhythms' or the 'patterns' ": Popper (1957), p. 3.

196   "historicist superstitions": Popper (1957), p. vii.

197   "the constraints of geography": Waters (1995).

197   "resides in the ideas": Gilder (1989); see Wright (1989b).

198   †higher ratios of value to mass: What's more, this specific incarnation of that trend is far from a product of the information age. Books have long been bought and sold, even over long distances. And it misses the point to say that these books were bulky, hence expensive to ship, whereas today's forms of data flit through optical fibers at almost no cost. You don't have to ship books from the United States to England to sell them there. You can just print them in England.

198   "In this global information age": radio address, June 27, 1998.

199   "world's largest thought warehouse": *Washington Post,* April 19, 1998, p. A26.

199   "glorification of the present": Butterfield (1965), p. v.

199   †centralized hierarchies: According to some management theorists, we are entering an age of "flat organizations" and "networks." It is possible, of course, that these pronouncements will turn out to have been greatly exaggerated. But certainly such a trend would have historical precedent. One way to describe the effect of the invention of money and writing in ancient times is to say that they undermined the logic of statist economic hierarchies.

200   American magazine history: *Encyclopaedia Britannica* (1989), vol. 26, pp. 483–85.

200   overemphasizing differences among media: Ithiel de Sola Pool (1983), in contrast to McLuhan and so many others, did stress and acutely assess the similarities in the social effects of different information technologies. He wrote (p. 5): "Freedom is fostered when the means of communication are dispersed, decentralized, and easily available, as are printing presses or microcomputers. Central control is more likely when the means of communication are concentrated, monopolized, and scarce, as are great [television] networks." See also Huber (1994) and Wright (1985).

203   "spinning out of control": Barber (1995), pp. 4–5.

203   "fragmegration": Rosenau (1983). See also Rosenau (1990).

203   neither new nor contradictory: Barber (1995) acknowledges that "neither Jihad nor McWorld is in itself novel" (p. 5). But he turns out not to be referring to the kinds of antecedents I cite; he simply means that neither the notion of the world dissolving in chaos (Jihad) or reaching a spookily rational order (McWorld) is new.

203   tribalism also lubricated by information: Barber (1995, p. 17) recognizes the irony that, even as McWorld imposes its culture via video and other media, "the information revolution's instrumentalities are also Jihad's favored weapons." But he doesn't seem to see the information revolution as being *causal*. And he doesn't seem to see that this irony has precedent in the past. What's more, he seems to see the irony as probably transient, imagining that either Jihad or McWorld will "win" in the end. Thus the scenario I lay out in the next chapter, in which globalization and tribalism are two sides of the same coin, doesn't fit easily into his worldview.

204   "the main lines of cleavage": Eisenstein (1993), p. 155.

205   "When England sneezes": Bradford De Long, personal communication.

205   Baring Brothers: Kindleberger (1993), p. 219.

205   Hundred Years' War: Kindleberger (1993), p. 45.

207   slain foes: Drews (1993), p. 49.

207   Athens and slaughter: Starr (1991), p. 276; McNeill (1963), p. 287.

208   "as though they were plants or animals": Starr (1991), p. 403.

208   †synergy rulers can't ignore: The attendant broadening of the power base over the millennia may explain a long-term, worldwide trend: powerful males are less and less able to get away with hoarding women. An emperor in the Chou dynasty had thirty-seven wives and eighty-one concubines. Inca nobility had "houses of virgins," each stocked with hundreds of women, and if a commoner dared seduce one of them, he would be killed, along with his family. Betzig (1993), p. 37, writes of the six "pristine" civilizations: "In every one of them—in ancient Mesopotamia and Egypt, Aztec Mexico and Inca Peru, and imperial India and China—powerful men mate with hundreds of women, pass their power on to a son by one legitimate wife, and take the lives of men who get in their way." It is Betzig's belief that this sort of hoarding of women is strongly correlated with authoritarian government, in which case the world's general drift away from polygamy toward monogamy may reflect a broadening of the power base beyond a centralized elite. See also Wright (1994).

208   †tougher moral standards: Other examples of the moral superiority of the present to the past: Of all known agrarian societies, 45 percent have practiced slavery. Of all known "advanced horticultural societies," 83 percent have practiced slavery (Lenski, Nolan, and Lenski, 1995, p. 166). In the sixteenth century, when the Portuguese, one of the most advanced civilizations of the day, subdued India for commercial purposes, they severed the hands, ears, and noses of recalcitrant natives. (Palmer and Coulton, 1965, p. 91). As Palmer and Colton note (p. 556), "Torture went out of use about 1800, even in the illiberal European states; and legalized caste and slavery in the course of the nineteenth century."

15. NEW WORLD ORDER

209   epigraph: *Wall Street Journal,* January 31, 1997.

209   600,000 polities: Carneiro (1978).

209   "grid of nation-states": Kaplan (1994). Kaplan (1997) concedes that eventually solutions to these problems may be found, but he considers the coming decades almost assuredly chaotic, and he stresses that throughout history, golden ages of peace and tranquility have been few and far between.

210   Antichrist: An advertisement in the October 14, 1997, *New York Times* reads, in part: "The New World Order. Increased centralization of world financial and political power is a prelude to the soon-coming world power system in the hands of 'Antichrist,' who will be the incarnation of Satan and who will deceive most of the world (Daniel 7–12, Matthew 24:15, Revelation 13)."

210   government versus goverance: See, e.g., Rosenau and Czempiel, eds. (1992).

211   emerging supranational non-zero-sum problems: One classic text on this subject is Keohane and Nye (1989).

212   non-wild-eyed people: *New York Times,* May 11, 1998.

212   "only the most dogmatic": *New York Times,* February 12, 1998, p. D2.

212   mainstream IMF reform plan: See, e.g., Soros (1999).

212   "contingent credit line": *Wall Street Journal,* April 26, 1999. This IMF initiative was not as forceful as some had hoped; though nations would have to qualify for eligibility in advance, they would still have to pass a final review at the time of the emergency need. Thus they wouldn't have *automatically* "pre-qualified"—a status that would presumably offer a stronger incentive for a nation's participation in the program. (Barry Eichengreen, personal communication.) As this book went to press, the details of the IMF plan were still being worked out.

213 "dooming the proposal": *Wall Street Journal,* Asian online edition, April 17, 1998.

213 Italian tariff harmonization: German states did the same in the nineteenth century in moving toward unification. See Kindleberger (1993), pp. 120, 137.

214 Viagra: *New York Times,* September 24, 1998.

215 U.S. respect for WTO ruling: The U.S. appealed the ruling, and the ruling was partly affirmed. (Though the U.S. environmental law itself was deemed legitimate, the U.S. had not applied it equally to all nations.) The U.S., as of the end of 1998, had reportedly decided to abide by the ruling and permit shrimp imports from the Asian nations in question. See *The Statesman* (India), December 27, 1998.

215 food laws as trade barriers: To be sure, once food regulation is at the supranational level, interest groups in each nation may continue to try to subvert it. French farmers (for example) would still like to see Portuguese foods deemed substandard, or non-French cheeses deemed non-cheeses. But such maneuvers will now be harder to pull off. And even when a law that is nationally protectionist *is* passed at the supranational level, complying with it will be cheaper for pan-European companies than complying with a patchwork of different national laws would be.

216 outright confederacy: Elazar (1998), p. 14, classifies the EU as a confederation and the WTO as a "confederal arrangement."

216 world monetary unification: "One World, One Currency": *The Economist,* September 26–October 2, 1998. For skepticism about transnational currency unification, see Krugman (1999). For bullishness, see Beddoes (1999).

217 escaping calamity: Actually, some forms of economic non-zero-sumness have this same "pushing" property, as when the IMF makes loans to countries to avoid a global contagion of collapse. As noted earlier, the distinction between "pushing" and "pulling" forces blurs on close inspection but does have expository value.

217 a future, stricter CWC: This doesn't mean every American house will be fair game for searches initiated by distant bureaucratic busybodies in the Hague. The inspectorate could have real power yet be constrained in various ways. For example, American officials could accompany international inspectors to ensure that no evidence is planted. And there could be limits on how many surprise searches a given nation can be subjected to each year. And so on.

218 "as Christians": McNeill (1989), pp. 183–84.

219 Gaelic Channel: *New York Times,* October 21, 1996.

220 Nebraska's Original Betty: *New York Times,* April 28, 1998.

220 European Headache Federation: *Wall Street Journal,* June, 17, 1996.

220 NGOs: See Mathews (1997).

220 Rainforest Action Network: *New York Times,* May 24, 1997.

221 "tribes" and governance: See Mathews (1997), Wright (1997), and Rosenau and Czempil (1992).

221 code governing clothing factories: *New York Times,* November 5, 1998.

221 sporting goods factories: *New York Times,* December 25, 1996.

221 "Rugmark": *Washington Post,* December 8, 1996.

222 "This is a case": *New York Times,* May 15, 1999.

223 lobbying WTO: *New York Times,* December.13, 1997.

223 World Intellectual Property Organization: *New York Times,* December 21, 1996.

223 disputed islands: *Wall Street Journal,* October 18, 1996.

223 Rwandan mayor: Associated Press, August 2, 1998.

224 Irish support for EU: *Wall Street Journal,* online edition, September 18, 1998.

225 "great political body of the future": Kant (1784), p. 51.

225 Congress of Vienna: See Palmer and Colton (1965), pp. 447–48.

226 †UN's glimmers of success: The Persian Gulf War was in some ways a text-book exercise of collective security as outlined in the UN charter. The Security Council authorized the forceful rollback of Iraqi transborder aggression, and a UN-sanctioned multinational force accomplished that rollback. The intervention in Bosnia that led to the Dayton accords was also—if in a more tenuous sense—a UN-backed intervention to stem transborder aggression. (Bosnia had been recognized as a sovereign nation.)

226 †creeping toward supranational level: The ability of multinational corporations to use differing nations' tax laws to dodge taxes has long been an argument for harmonization of tax law, something that is starting to happen within the European Union. Cyberspace only adds to the logic, by making it even easier to dodge taxes. Meanwhile, the World Bank is encouraging the adoption of global-ized accounting standards by asking the "big five" accounting firms not to sign audits unless they conform to those standards (Friedman, 1999). As for policing: in addition to such incremental steps toward supranational policing as the Chemical Weapons Convention, there are improvements in data sharing and coordination among national police agencies, particularly as effected by INTERPOL. And in international environmental law, in addition to a growing number of treaties, there is growing recognition of the need for sanctions to enforce the treaties.

226 Taliban telephones: *Washington Post,* May 8, 1999, p. A13.

228 "full civil war": Service (1975), p. 158.

16. DEGREES OF FREEDOM

229 epigraph: Spencer (1851), p. 456.

230 †experiments in governance conducted in serial: This growing fragility is a product not just of world governance, but of world interdependence. In an inter-dependent world, mistakes in one nation can sink a whole world into depression. This interdependence has developed gradually, and in that sense, world history has been growing less insulated from the errors of individual leaders for some time. (As we saw in chapter 12, one scholar believes Eurasia was significantly interde-pendent even during the Mongol era.) Still, the problem has reached great extent only in the past century.

231 †Denying self-determination: This isn't to say that all these groups will *want* sovereignty. It's possible that the forces of "McWorld" will so integrate some such groups into the larger economy and culture that nationalist sentiment will fade. The point is just that those coherent groups that remain bent on sovereignty will be denied it only at some peril.

231 Med TV: *New York Times,* April 29, 1999. The occasion for revoking the license was the broadcast of interviews with people who advocated violent uprising against Turkey.

232 "as quickly as possible": Ogburn (1950), p. 201.

233 †slow material change *by* hastening adaptive change: This may seem like double counting, but in fact these are two separate effects. Suppose supranational labor accords raise Mexican wages at a factory that makes microprocessors. This can (a) slow the exodus of jobs from more developed countries; and (b) slow the rate at which the price of microprocessors drops, thus slowing the global diffusion of computer technology. The first effect is part of the adaptive change, and the second effect is a slowing of material change. (Of course, the slowed material change may in turn *further* slow the migration of jobs from one country to another, since this migration is ultimately driven by the global diffusion of technology.)

235 †the life of a cell: Actually, we may be choosing the life of a cell in more ways than one—by choosing vocations that involve a loss of liberty. Bosses increasingly have the technological capacity to monitor the activities of workers minutely. What's more, there's no getting away from the boss; it's simply not true that you can't be reached on an airplane, or that a memo can't be transmitted to you while you're on a remote island. Of course, when people "choose" to work for such bosses, they're doing so under some constraints—they need the job, and it may be increasingly hard to find jobs without these forms of oppression.

235 "To say 'love' ": Teilhard de Chardin (1969), p. 140.

236 Gandhi and King: In Gandhi's case the relatively nonviolent "transformation" I'm referring to is independence from the British Empire. Obviously, the subsequent partition of India and Pakistan (which Gandhi opposed) involved considerable violence.

238 †mindless materialism: The concern here isn't about "finite resources" per se. There is good reason to believe that world population will level off in the twenty-first century, as developing nations, having developed, adopt such bourgeois values as bearing fewer children. And, though we'll still have another few billion mouths to feed, that should be doable. The concern, rather, is with the generally unpredictable and often unpleasant environmental effects of large-scale disruption: ozone depletion, global warming, disruption of marine ecosystems, and various unpleasant surprises that are no doubt in store.

238 "no further benefits": Carneiro (1992), p. 133.

238 frugality as an affordable luxury: On various reasons to reign in rampant material acquisition, see Frank (1999).

239 "made the subject of reasoning": Boas (1940), p. 288.

17. THE COSMIC CONTEXT

243 epigraph: Mackay, ed. (1981), p. 77.

244 "inert lump of matter": Schrödinger (1967), p. 74.

244 "appears so enigmatic": Schrödinger (1967), p. 75.

245 "the essential thing": Schrödinger (1967), p. 76.

245 †folded into more complex forms: Order and complexity are not the same. See appendix 2 and Wright (1988), chapter 18.

247 †not the other way around: I don't mean that information isn't in any sense *dependent* on matter and energy. For one thing, information always *consists* of either matter (ink, say) or energy (radio waves, say). Second, whenever information (or for that matter plain old matter) is sent somewhere it depends for its propulsion on energy. But there is a difference between raw propulsion and guidance, and whenever energy or matter in a living system is *guided* to a particular place, information does the guiding. It is in charge of putting things in their place.

247   "informal headmen": Flannery (1972), p. 411.

247   "one of the main trends": Flannery (1972), p. 411.

247   cyclic AMP as a symbol: Tomkins (1975).

248   pragmatic conception of meaning: See Wright (1988), chapter 10, and Peirce (1878).

248   "microscopic discriminative . . . faculty": Monod (1971), p. 46.

248   "appear to escape the fate": Monod (1971), p. 59.

250   "we do not know": White (1959), p. 206. Clearly, White had failed to update his thinking in light of the birth during the 1950s of molecular biology.

18. THE RISE OF BIOLOGICAL NON-ZERO-SUMNESS

251   epigraph: Ridley (1996), p. 17.

252   "subservient to the needs": Dulbecco (1987), p. 54.

252   "encapsulated slaves": Maynard Smith and Szathmáry (1995), p. 141.

254   Margulis: For a taste of Margulis's theory, see chapter 5 of Margulis and Sagan (1995). Her treatise on the subject is called *Symbiosis in Cell Evolution.*

255   raised cell's efficiency: This is the surmise of Maynard Smith and Szathmáry (1995), p. 138. Some biologists might speculate that this genetic migration served the Darwinian interest of the nucleus more than that of the mitochondrion. (As we'll see later in this chapter, there is *some* divergence of interest between the two.) On the other hand, one can also imagine the prime beneficiary being the genes remaining in the mitochondrion. Maybe the mitochondrion, by jettisoning a few genes, became a leaner and meaner reproductive machine; thus "the mitochondria that lost genes to the nucleus are the ones that out-reproduced their siblings, and survive today." (James Gould, personal communication.) In any event, this whole category of analysis seems alien to Dulbecco; he apparently doesn't see that in order to make meaningful judgments about "subservience," we need to talk about comparative Darwinian interests, and not just the mechanics of control.

255   "metabolic exploitation": Maynard Smith and Szathmáry (1995), p. 141.

256   "restorer" genes: See Frank (1989).

256   †leaping to conclusions of tension: Maynard Smith and Szathmáry, unlike Dulbecco, do at times recognize that the question of "exploitation" hinges ultimately on Darwinian interest. Thus, they suggest (p. 141), if there were a free-living variety of mitochondrion, then we could compare its reproductive rate with the reproductive rate of encapsulated mitochondria to see whether mitochondrial DNA benefits from cellular encapsulation. But really we couldn't. For a free-living variety, in order to be adapted to free life, would have appreciably different DNA. Those genes peculiar to the encapsulated version obviously benefit from encapsulation (without which they presumably would not exist), and those genes peculiar to the non-encapsulated version would obviously have benefited from freedom (ditto). To be sure, we could ask whether *those genes common to* the encapsulated and free-living variants were better off in encapsulated form. But the way to answer *that* question would be by retrospective analysis: Go back and find the fork in the road at which some of the mitochondria's ancestors became encapsulated and others didn't, and compare the relative fruitfulness of the two paths—compare the number of free-living descendants (which for all we know do still exist in much-altered form) with the number of encapsulated descendants. This isn't a practical analysis, of course, but it at least makes sense in principle.

In short: Notwithstanding the convenient shorthand that we all use when talking about "who's exploiting whom," it seems to me ultimately meaningless to ask whether "the mitochondrion" is being exploited, since the answer may differ for different mitochondrial genes. For some genes the answer is, obviously not—they owe their existence to encapsulation—and for other genes the answer, though not practically obtainable, is a historical question. And, as I've suggested in the text, if the answer to that historical question is indeed the one that Maynard Smith and Szathmáry imply by comparing mitochondria to captive pigs, then their conclusion that mitochondria are exploited is assuredly wrong; pig-human symbiosis, in Darwinian terms, is mutualistic, not parasitic.

258   cellular slime mold: Bonner (1993), pp. 3–5.

259   relatedness of nearby slime mold cells: Bonner (1988), p. 163. See also Maynard Smith and Szathmáry (1995), p. 214. As they note, the chances that a cell is genetically identical to a neighbor are less than 100 percent. Still, kin selection can operate amid uncertainty, so long as the average degree of relatedness is high.

259   "vital force": See Bonner (1993), p. 13.

259   proteins sent to mitochondrion: Dulbecco (1987), p. 54.

260   †communication within *E. coli:* See Tomkins (1975). The communication also involves other elements in what turns out to be a roundabout process. The genes that I've said are responsible for "sensing" the carbon actually create an enzyme that will do the sensing; the enzyme automatically builds cyclic AMP molecules *unless* inhibited by such carbon-containing molecules as sugars. Thus in the absence of carbon, the enzyme builds a cyclic AMP molecule, which then attaches itself to a protein, and the resulting complex attaches itself to the DNA, inducing the construction of the flagellum. In the end, then, the "communication" between the genes for sensing a carbon shortage and the genes for building the flagellum has been mediated by a third party; the former genes in effect build a "proxy sensor" and charge it with the task of executing the communication when appropriate. This illustrate that, in breaching the "information barrier" to non-zero-sum gain, players needn't always directly communicate. Sometimes, for example, people coordinate their behavior not by explicitly communicating about coordination, but by drawing on past knowledge about the likely behavior of each other. Here the information has been exchanged well in advance—and, in some cases it may have been exchanged not via active communication, but via passive observation.

260   †biological "trust" technologies: "Trust" is important not just between cells but within them. Remember that odd reproductive fact about your mitochondria—that the DNA is always passed on by the mother? Some biologists think this "uniparental inheritance" is a way of preventing conflict among mitochondria. Conventional, biparental inheritance might mean that the various mitochondria in each cell were not genetically identical. And this lack of complete kinship could lead to conflict among them over whose genes get into the next generation. Such conflict might be so bad for the organism that natural selection weighed against it—and thus in favor of conflict-avoidance mechanisms, such as uniparental inheritance. (Steve Frank, personal communication.) For more on anti-cheating devices of both a biological and cultural sort, see Ridley (1996). On the segregation of the germ line (discussed by Ridley), see also Michod (1997).

261   degree of altruism and degree of relatedness: Note that sometimes kin selection doesn't evolve, even though close relatives live near one another, because the required fortuitous mutation doesn't happen.

263   knotty technical problems: See Maynard Smith and Szathmáry (1995) for a good rundown of the major thresholds crossed by evolution, and a good sense for how many logistical feats were required to cross them.

263   "Is Nature Motherly?": Sapp (1994).

263   Dolphin coalitions: See Ridley (1996), pp. 160–63.

264   "So careless of the single life": from *In Memoriam*. In the next part of the poem, having reflected on the many extinct species, Tennyson took back the part about nature being "careful of the type."

19. WHY LIFE IS SO COMPLEX

265   epigraph: Bonner (1988), pp. 5–6.

265   bombardier beetle: Keeton and Gould (1986), p. 869.

266   social Darwinism: See Hofstadter (1955).

266   †confusions underlying social Darwinism: Among the obvious problems with this variant of the logic of social Darwinism is the key assumption that nature is entirely the handiwork of a good God, notwithstanding the much-discussed "problem of evil." (See final chapter.) In a more general, abstract sense, the fallacy underlying social Darwinism is known as the "naturalistic fallacy"—inferring "ought" (moral prescription) from "is" (objective description).

267   "that progress defines history": Gould (1996), p. 4.

267   "inherently progressive process": Gould (1996), pp. 18–20.

267   "regrown under similar conditions": Gould (1996), p. 18.

267   "extremely small": Gould (1996), p. 214.

267   "the most complex creature": Gould (1996), p. 148.

267   species getting less complex: For example, species (such as bats) may lose their eyes in the course of adapting to a new, lightless habitat such as a cave. Interestingly, most examples of declining complexity seem to come from parasites, such as single-celled stomach parasites, or ants that parasitize the nests of other ants and can't survive on their own. See Keeton and Gould (1986), p. 880. In these cases, the decline in complexity is only feasible because of the existence of some more complex organism that can serve as host.

268   "On any possible, reasonable": Gould (1996), p. 176.

268   "myopic focus": Gould (1996), p. 168.

268   Gould's pet peeve: See Gould (1989) and my review of it—Wright (1990).

269   "vaunted progress": Gould (1996), p. 173.

270   "eat the beetles": Keeton and Gould (1986), p. 869.

270   predator-prey brain growth: Jerison (1973), pp. 320–39. See also Mark Ridley (1993), pp. 570–71; Dawkins (1986), p. 191.

271   chimp colony arms races: See, e.g., de Waal (1982).

271   "effectively random": Gould (1996), pp. 139–40.

271   †organisms as environments: Gould subsequently evinces awareness of this argument—that part of an organism's environment may consist of other organisms, and that this situation could drive a growth in complexity. But, rather than confront the argument—either dispose of it or concede its power—he sidesteps it by treating it as a curiosity of intellectual history. This allows him to obliquely (and misleadingly) cast aspersions on the argument without ever taking an explicit stance on it. Indeed, this passage—pp. 142–44 of Gould (1996)—is well worth reading; it is a masterpiece of evasion. Gould begins by noting that Darwin made a distinction between a species' "struggle" against its inanimate environment ("abiotic" competition) and against its animate environment ("biotic" competition) and argued that the latter dynamic could impart directionality to evolution. Then (here come the aspersions) Gould tries to make it sound as if two uncertainties militate against Darwin's argument. First, Gould writes as if the logic holds only if "biotic competition is much more important than abiotic competion"— whereas, in fact, any degree of biotic competition whatsoever could sustain some degree of directionality in the evolution of the species involved (and, in any event, biotic competition probably *has* been more important than abiotic). Second, Gould claims the world would have to be "persistently full" of species for biotic competition to sponsor directional evolution—and again the claim is logically false; isolated instances of interspecies crowding could bring enough arms races for *some* degree of directionality to ensue. (And, once again, the claim's falseness is in a sense irrelevant; the world *does* seem to have been pretty damn full of life for a long time now.) Then finally, perhaps aware that he hasn't provided any valid reason to doubt Darwin's argument, Gould admits that the argument isn't necessarily implausible—and then, before the significance of this concession has time to sink in, he changes the subject to the pressing question of Darwin's political views and cultural milieu. Here is the passage: "I do not say that any obvious error pervades tne logic of this argument, but we do need to inquire why Darwin bothered, and why the issue seemed important to him." Gould then embarks on a psychoanalysis of Darwin: "I do feel his strained and uncomfortable argument for progress arises from a conflict between two of his beings—the intellectual radical and the cultural conservative."

Let me now summarize the situation: Gould (a) writes a book whose central thesis is that non-random directionality doesn't exist; (b) defends his thesis on grounds that the environment of evolution is "effectively random" over time; (c) concedes that no less an authority on evolution than Charles Darwin took a different view of the matter; (d) employing flagrant illogic, tries to obliquely cast doubt on Darwin's argument; (e) in a brief fit of candor, concedes that Darwin's argument has no obvious flaw; then (f) rather than do what is obviously in order— either directly confront the argument and provide plausible and explicit grounds for rejecting it, or retract his earlier assertion that evolution's environment is "effectively random"—he changes the subject in mid-sentence, switching to a claim that the person *making* the argument was motivated by unconscious biases. This is vintage Gould.

272   human brain expansion: Cronin et al. (1981). The authors argue that the fossil record shows growth in cranial capacity to be a case of "gradual change with periods of varying rates of evolution," and not a case where periods of stasis are punctuated by rapid change.

272   †most serendipitous drunken walk: By the way, it won't do to try to save Gould's model of evolution by depicting *Homo habilis* and other human ancestors as the drunk bouncing off a wall. Yes, there's a wall at "zero complexity," but there isn't one at "*Homo-habilis*-level complexity." A Gouldian might be tempted to posit the existence of a wall of a different sort—to say that the niches just below our ancestors' level of complexity were "occupied," whereas the niches above their level of complexity weren't. And it's true that our ancestors do seem to have sometimes lived alongside species that were a bit less cognitively sophisticated—a bit less complex—than they. But there is reason to believe that when this happened, our ancestors tended to extinguish these rivals (directly or indirectly), leaving no wall to bounce off. (See, e.g., Pinker, 1997, p. 200.) If evolution were just a matter of species drifting away from walls, you might expect that these less complex species would have evolved toward even lower complexity. But instead they died out. Why? Probably because they were involved in an arms race, and they weren't well equipped.

273   Gould's interpretation challenged: See, e.g., Fortey (1998), pp. 94–98, and Morris (1998).

274   skirting the intelligent-life question: In *Wonderful Life,* Gould answered the question of whether the evolution of some form of self-conscious intelligence was highly likely from the beginning, "I simply do not know." In *Full House* (p. 214) Gould says the chances that evolution, if started anew, would produce some form of self-conscious intelligence are "extremely small." It's odd that Gould's stance against the evolution of great intelligence should have become firmer between these two books, for his stance against the evolution of complexity had gotten weaker. He went from refusing to admit that evolution evinces movement toward complexity to acknowledging the movement but calling it random drift. One might ask: If indeed this random drift accounts (in Gould's view, at least) for the great diversity of fairly complex, fairly intelligent nonhuman species, why should we think that, given long enough, it wouldn't push one of them up to a roughly human level of intelligence?

275   [†]sensory technologies: My focus on information technologies isn't meant to imply that other technologies have been neglected by natural selection. For example, natural selection has invented lots of different kinds of antifreeze for the sake of animals in chilly climes. In one case, it even happened on the same antifreeze formula—the amino acids threonine, alanine, and proline—on two separate occasions. It invented the stuff for fish in the Antarctic more than 7 million years ago, and then, millions of years later, for cod in the Arctic. See *New York Times,* December 15, 1998.

275   "a particular biological property": Morris (1998), p. 139. The same point was made, also in the course of critiquing Gould's analysis, in Wright (1990).

276   "luck of the draw": Gould (1996), p. 175.

276   2 *billion* years: Maynard Smith and Szathmáry (1995), p. 145, in a fairly conventional estimate, say the oldest prokaryote fossils are 3.5 billion years old and the oldest eukaryote fossils 1.5 billion years old. But, as this book was going to press, a group of Australian scientists had reported evidence suggesting some type of eukaryotic cell could have existed as early as 2.7 billion years ago. See *New York Times,* August 13, 1999.

277   photophosphorylation: See Keeton and Gould (1986), pp. 1007–14.

277   Gooey mats: Fortey (1998), p. 58.

278   scarlet gilia, orchids: Keeton and Gould (1986), pp. 867–68.

278   bee tongues: Keeton and Gould (1986), p. 867.

278   threshold to multicelled animals: Various people have argued that multicelled animals may have existed long before the immediately pre-Cambrian period. See, e.g., *New York Times,* October 25, 1996. And note, in any event, that very crude multicellular *plants* show up long before the Cambrian.

279 trilobite's tracks: Fortey (1998), p. 92.

279 natural selection's ingenuity: See Maynard Smith and Szathmáry (1995) for an account that emphasizes the difficulties involved in life's passage through various thresholds of complexity but, at the same time, often concludes that such passage was not highly improbable, given the way natural selection works.

280 independent inventions of multicellular life: Whittaker (1969), for example, estimates that multicellular life evolved more than 10 times. Other estimates run as high as 17, according to Richard Michod (personal communication).

20. THE LAST ADAPTATION

282 epigraph: Huxley (1959), p. 13.

282 "drier, less forested": Hayden (1998). This view is "almost universal," says Hayden.

283 near-tripling of cranial capacity: Cronin et al. (1981). On co-evolution see also Wilson (1978). The identity of the first stone-tool-using hominid has undergone revision. This hominid was once thought to be *Homo habilis,* which lived around 2 million years ago, but now evidence of stone-tool use extends back to 2.5 million years. One candidate for user of these early tools is a recently discovered species labeled *Australopithecus garhi,* but that label may be misleading, as the species seems to have been about intermediate in development—including cranial development—between the most recent previously discovered australopithecines and the earliest members of the *Homo* genus. (See *Time,* August 23, 1999, pp. 50–58).

284 sharp "sexual dimorphism": See Wright (1994). The quantitative index of polygyny is the degree of variation in reproductive success among males.

285 †the favor of females means more progeny: This isn't to say that the males would have to be conscious of the reproductive payoff—any more than rams or peacocks are. It is just to say that the males might be genetically inclined to warm up to tools and techniques that helped them reach goals that were conducive to the spreading of their genes—such as gaining sexual access to females.

285 human and nonhuman primate sex-for-meat swaps: Pinker (1997), p. 197.

285 hunting, social status, and sex: See Wright (1994).

285 thick skulls: Ridley (1996), p. 165.

285 elephants: Bonner (1980), p. 177.

286 titmouse: Bonner (1980), p. 183.

286 baboons and tainted fruit: Nishida (1987), p. 472.

287 bears: See Gould (1982), p. 23.

287 chimpanzee feats: See Wilson (1975), p. 173.

287 "wiped her foot": Wilson (1975), p. 173.

287 chimps and water fountain: Bonner (1980), p. 172.

288 †language is a must: Among the culture-boosting properties of language is the way it aids the intellectual division of labor, turning society into a social brain that brings good ideas together, across space and time, to build even better ideas. The social brain may not be very complex at first, but eventually it can reach a point where the actual *invention* of a water fountain is possible. Or the discovery of laws of physics. Newton crystallized the collaborative nature of knowledge: "If I have seen farther than others, it is because I stand on the shoulders of giants."

288   bird warning calls: Wilson (1975), p. 183.

288   squeaking "Help!": Wilson (1975), p. 211.

288   vervet monkey vocabulary: Seyfarth (1987), p. 444.

288   young vervet learning: Nishida (1987), p. 473.

288   ten to forty messages: Wilson (1975).

289   neocortex and large social groups: Dunbar (1992).

289   Vampire bats: Ridley (1996), p. 69.

290   coalition-forming and intelligence: Ridley (1996), p. 160.

290   "feasted in peace": Pinker (1997), p. 193.

290   "mind blind": Pinker (1997), pp. 331–33.

290   †our evolved *social* intelligence: The properties of a specifically social intelligence may have helped shape the nature of science. The tendency of the scientific mind to distinguish between "cause" and "effect" may indeed, as the mystics say, misrepresent the ultimate nature of a seamless reality, attributing independent agency to parts of an interdependent whole. But this tendency has certainly had its practical payoffs. And it may owe much to our instinct for social analysis—thinking about who has caused whom to do what to whom, and why. The Darwinian importance of answering such questions probably shaped the human penchant for formulating theories about cause and effect, assessing them in light of new evidence, and refining them.

291   collaborative hunting by chimps: Wrangham and Peterson (1996), p. 216.

291   "even hit him": de Waal (1982), p. 207.

292   bonobos: Though bonobos have reciprocal altruism, and females use this faculty to sustain coalitions, the coalitional dimension seems to be generally weaker in bonobos than in chimps. See Wrangham and Peterson (1996).

292   chimps and bonobos "held back": James Gould (personal communication).

292   macaques on islands: Bonner (1980), p. 184.

293   dinosaur brains: Wills (1993), pp. 265–66.

293   "panda's true thumb": Gould (1982), p. 24.

293   "for other purposes": Gould (1982), p. 20.

294   "improbable foundations": Gould (1982), p. 24.

294   reciprocal altruism: For an overview, see Dugatkin (1997). The evidence that mutual grooming among impalas qualifies as reciprocal altruism isn't definitive but, as Dugatkin argues (pp. 90–94), is strong. Cooperative hunting in hawks (pp. 77–78) and other birds may also be reciprocally altruistic.

295   dolphin air art: Marten et al. (1996).

297   "Every epic": Wilson (1978), p. 203. See Rue (1999), Goodenough (1998), Barlow (1997), and Swimme and Berry (1994) for examples of evolutionary epics in one sense or another.

21. NON-CRAZY QUESTIONS

301   "brain of brains": Teilhard de Chardin (1969), p. 173.

301   Teilhard's tensions with the church: Hefner (1970), pp. 13–14.

302   †superorganisms and fascism: The stigma associated with comparisons between societies and organisms has, to some extent, endured to this day. There are several problems with the logic behind this stigma.

First, it should go without saying that to compare a human society to *any-thing*—an organism, an ecosystem, a grandfather clock—is not to advocate that we instill all the properties of that thing in that society. We are describing, not prescribing.

Second, any attempt to attribute fascist or Nazi beliefs to people who make the society-organism comparison has to contend with the inconvenient fact that it isn't only fascists and Nazis who make the comparison. Leslie White, for example, was a Marxist. Of course, Marxism, in some real-world manifestations, is totalitarian, so you might argue that here, too, we see organism-society comparisons giving aid and comfort to those who would build tight, oppressive societies.

Unfortunately for this argument, another big champion of organism-society comparisons was Herbert Spencer, who liked very loose societies—laissez-faire. (See Morris, 1987, p. 96.) Further, one recent wave of enthusiasm for organism-society comparisons comes from Silicon Valley libertarians. They see that nature, lacking a central government, nonetheless produces elegant structure through evolution, and infer that centralized control is often superfluous. (See, e.g., Kelly, 1995.)

Speaking as someone who is neither a Nazi, a fascist, a Marxist, a fan of laissez-faire government, or a libertarian, but who does like comparing societies to organisms, I propose that we just get on with the comparisons and quit trying vainly to stereotype them. The question should be: How intellectually productive are they?

303 "be called an organism": Wilson (1975), p. 379.

303 Portuguese man-of-war: Wilson (1975), pp. 383–84; Ridley (1996), p. 15; Bonner (1988), p. 119.

304 "segregation distorter": Ridley (1996), pp. 32–33; Crow (1999). The gene stacks the deck by destroying sperm that don't contain a copy of it, thus ending their chances of making it into the gamete that will sail to the next generation.

304 B chromosomes: Ridley (1996), pp. 31–32.

304 "no such thing": Ridley (1996), pp. 15, 33.

305 superorganism's comeback: See Queller and Strassman (1998). See Wilson (1971) for a discussion of Wheeler's superorganism concept. For a biologist's view of the sense in which it may be legitimate to speak of a "group mind" in both insect societies and human societies, see D. S. Wilson (1997), pp. S128–S133. See also Kelly (1995).

305 [†]interdependence in ants: There is one other form of dependence that might sharply distinguish ant societies from human societies—reproductive dependence. Most of the ants you see have no chance of being parents. They are sterile, and their only hope of getting genes into the next generation lies with their more privileged relatives, the tiny reproductive elite. That is why worker ants slave so tirelessly on behalf of these pampered few. The same can be said of our skin cells, which, lacking reproductive power, sacrifice on behalf of their fertile relatives, the germ cells. But the same cannot be said of entire human beings; any young man or woman carries sperm or eggs. Because of this reproductive autonomy, our biological evolution has not inclined us to efface ourselves for the sake of an elite. So human society doesn't naturally possess the authoritarian cohesion of an ant society.

Still, as the biologist David Queller has pointed out, the reproductive independence of humans doesn't by itself deny them superorganism status. As we've seen, a eukaryotic cell comprises organelles that take separate genetic routes to the next generation, yet nobody accuses it of being a mere society. (Queller, personal communication. See Queller, 1997.)

Moreover, even if you insisted that reproductive independence *is* disqualification from superorganism status, there would remain a (fairly tenuous but interesting) rationale for not disqualifying the human species on these grounds. See the explanatory note below labeled "ironing out kinks in the Crickian meta–natural-selection scenario."

306 †underlies mainstream behavioral science: I'm not saying that the "epiphenomenal" view of consciousness I describe in this chapter is often made explicit in the behavioral sciences. But it is often implicit in assumptions governing research in the behavioral sciences, such as the assumption that all forces that influence behavior are physical. To be sure, when the epiphenomenal view of consciousness in made explicit, some people who consider themselves mainstream behavioral scientists take issue with it. Certainly there are psychology professors at major universities who will say they see "mental" forces, and not just physical forces, as causal. But if that's true, then they can't call themselves behavioral *scientists* in the strict sense, since truly scientific models must invoke only "publicly observable" causal phenomena (neurotransmitters, etc.).

307 †all causality is physical: To be sure, behavioral scientists may talk as if subjective states had causal effect. In fact, I myself, in discussing evolutionary psychology, have written that "the function of anxiety" or "the function of love" is to cause such and such behavior. But this is just a shorthand, an efficient way of talking. Strictly speaking, when I write such things, I'm referring to the function of the biochemical information flow that gives rise to the anxiety or the love. As observed in the previous note, scholars who speak *literally* of things like anxiety having causal effect aren't behavioral *scientists* in the strict sense.

307 Nagel: See Nagel (1974).

308 †Dennett: Actually, Dennett and like-minded scholars might claim that consciousness does have a function. But they mean this in a highly restricted, even (to me) incomprehensible sense. By their definition, consciousness is *nothing more than* the informational processes underlying it. They don't just mean that consciousness is entirely a product of those processes (something I and other epiphenomenalists assume to be true) but rather that consciousness is *identical* to those processes. And, since those processes clearly have a function, then consciousness itself, being identical to them, must have a function. In my view, the problem here is with the claim that consciousness is "identical" to physical brain states. The more Dennett et al. try to explain to me what they mean by this, the more convinced I become that what they really believe is that consciousness doesn't exist. This position, associated with behaviorism, was common several decades ago, and is not without its merits. One of these merits is its simple clarity: it holds that consciousness doesn't exist and therefore has no function. Still, it is a position that makes no sense to me.

308 †robots without sentience: You may object that perhaps these machines *do* have sentience. That's my point. For all we know, they do—which means this issue is superfluous to the issue of how they function behaviorally, which we understand entirely by reference to the strictly physical parts of the machines.

308 Chalmers: Chalmers (1996). A similar view is found in Wright (1988), part 5.

309 Teilhard and sentient bacteria: Actually, Teilhard believed that even inorganic matter had some tiny increment of consciousness; the whole physical world, he believed, had an "outer" manifestation—the physical part—and an "inner" manifestation. And the "inner" manifestation—consciousness, sentience—deepened as outer complexity grew. All of this meshes with his view that "complexification" was an evolutionary trend predating organic evolution, manifested in the evolution of the universe from an early homogenous state into a complex system of galaxies.

309 " 'etherized' universal consciousness": Teilhard de Chardin (1969), p. 174.

309 "suddenly completed": Teilhard de Chardin (1969), p. 182.

310 Paley: quoted in Dawkins (1986), p. 5.

312 "persistence towards the goal": Braithwaite (1953), p. 329. See Beckner (1967).

312 purpose and information processing: Rosenblueth, Wiener, and Bigelow (1943) assert "all purposeful behavior may be considered to require negative feedback." One thing I like about information processing being a prerequisite for purposive behavior is that, as we've seen, information processing may be the foundation of consciousness; and the possibility of some sort of "metaphysical" link between consciousness and purposive behavior is intriguing and in some ways (to me, at least) intuitively plausible.

314 †Teilhard's scientific vision: There is a tendency to dismiss Teilhard de Chardin's view of life as one giant heap of muddle: he had muddled ideas about how natural selection works, and the muddled notion that it has purpose. Daniel Dennett, for example, has written (1995, p. 320): "The problem with Teilhard's vision is simple. He emphatically denied the fundamental idea: that evolution is a mindless, purposeless, algorithmic process." But that's not a single "problem"— it's at least two separate issues, and only one of them is demonstrably problematic. Yes, Teilhard was muddled in thinking that natural selection works in a mushy, mystical way, instead of a nuts and bolts, algorithmic way. But that doesn't mean he was wrong to say evolution has purpose. A process can work in a nuts and bolts, algorithmic way and still serve a larger purpose. The growth of an organism is a perfect example—a nuts and bolts process subordinate to the goal that animals are designed to pursue: genetic proliferation.

314 Dobzhansky: Dobzhansky (1968), pp. 248–49.

314 "natural selection is a process": Dobzhansky (1968), p. 248.

316 †ironing out kinks in the Crickian meta–natural-selection scenario: One objection to seeing biological evolution as the unfolding of a planet-wide superorganism is that organic evolution—the supposed maturation of our meta-organism, the biosphere—is a wild, hurly-burly, unpredictable process. Aside from a general growth in complexity and intelligence, it is quite unpredictable. The maturation of *real* organisms, in contrast, is orderly—a tidy unfolding of a genetic blueprint.

In truth, though, genetic "blueprint" is not a very apt word, and biologists tend not to use it. One reason is that the maturation of an organism is a fairly contingent, open-ended process, full of if–then junctures. At every stage, and especially early, environmental influences matter. (That's why maternal nutrition can be pivotal during gestation.) Indeed, according to the theory of "neural Darwinism," organic structure can depend heavily on a process quite like natural selection; neural connections proliferate randomly, and only the adaptive survive. All told, if you viewed an unfolding organism from as close a range as you view the biosphere, you might well remark to yourself: It's a jungle in there.

Another possible objection to the notion of meta–natural selection gets back to our wondering, at the beginning of this chapter, whether the human species could be an organism. One source of reluctance to say yes (articulated above, in explanatory note labeled "Teilhard's scientific vision") had to do with the reproductive independence of human beings. In a real organism—a dog, say—almost none of the cells have this sort of autonomy; the skin cells and brain cells and kidney cells are all sterile, and are programmed to sacrifice on behalf of the elite sperm or egg cells, whose genes the sterile cells share. Well (we asked above), since human beings *aren't* sterile, how can we attribute a truly organic unity to human society?

From a cosmic, Crickian vantage point, this question disappears. Once you view the human species as a mere organ in a larger superorganism—the biosphere's superbrain—individual human beings *don't* have reproductive autonomy. For, in this view, when a human being procreates, he or she is just making more brain cells. And, of course, even a dog's "sterile" brain cells get to do *that* sort of procreation during the maturation of the organism. It's the procreation of the whole dog that an individual brain cell doesn't contribute any genetic information to. Analogously, it's the procreation of our biosphere—the launching of that rocket full of bacteria—that, in a Crickian scenario, human beings wouldn't contribute any genetic information to. People get to build the rocket, yes, but not ride in it.

Actually, in one sense, viewing individual people as brain cells only leads to more doubts about the meta–natural-selection scenario. After all, in an ordinary organism, don't all the different cells—skin cell, brain cell, muscle cell, germ cell—have the *same* genetic information? Indeed, aren't these cells united precisely by the goal of propelling their *common* genetic heritage—the distinctive genetic information of their organism—into posterity via the germ cells? If so, then how can you refer to people and bacteria as being cells in a superorganism, given the vast difference in their genetic information?

The only answer I can imagine is this: In meta-evolution, the essence of the information being propelled into posterity by rocket isn't the particular genotype of a bacterium. You could just as well have sent an amoeba instead; a premise of Crick's scenario (and of this book) is that whatever particular tiny creature you start out with, natural selection stands a good chance of eventually producing lots of species, including a very smart one. Rather, the essence of the information being sent—the thing that is truly essential to natural selection's getting started—is the genetic *code,* the code that underlies the biosphere on the planet in question. On our planet, we use a code of four bases; they are the "letters" in our genetic alphabet. Maybe that is a better code than some other codes, more likely to flourish on a distant planet. Maybe not. But, in any event, this is what is shared by all organisms on earth; this is the planet's common genetic heritage. And this is what, if we opted for panspermia, we would be propelling into the next metageneration.

As for the objection cited in the text—that there hasn't been enough time during the life of this universe for meta–natural selection to do much work—I suppose you could always concoct scenarios involving the births of new universes (via black holes or some such), with the birthing process miraculously leaving key organic molecules from the parent universe intact. But here we threaten to exceed even my tolerance for wild speculation.

22. YOU CALL THIS A GOD?

318   Hegel: Tucker, ed. (1978), p. xxi, and Williamson (1984), p. 168.

318   Bergson: Goudge (1967), p. 293.

319   a benign, almighty God: Actually, the biblical phrase "God almighty" is a Greek mistranslation of a Hebrew phrase that meant something closer to "God of the mountain" (Oden, 1995).

319   Zoroastrians: As it happens, the evil spirit is the son of the good God, Ahura Mazda, who had twin sons, one of whom is a good spirit and one of whom is a bad spirit (Angra Mainyu).

320   design constraints: My premise that the creator had good intentions is prominent in Christianity but is far from being universally shared. A. N. Wilson (1997, p. 58) notes how few religious traditions view the first cause as wholly benign. In the Christian tradition, Marcion is one of the few thinkers to have seen the creator as evil.

321   †a world in which "right" and "wrong" made sense: This isn't to say these robots wouldn't *use* such terms as "right" and "wrong." People often use moralistic language in pursuing their self-interest, and presumably natural selection would have left these insentient robots with the same tendency.

321   "to put consciousness in": *Time,* March 25, 1996.

323   †awe-inspiring reality: Actually, if the weirdness of consciousness, and the specialness of consciousness, were more widely appreciated, that might also cut down on the number of people who consider evolution directionless. As you may recall, one of the grounds for denying progress to biological evolution is the alleged absence of a gauge by which to measure progress. After all (the argument goes), one can imagine any number of measures by which to rank organisms. As one scholar has put it, the measure could, on the one hand, be the organism's "ability to synthesize its own biological material from inorganic sources"—in which case simpler life forms would rank above complex ones. Or, on the other hand, the measure could be "ability to gather and process information about the environment." Who's to say? "There is, of course, no reason to prefer one criterion over another." (Mokyr, 1992, p. 288.) That's one opinion. Personally, I'd go with information processing—not because it's what my own species is good at, but because, of all objective criteria for ranking, it is the only one correlated with meaning in life.

325   "between-group enmity": See Alexander (1979).

325   conservation of antipathy: Wright (1989a).

327   "will find its *heart*": Teilhard de Chardin (1969), p. 184.

329   †religious, national, ethnic boundaries: A big question is whether boundaries of social class will be so easily crossed—or whether, on the other hand, differences of social class within a society might sharpen as people invest more of their energy in virtual communities consisting of like-minded people.

330   God becoming nicer: See Easterbrook (1998) on the theme of the biblical God undergoing a kind of moral education.

330   "Should I not pity Nineveh": Jonah 4:11.

330   "the soul of religion": Durkheim (1912), p. 191.

330   "In the last analysis": Aron (1967), vol. 2, p. 44.

332   "less Alpha than Omega": Haught (1999).

333   "point," "purpose": Robinson (1997).

333   Philo as the link between *Logos* and Christianity: Latourette (1975), vol. 1, p. 141.

333   "divine *Logos*": See Wolfson (1967), p. 154.

333   "The whole world": See Wolfson (1967), p. 154.

334   "consummation of the whole world": Philo (1927), pp. 73–75.

APPENDIX 2. WHAT IS SOCIAL COMPLEXITY?

344   "Culture develops when": White (1943) pp. 338, 346.

346   Anglo-Saxon England: Carneiro (1968).

347   Naroll's and Carneiro's measures compared: Carneiro (1969). The coefficient of correlation was .891.

347   †"levels of hierarchical control": Scholars differ over the extent to which hierarchy is inherent in, and proportional to, complexity. For example, some management theorists have recently argued that in a modern technological milieu, corporations can grow more efficient by becoming *less* hierarchical, as decision-making power is moved from higher levels in the hierarchy to lower levels; and as, in some cases, whole levels in the management hierarchy are eliminated. On the other hand, sometimes these levels of hierarchy are "eliminated" through automation; so the level of control still exists; it is just occupied by a computer instead of a human.

347   "cell types": Bonner (1988).

347   no consensus on the essence of complexity: See Rothman (1994) on various approaches to measuring social complexity. See Wright (1988), chapter 18, on attempts by physicists to define complexity.

BIBLIOGRAPHY

Abu-Lughod, Janet L. (1989) *Before European Hegemony: The World System*, A.D. *1250–1350*. Oxford University Press.

Acheson, James M. (1989) "Management of Common-Property Resources," in Plattner, ed. (1989).

Adams, Richard E. W. (1997) *Ancient Civilizations of the New World*. Westview Press.

Adams, Robert McC. (1966) *The Evolution of Urban Society*. Aldine.

——— (1996) *Paths of Fire: An Anthropologist's Inquiry into Western Technology*. Princeton University Press.

Albrecht-Carrie, René (1981) "Liberation Movements in Europe," in Garraty and Gay, eds. (1981).

Alexander, Richard (1979) *Darwinism and Human Affairs*. University of Washington Press.

Allardyce, Gilbert (1990) "Toward World History: American Historians and the Coming of the World History Course." *Journal of World History* 1:23–76.

Alston, Richard (1998) *Aspects of Roman History, A.D. 14–117*. Routledge.

Anderson, Benedict (1991) *Imagined Communities: Reflections on the Origin and Spread of Nationalism*. Verso.

Armstrong, Karen (1994) *A History of God*. Ballantine Books.

Arnold, Jeanne (1996) "The Archaeology of Complex Hunter-Gatherers." *Journal of Archaeological Method and Theory* 3:77–126.

Aron, Raymond (1967) *Main Currents in Sociological Thought*. Basic Books.

Augustine (1958) *City of God*. Image Books.

Axelrod, Robert (1984) *The Evolution of Cooperation*. Basic Books.

——— (1987) "Laws of Life." *The Sciences* 27:44–51.

Bagley, Robert (1999) "Shang Archaeology," in Michael Loewe and Edward L. Shaughnessy, eds., *The Cambridge History of Ancient China*. Cambridge University Press.

Baines, John (1988) "Literacy, Social Organization, and the Archaeological Record: The Case of Early Egypt," in Gledhill, Bender, and Larsen, eds. (1995).

Barber, Benjamin (1995) *Jihad vs. McWorld*. Times Books.

Barlow, Connie (1997) *Green Space, Green Time*. Copernicus.

Beckner, Morton (1967)   "Teleology," in Edwards, ed. (1967).

Beddoes, Zanny Minton (1999)   "From EMU to AMU." *Foreign Affairs,* July–August, pp. 8–13.

Bellah, Robert N., ed. (1973)   *Emile Durkheim: On Morality and Society.* University of Chicago Press.

Bender, Barbara (1985)   "Emergent Tribal Formations in the American Midcontinent." *American Antiquity* 50:52–62.

Benedict, Ruth (1959)   *Patterns of Culture.* Houghton Mifflin Sentry. (Originally published, 1934.)

Beniger, James (1986)   *The Control Revolution: Technological and Economic Origins of the Information Society.* Harvard University Press.

Berdan, Frances (1989)   "Trade and Markets in Precapitalist States," in Plattner, ed. (1989).

Berlev, Oleg (1997)   "Bureaucrats," in Sergio Donadoni, ed. (1997), *The Egyptians.* University of Chicago Press.

Berlin, Isaiah (1954)   *Historical Inevitability.* Oxford University Press.

Betzig, Laura (1993)   "Sex, Succession, and Stratification in the First Six Civilizations," in Lee Ellis, ed., *Social Stratification and Socioeconomic Inequality.* Praeger.

Bielenstein, Hans (1981)   "The Chinese Empire: Foreign Rulers and National Restoration," in Garraty and Gay, eds. (1981).

Binford, Lewis R. (1968)   "Post-Pleistocene Adaptations," in Sally R. Binford and Lewis R. Binford, eds., *New Perspectives in Archeology.* Aldine.

Blackmore, Susan (1999)   *The Meme Machine.* Oxford University Press.

Blaut, J. M. (1993)   *The Colonizer's Model of the World: Geographical Diffusionism and Eurocentric History.* Guilford Press.

Boas, Franz (1899)   "The Northwestern Tribes of British Columbia." *Report of the Sixty-eighth Meeting of the British Association for the Advancement of Science.* London.

——— (1929)   *The Mind of Primitive Man.* Macmillan.

——— (1940)   *Race, Language and Culture.* Macmillan.

Boltz, William G. (1999)   "Language and Writing," in Michael Loewe and Edward L. Shaughnessy, eds., *The Cambridge History of Ancient China.* Cambridge University Press.

Bonner, John (1980)   *The Evolution of Culture in Animals.* Princeton University Press.

——— (1988)   *The Evolution of Complexity.* Princeton University Press.

——— (1993)   *Life Cycles.* Princeton University Press.

Boone, Elizabeth Hill, and Walter D. Mignolo, eds. (1994)   *Writing Without Words: Alternative Literacies in Mesoamerica and the Andes.* Duke University Press.

Boulding, Kenneth (1953)   *The Organizational Revolution.* Harper and Brothers.

——— (1985)   *The World as a Total System.* Sage.

Bowersock, G. W. (1988)   "The Dissolution of the Roman Empire," in Yoffee and Cowgill, eds. (1988).

Boyd, Robert, and Peter J. Richerson (1985)   *Culture and the Evolutionary Process.* University of Chicago Press.

Braidwood, Robert J. (1960)   "The Agricultural Revolution," *Scientific American,* September, pp. 131–48.

Braithwaite, Richard B. (1953)   *Scientific Explanation.* Cambridge University Press.

Braudel, Fernand (1981)   *Civilization and Capitalism, 15th–18th Century.* Vol. 1, *The Structures of Everyday Life.* Harper and Row.

Bray, Warwick (1991)   *Everyday Life of the Aztecs.* Peter Bedrick Books.

Brogan, Hugh (1989)   "The Forms of Government," in *Encyclopaedia Britannica,* 15th ed., vol. 20, pp. 189–95.

Bronson, Bennet (1988)   "The Role of Barbarians in the Fall of States," in Yoffee and Cowgill, eds. (1988).

Brown, James A., and T. Douglas Price (1985)   "Complex Hunter-Gatherers: Retrospect and Prospect," in Price and Brown, eds. (1985).

Brumfiel, Elizabeth (1992)   "Breaking and Entering the Ecosystem—Gender, Class, and Faction Steal the Show." *American Anthropologist* 94:551–67.

Butterfield, Herbert (1965)   *The Whig Interpretation of History.* W. W. Norton.

Cahill, Thomas (1995)   *How the Irish Saved Civilization.* Anchor.

Camerer, Colin F. (1997)   "Progress in Behavioral Game Theory." *Journal of Economic Perspectives* 11:167–88.

Camerer, Colin, and Richard H. Thaler (1995)   "Ultimatums, Dictators and Manners." *Journal of Economic Perspectives* 9:209–19.

Cameron, Euan (1999)   "The Power of the Word: Renaissance and Reformation," in Euan Cameron, ed., *Early Modern Europe.* Oxford University Press.

Campbell, Jeremy (1982)   *Grammatical Man: Information, Entropy, Language and Life.* Simon and Schuster.

Cantor, Norman F. (1993)   *The Civilization of the Middle Ages.* HarperPerennial.

Carneiro, Robert, ed. (1967)   *The Evolution of Society: Selections from Herbert Spencer's Principles of Sociology.* University of Chicago Press.

Carneiro, Robert (1968)   "Ascertaining, Testing, and Interpreting Sequences of Cultural Development." *Southwestern Journal of Anthropology* 24:354–74.

—— (1969)   "The Measurement of Cultural Development in the Ancient Near East and in Anglo-Saxon England." *Transactions of the New York Academy of Sciences* 31:1013–23.

—— (1970)   "A Theory of the Origin of the State." *Science* 169:733–38.

—— (1972)   "The Devolution of Evolution." *Social Biology* 19:248–58.

—— (1974)   "A Reappraisal of the Roles of Technology and Organization in the Origin of Civilization." *American Antiquity* 39:179–86.

—— (1978)   "Political Expansion as an Expression of the Principle of Competitive Exclusion," in Elman Service, ed., *Origins of the State,* pp. 205–23. Institute for the Study of Human Issues.

—— (1990)   "Chiefdom-Level Warfare as Exemplified in Fiji and the Cauca Valley," in Jonathan Haas, ed. (1990), *The Anthropology of War.* Cambridge University Press.

—— (1991)   "The Nature of the Chiefdom as Revealed by Evidence from the Cauca Valley of Colombia," in A. Terry Rambo and Kathleen Gillogly, eds. (1991), *Profiles in Cultural Evolution.* University of Michigan Press.

—— (1992)   "The Calusa and the Powhatan, Native Chiefdoms of North America." *Reviews in Anthropology* 21:27–38.

Cashdan, Elizabeth (1980)   "Egalitarianism Among Hunters and Gatherers." *American Anthropologist* 82:116–20.

Cavalli-Sforza, L. Luca, et al. (1994)   *The History and Geography of Human Genes.* Princeton University Press.

Chagnon, Napoleon (1968)   *Yanomamo: The Fierce People.* Holt, Rinehart and Winston.

——— (1973)   "The Culture-Ecology of Shifting (Pioneering) Cultivation Among the Yanomamo Indians," in Daniel R. Gross, ed. (1973), *Peoples and Cultures of Native South America.* Doubleday/Natural History Press.

——— (1979)   "Is Reproductive Success Equal in Egalitarian Societies?," in Napoleon Chagnon and William Irons, eds., *Evolutionary Biology and Human Social Behavior: An Anthropological Perspective.* Duxbury.

Chalmers, David (1996)   *The Conscious Mind.* Oxford University Press.

Chang, K. C. (1983)   "Sandai Archaeology and the Formation of States in Ancient China," in Keightley, ed. (1983).

Childe, V. Gordon (1954)   "Early Forms of Society," in Singer et al., eds. (1954–56).

Childers, Thomas (1998)   "Europe and Western Civilization in the Modern Age," audiotape lecture series. The Teaching Company.

Chirot, Daniel (1985)   "The Rise of the West." *American Sociological Review* 50:181–95.

Cipolla, Carlo M. (1994)   *Before the Industrial Revolution: European Society and Economy, 1000–1700.* W. W. Norton.

Clark, Gregory (1996)   "The Political Foundations of Modern Economic Growth: England, 1540–1800." *Journal of Interdisciplinary History* 26:563–88.

Clark, Kenneth (1969)   *Civilisation.* John Murray.

Clough, Shepard (1981a)   "The Industrial Revolution," in Garraty and Gay, eds. (1981).

——— (1981b)   "A World Economy," in Garraty and Gay, eds. (1981).

Coe, Michael (1992)   *Breaking the Maya Code.* Thames and Hudson.

Cohen, Mark Nathan (1977)   *The Food Crisis in Prehistory.* Yale University Press.

——— (1985)   "Prehistoric Hunter-Gatherers: The Meaning of Social Complexity," in Price and Brown, eds. (1985).

Coon, Carleton S. (1971)   *The Hunting Peoples.* Little, Brown.

Cotterell, Arthur (1995)   "The Unification of China," in Cotterell, ed. (1995).

Cotterell, Arthur, ed. (1995)   *The Penguin Encyclopedia of Classical Civilizations.* Penguin Books.

Craig, Albert M., et al. (1990)   *The Heritage of World Civilizations.* Macmillan.

Cronin, Helena (1991)   *The Ant and the Peacock.* Cambridge University Press.

Cronin, J. E., N. T. Boaz, C. B. Stringer, and Y. Rak (1981)   "Tempo and mode in hominid evolution." *Nature* 292:113–22.

Crow, James F. (1999)   "Unmasking a Cheating Gene." *Science* 283:1651.

Damerow, Peter (1996)   "Food Production and Social Status as Documented in Proto-Cuneiform Texts," in Wiessner and Schiefenhövel, eds. (1996).

Daniels, Peter T., and William Bright, eds. (1996)   *The World's Writing Systems.* Oxford University Press.

Davie, Maurice R. (1968)   *The Evolution of War.* Kennikat Press.

Davis, Morton D. (1989)   "Game Theory," in *Encyclopaedia Brittanica,* 15th ed., vol. 19, pp. 643–49.

Dawkins, Richard (1976)   *The Selfish Gene.* Oxford University Press.

——— (1986)   *The Blind Watchmaker.* W. W. Norton.

——— (1997)   "Human Chauvinism." *Evolution* 51:1015–20.

De Long, J. Bradford, and Andrei Shleifer (1993) "Princes and Merchants: European City Growth Before the Industrial Revolution." *Journal of Law and Economics* 36:671–702.

Dennett, Daniel (1991) *Consciousness Explained.* Little, Brown.

—— (1995) *Darwin's Dangerous Idea.* Simon and Schuster.

—— (1997) "Appraising Grace." *The Sciences,* January/February, pp. 39–44.

de Sola Pool, Ithiel (1983) *Technologies of Freedom.* Harvard University Press.

Deur, Douglas E. (1999) "Salmon, Sedentism, and Cultivation: Towards an Environmental Prehistory of the Northwest Coast," in Paul Hirt and Dale Goble, eds., *Northwest Lands and Peoples.* University of Washington Press.

Deur, Douglas E., and Nancy J. Turner, eds. (in press) *Keeping It Living: Traditional Plant Tending and Cultivation on the Northwest Coast.* University of Washington Press.

de Waal, Frans (1982) *Chimpanzee Politics.* Johns Hopkins University Press.

Diamond, Jared (1992) *The Third Chimpanzee: The Evolution and Future of the Human Animal.* HarperPerennial.

—— (1997) *Guns, Germs, and Steel: The Fates of Human Societies.* W. W. Norton.

Dobzhansky, Theodosius (1968) "Teilhard de Chardin and the Orientation of Evolution." *Zygon* 3:242–58.

Dray, W. H. (1967) "Philosophy of History," in Edwards, ed. (1967).

Drennan, Robert (1991) "Pre-Hispanic Chiefdom Trajectories in Mesoamerica, Central America, and Northern South America," in Earle, ed. (1991).

Drews, Robert (1993) *The End of the Bronze Age: Changes in Warfare and the Catastrophe ca. 1200 B.C.* Princeton University Press.

Drummond, Andrew (1995) "The World of Rome," in Cotterell, ed. (1995).

Dugatkin, Lee Alan (1997) *Cooperation Among Animals: An Evolutionary Perspective.* Oxford University Press.

Dulbecco, Renato (1987) *The Design of Life.* Yale University Press.

Dunbar, R. I. M. (1992) "Neocortex Size as a Constraint on Group Size in Primates." *Journal of Human Evolution* 20:469–93.

Durkheim, Émile (1912) "Elementary Forms of Religious Life," in Robert N. Bellah, ed. (1973), *Émile Durkheim: On Morality and Society.* University of Chicago Press.

Duus, Peter (1993) *Feudalism in Japan.* McGraw-Hill.

Earle, Timothy (1997) *How Chiefs Come to Power: The Political Economy in Prehistory.* Stanford University Press.

Earle, Timothy, ed. (1991) *Chiefdoms: Power, Economy, and Ideology.* Cambridge University Press.

Easterbrook, Gregg (1998) *Beside Still Waters: Searching for Meaning in an Age of Doubt.* Morrow.

Edwards, Paul, ed. (1967) *The Encyclopedia of Philosophy.* Macmillan.

Eisenstein, Elizabeth L. (1993) *The Printing Revolution in Early Modern Europe.* Cambridge University Press/Canto.

Elazar, Daniel J. (1998) *Constitutionalizing Globalization.* Rowman and Littlefield.

Elvin, Mark (1973) *The Pattern of the Chinese Past.* Stanford University Press.

Embree, Ainslee (1981) "India: 1500–1750," in Garraty and Gay, eds. (1981).

Engels, Friedrich (1884) "The Origin of Family, Private Property, and State," in Tucker, ed. (1978).

Evans-Pritchard, E. E. (1969)   *The Nuer.* Oxford University Press.

Fagan, Brian M. (1995)   *People of the Earth: An Introduction to World Prehistory.* HarperCollins.

Fairbank, John King (1992)   *China: A New History.* Harvard University Press.

Farb, Peter (1969)   *Man's Rise to Civilization.* Avon.

Ferguson, Yale (1991)   "Chiefdoms to City-States: The Greek Experience," in Earle, ed. (1991).

Fisher, H. A. L. (1936)   *A History of Europe.* Edward Arnold and Co.

Flannery, Kent (1972)   "The Cultural Evolution of Civilizations." *Annual Review of Ecology and Systematics* 3:399–426.

——— (1973)   "The Origins of Agriculture." *Annual Review of Anthropology* 2:271–310.

——— (1995)   "Prehistoric Social Evolution," in Carol R. Ember and Melvin Ember, eds. (1995), *Research Frontiers in Anthropology.* Prentice-Hall.

Fortey, Richard (1998)   *Life: A Natural History of the First Four Billion Years of Life on Earth.* Alfred A. Knopf.

Frank, André Gunder (1998)   *ReOrient: Global Economy in the Asian Age.* University of California Press.

Frank, Robert (1999)   *Luxury Fever.* Free Press.

Frank, Steven A. (1989)   "The Evolutionary Dynamics of Cytoplasmic Male Sterility." *American Naturalist* 133:345–76.

——— (1997)   "Models of Symbiosis." *American Naturalist* 150:S80–S99.

Fried, Morton (1983)   "Tribe to State or State to Tribe in Ancient China," in Keightley, ed. (1983).

Friedman, Thomas (1999)   *The Lexus and the Olive Tree.* Farrar, Straus and Giroux.

Friedrich, Otto (1986)   *The End of the World: A History.* Fromm International.

Fromkin, David (1981)   *The Independence of Nations.* Praeger.

Fukuyama, Francis (1993)   *The End of History and the Last Man.* Avon.

Gaddis, John L. (1999)   "Living in Candlestick Park." *The Atlantic,* April, pp. 65–74.

Garraty, John A., and Peter Gay, eds. (1981)   *The Columbia History of the World.* Harper and Row.

Garsoian, Nina (1981)   "Early Byzantium," in Garraty and Gay, eds. (1981).

Gernet, Jacques (1962)   *Daily Life in China on the Eve of the Mongol Invasion, 1250–1276.* Macmillan.

——— (1996)   *A History of Chinese Civilization.* Cambridge University Press.

Gies, Frances, and Joseph Gies (1995)   *Cathedral, Forge, and Waterwheel: Technology and Invention in the Middle Ages.* HarperPerennial.

Gilder, George (1989)   *Microcosm: The Quantum Revolution in Economics and Technology.* Simon and Schuster.

Gledhill, J., B. Bender, and M. T. Larsen, eds. (1995)   *State and Society: The Emergence and Development of Social Hierarchy and Political Centralization.* Routledge.

Goffart, Walter (1980)   *Barbarians and Romans, A.D. 418–584: The Techniques of Accommodation.* Princeton University Press.

Goodenough, Oliver (1995)   "Mind Viruses: Culture, Evolution and the Puzzle of Altruism." *Biology and Social Life* 34:287–320.

Goodenough, Ursula (1998)   *The Sacred Depths of Nature.* Oxford University Press.

Goody, Jack (1986) *The Logic of Writing and the Organization of Society*. Cambridge University Press.

Goudge, T. A. (1967) "Henri Bergson," in Edwards, ed. (1967).

Gould, Stephen Jay (1982) *The Panda's Thumb*. W. W. Norton.

—— (1989) *Wonderful Life: The Burgess Shale and the Nature of History*. W. W. Norton.

—— (1996) *Full House: The Spread of Excellence from Plato to Darwin*. Harmony Books.

Grew, Raymond (1993) "On the Prospect of Global History," in Mazlish and Buultjens, eds. (1993).

Hamilton, William D. (1964) "The Genetical Evolution of Social Behaviour," parts 1 and 2. *Journal of Theoretical Biology* 7:1–52.

Hammond, N. G. L. (1989) *Alexander the Great*. Bristol Press.

Harris, Marvin (1991) *Cannibals and Kings*. Vintage Books.

Haught, John (1999) *God After Darwin: A Theology of Evolution*. Westview Press.

Hawkes, Kristen, and James F. O'Connell (1981) "Affluent Hunters?" *American Anthropologist* 83:622–26.

Hayden, Brian (1993) *Archaeology: The Science of Once and Future Things*. W. H. Freeman.

—— (1995) "A New Overview of Domestication," in Price and Gebauer, eds. (1995).

—— (1998) "Practical and Prestige Technologies: The Evolution of Material Systems." *Journal of Archaeological Method and Theory* 5:1–55.

Hays, H. R. (1958) *From Ape to Angel: An Informal History of Social Anthropology*. Alfred A. Knopf.

Hefner, Philip (1970) *The Promise of Teilhard*. Lippincott.

Henry, Donald O. (1985) "Preagricultural Sedentism: The Natufian Example," in Price and Brown, eds. (1985).

Herskovits, Melville J. (1938) *Dahomey: An Ancient West African Kingdom*. J. J. Augustin.

Hill, Kim, et al. (1985) "Men's Allocation to Subsistence Work among the Ache of Eastern Paraguay." *Human Ecology* 13:29–47.

von Hippel, Arndt (1994) *Human Evolutionary Biology*. Stone Age Press.

Ho, Ping-Ti (1975) *The Cradle of the East*. University of Chicago Press.

Hoebel, E. Adamson (1983) *The Law of Primitive Man*. Atheneum.

Hofstadter, Richard (1955) *Social Darwinism in American Thought*. Beacon Press.

Hollister, C. Warren (1974) *Medieval Europe*. John Wiley and Sons.

Hooker, J. T. (1995) "Hellenic Civilization," in Cotterell, ed. (1995).

Houston, Stephen (1994) "Literacy Among the Pre-Columbian Maya: A Comparative Perspective," in Boone and Mignolo, eds. (1994).

Hsu, Cho-yun (1988) "The Roles of the Literati and of Regionalism in the Fall of the Han Dynasty," in Yoffee and Cowgill, eds. (1988).

—— (1989) "The Chou and Ch'in Dynasties," in *Encyclopaedia Britannica*, 15th ed., vol. 16, pp. 70–75.

Huber, Peter (1994) *Orwell's Revenge*. Free Press.

Hucker, Charles O. (1975) *China's Imperial Past: An Introduction to Chinese History and Culture*. Stanford University Press.

Huntington, Samuel (1996)   *The Clash of Civilizations and the Remaking of World Order.* Simon and Schuster.

Hutchinson, John, and Anthony D. Smith, eds. (1994)   *Nationalism.* Oxford University Press.

Huxley, Julian (1959)   *New Bottles for New Wine.* Chatto and Windus.

Insko, Chester A., et al. (1983)   "Trade Versus Expropriation in Open Groups: A Comparison of Two Types of Social Power." *Journal of Personality and Social Psychology* 44:977–99.

Irwin, Douglas (1996)   *Against the Tide: An Intellectual History of Free Trade.* Princeton University Press.

Jacob, François (1982)   *The Logic of Life: A History of Heredity.* Pantheon Books.

Jastrow, Robert (1981)   *The Enchanted Loom.* Touchstone.

Jerison, Harry J. (1973)   *Evolution of the Brain and Intelligence.* Academic Press.

Johnson, Allen (1989)   "Horticulturalists: Economic Behavior in Tribes," in Plattner, ed. (1989).

Johnson, Allen, and Timothy Earle (1987)   *The Evolution of Human Societies: From Foraging Group to Agrarian State.* Stanford University Press.

Johnson, Gary (1995)   "The Evolutionary Origins of Government and Politics," in Albert Somit and Joseph Losco, eds. (1995), *Human Nature and Politics.* JAI Press.

Jones, A. H. M. (1986)   *The Later Roman Empire, 284–602.* Johns Hopkins University Press.

Jones, E. L. (1987)   *The European Miracle.* Cambridge University Press.

Kant, Immanuel (1784)   "Idea for a Universal History with a Cosmopolitan Purpose," in Hans Reiss, ed. (1991), *Kant: Political Writings.* Cambridge University Press.

Kaplan, Robert (1994)   "The Coming Anarchy." *The Atlantic,* February.

——— (1997)   *The Ends of the Earth.* Vintage Books.

Kaufman, Herbert (1988)   "The Collapse of Ancient States and Civilizations as an Organizational Problem," in Yoffee and Cowgill, eds. (1988).

Keeley, Lawrence H. (1995)   "Protoagricultural Practices Among Hunter-Gatherers," in Price and Gebauer, eds. (1995).

——— (1996)   *War Before Civilization.* Oxford University Press.

Keeton, William T., and James L. Gould (1986)   *Biological Science,* 4th ed. W. W. Norton.

Keightley, David N. (1983)   "The Late Shang State: When, Where, and What," in Keightley, ed. (1983).

——— (1989a)   "The Origins of Writing in China: Scripts and Cultural Contexts," in Senner, ed. (1989).

——— (1989b)   "The First Historical Dynasty: The Shang," in *Encyclopaedia Britannica,* 15th ed., vol. 16, pp. 68–70.

——— (1995)   "A Measure of Man in Early China: In Search of the Neolithic Inch." *Chinese Science* 12:18–40.

——— (1996)   "Art, Ancestors, and the Origins of Writing in China." *Representations* 56:68–94.

Keightley, David N., ed. (1983)   *The Origins of Chinese Civilization.* University of California Press.

Kelly, Kevin (1995)   *Out of Control.* Perseus.

Kennedy, Paul (1987) *The Rise and Fall of the Great Powers.* Random House.

Kenoyer, Jonathan Mark (1998) *Ancient Cities of the Indus Valley Civilization.* Oxford University Press.

Keohane, Robert, and Joseph Nye (1989) *Power and Interdependence.* Harper-Collins.

Kindleberger, Charles P. (1993) *A Financial History of Western Europe.* Oxford University Press.

Kirch, Patrick V. (1989) *The Evolution of the Polynesian Chiefdoms.* Cambridge University Press.

Klein, Richard G. (1989) *The Human Career.* University of Chicago Press.

Kohn, Hans (1973) "Nationalism," in Philip P. Wiener, ed., *Dictionary of the History of Ideas,* vol. 3. Scribner's.

——— (1994) "Western and Eastern Nationalism," in Hutchinson and Smith, eds. (1994).

Kremer, Michael (1993) "Population Growth and Technological Change: One Million B.C. to 1990." *The Quarterly Journal of Economics* 103:681–716.

Krugman, Paul (1999) "Monomoney Mania," *Slate,* April 15.

Kuran, Timur (1997) "Islam and Underdevelopment: An Old Puzzle Revisited." *Journal of Institutional and Theoretical Economics* 153:41–71.

Lamberg-Karlovsky, C. C., and Jeremy A. Sabloff (1995) *Ancient Civilizations: The Near East and Mesoamerica.* Waveland Press.

Landes, David S. (1998) *The Wealth and Poverty of Nations.* W. W. Norton.

Larsen, Mogens Trolle (1988) "Introduction: Literacy and Social Complexity," in Gledhill, Bender, and Larsen (1995).

Latourette, Kenneth Scott (1975) *A History of Christianity.* HarperCollins.

Leakey, Richard (1994) *The Origin of Humankind.* Basic Books.

Lee, Richard B. (1979) *The !Kung San.* Cambridge University Press.

Lenski, Gerhard, Patrick Nolan, and Jean Lenski (1995) *Human Societies: An Introduction to Macrosociology.* McGraw-Hill.

Loewenstein, Werner R. (1999) *The Touchstone of Life: Molecular Information, Cell Communication, and the Foundations of Life.* Oxford University Press.

Lopez, Robert S., and Raymond W. Irving (1990) *Medieval Trade in the Mediterranean World.* Columbia University Press.

Lounsbury, Floyd (1989) "The Ancient Writing of Middle America," in Senner, ed. (1989).

Lourandos, Harry (1985) "Intensification and Australian Prehistory," in Price and Brown, eds. (1985).

Lowie, Robert (1920) *Primitive Society.* Liveright.

——— (1946) "Evolution in Cultural Anthropology: A Reply to Leslie White." *American Anthropologist* 48:223–33.

——— (1966) *Culture and Ethnology.* Basic Books.

Mackay, Alan L., ed. (1981) *The Harvest of a Quiet Eye: A Selection of Scientific Quotations.* Crane, Russak.

McLuhan, Marshall (1966) *Understanding Media: The Extensions of Man.* Signet.

MacMullen, Ramsay (1988) *Corruption and the Decline of Rome.* Yale University Press.

McNeill, William H. (1963) *The Rise of the West: A History of the Human Community.* University of Chicago Press.

———— (1977)   *Plagues and People.* Anchor.

———— (1980)   *The Human Condition: An Ecological and Historical View.* Princeton University Press.

———— (1989)   *Arnold J. Toynbee: A Life.* Oxford University Press.

———— (1990)   "*The Rise of the West* after Twenty-five Years." *Journal of World History* 1:1–21.

Madina, Maan Z. (1981)   "The Arabs and the Rise of Islam," in Garraty and Gay, eds. (1981).

Margulis, Lynn, and Dorion Sagan (1995)   *What Is Life?* Simon and Schuster.

Marten, Ken, et al. (1996)   "Ring Bubbles of Dolphins." *Scientific American,* August, pp. 83–87.

Martin, Henri-Jean (1994)   *The History and Power of Writing.* University of Chicago Press.

Marx, Karl (1846)   "Society and Economy in History," in Tucker, ed. (1978).

Maschner, Herbert D. G. (1991)   "The Emergence of Cultural Complexity on the Northern Northwest Coast." *Antiquity* 65:924–34.

Mathews, Jessica (1997)   "The Rise of Global Civil Society." *Foreign Affairs,* January.

Maynard Smith, John, and Eörs Szathmáry (1995)   *The Major Transitions in Evolution.* W. H. Freeman.

Mazlish, Bruce (1993)   "An Introduction to Global History," in Mazlish and Buultjens, eds. (1993).

Mazlish, Bruce, and Ralph Buultjens, eds. (1993)   *Conceptualizing Global History.* Westview Press.

Michod, Richard E. (1997)   "Evolution of the Individual." *American Naturalist* 150:S80–S99.

Mill, John Stuart (1872)   "Of the Inverse Deductive, or Historical Method," in John Stuart Mill (1872), *A System of Logic.* Longmans, Green, Reader, and Dyer.

Mokyr, Joel (1992)   *The Lever of Riches: Technological Creativity and Economic Progress.* Oxford University Press.

Monod, Jacques (1971)   *Chance and Necessity.* Alfred A. Knopf.

Morgan, David (1990)   *The Mongols.* Blackwell.

Morgan, Lewis H. (1877)   *Ancient Society.* Charles H. Kerr.

Morris, Brian (1987)   *Anthropological Studies of Religion.* Cambridge University Press.

Morris, Simon Conway (1998)   *The Crucible of Creation: The Burgess Shale and the Rise of Animals.* Oxford University Press.

Mundy, John H. (1981)   "The Late Middle Ages," in Garraty and Gay, eds. (1981).

Murdock, George Peter (1934)   *Our Primitive Contemporaries.* Macmillan.

Nagel, Thomas (1974)   "What Is It Like to Be a Bat?" *Philosophical Review* 83:435–50.

Nasar, Sylvia (1999)   *A Beautiful Mind.* Touchstone.

Nash, John F., Jr. (1950)   "The Bargaining Problem." *Econometrica* 18:155–62.

Nef, John (1964)   *The Conquest of the Material World.* University of Chicago Press.

Nisbet, Robert A. (1969)   *Social Change and History.* Oxford University Press.

Nishida, Toshisada (1987)   "Local Traditions and Cultural Transmission," in Barbara B. Smuts et al., eds. (1987), *Primate Societies.* University of Chicago Press.

Nissen, Hans J. (1985) "The Emergence of Writing in the Ancient Near East." *Interdisciplinary Science Reviews* 10:349–61.

Nissen, Hans, Peter Damerow, and Robert Englund (1993) *Archaic Bookkeeping: Early Writing and Techniques of Economic Administration in the Ancient Near East.* University of Chicago Press.

Oates, Joan (1979) *Babylon.* Thames and Hudson.

Oden, Robert (1995) "The Old Testament: An Introduction," audiotape lecture series. The Teaching Company.

Ogburn, William (1950) *Social Change.* Viking.

Ogburn, William, and Dorothy Thomas (1922) "Are Inventions Inevitable?" *Political Science Quarterly* 37:83–98.

Oliver, Douglas L. (1955) *A Solomon Island Society.* Harvard University Press.

Pacey, Arnold (1990) *Technology in World Civilization.* MIT Press.

Painter, Sidney (1951) *Mediaeval Society.* Cornell University Press.

Palmer, R. R., and Joel Colton (1965) *A History of the Modern World.* Alfred A. Knopf.

Peace, William (1993) "Leslie A. White and Evolutionary Theory." *Dialectical Anthropology* 18:123–51.

Perdue, Peter C. (1999) "China in the Early Modern World." *Education About Asia.* Summer.

Peirce, Charles S. (1878) "How to Make Our Ideas Clear," in Justus Buchler, ed. (1955), *Philosophical Writings of Peirce.* Dover.

Philo (1927) "On the Cherubim," in *Philo in Ten Volumes.* Harvard University Press (1927).

Pinker, Steven (1997) *How the Mind Works.* W. W. Norton.

Plattner, Stuart, ed. (1989) *Economic Anthropology.* Stanford University Press.

Pohl, John M. D. (1994) "Mexican Codices, Maps, and Lienzos as Social Contracts," in Boone and Mignolo, eds. (1994).

Popper, Karl (1957) *The Poverty of Historicism.* Routledge and Kegan Paul.

Price, T. Douglas, and James A. Brown (1985) "Aspects of Hunter-Gatherer Complexity," in Price and Brown, eds. (1985).

Price, T. Douglas, and James A. Brown, eds. (1985) *Prehistoric Hunter-Gatherers: The Emergence of Cultural Complexity.* Academic Press.

Price, T. Douglas, and Anne Birgitte Gebauer (1995) "New Perspectives on the Transition to Agriculture," in Price and Gebauer, eds. (1995).

Price, T. Douglas, and Anne Birgitte Gebauer, eds. (1995) *Last Hunters, First Farmers.* School of American Research Press.

Pringle, J. W. S. (1951) "On the Parallel between Learning and Evolution." *Behaviour* 3:174–215.

Queller, David (1997) "Cooperators Since Life Began." *Quarterly Review of Biology* 72:184–88.

Queller, David C., and J. E. Strassmann (1998) "Kin Selection and Social Insects." *Bioscience* 48:165–75.

Ranum, Orest (1981a) "The Collapse of France," in Garraty and Gay, eds. (1981).

——— (1981b) "Forming National States," pp. 725–32, in Garraty and Gay, eds. (1981)

——— (1981c) "Elizabethans and Puritans," pp. 566–83, in Garraty and Gay, eds. (1981).

Rapaport, Anatol (1960)   *Fights, Games, and Debates.* University of Michigan Press.

Reischauer, Edwin O. (1981)   *Japan.* Charles E. Tuttle Co.

Rice, Eugene F., Jr. (1981a)   "The State System of the Italian Renaissance," in Garraty and Gay, eds. (1981).

—————— (1981b)   "The Reformation: Society," in Garraty and Gay, eds. (1981).

Riché, Pierre (1976)   *Education and Culture in the Barbarian West, Sixth Through Eighth Centuries.* University of South Carolina Press.

Richerson, Peter J., and Robert Boyd (1998)   "The Evolution of Human Ultrasociality," in Irenäus Eibl-Eibesfeldt and Frank Kemp Salter, eds. (1998), *Indoctrinability, Ideology, and Warfare.* Berghahn Books.

—————— (1999)   "Complex Societies: The Evolutionary Origins of a Crude Superorganism." *Human Nature,* vol. 10, no. 3.

Ridley, Mark (1993)   *Evolution.* Blackwell Scientific Publications.

Ridley, Matt (1994)   *The Red Queen.* Macmillan.

—————— (1996)   *The Origins of Virtue: Human Instincts and the Evolution of Cooperation.* Viking.

Roberts, J. M. (1993)   *History of the World.* Oxford University Press.

Robinson, Daniel N. (1997)   "The Great Ideas of Philosophy," audiotape lecture series. The Teaching Company.

Roccati, Alessandro (1997)   "Scribes," in Sergio Donadoni, ed. (1997), *The Egyptians.* University of Chicago Press.

Romer, Paul M. (1990)   "Endogenous Technological Change." *Journal of Political Economy* 98:S71–S102.

Romer, Paul (1993)   "Implementing a National Technology Strategy with Self-Organizing Industry Investment Boards." *Brookings Papers: Microeconomics 2,* pp. 345–99.

Rosenau, James (1983)   " 'Fragmegrative' Challenges to National Security," in Terry L. Heyns, ed. (1983), *Understanding U.S. Strategy: A Reader.* National Defense University.

—————— (1990)   *Turbulence in World Politics.* Princeton University Press.

Rosenau, James, and Ernst-Otto Czempiel, eds. (1992)   *Governance Without Government.* Cambridge University Press.

Rosenblueth, Arturo, Norbert Wiener, and Julian Bigelow (1943)   "Behavior, Purpose and Teleology." *Philosophy of Science* 10: 18–24.

Rossabi, Morris, ed. (1983)   *China Among Equals: The Middle Kingdom and Its Neighbors, 10th–14th Centuries.* University of California Press.

Rothman, Mitchell S. (1994)   "Evolutionary Typologies and Cultural Complexity," in Stein and Rothman, eds. (1994).

Rue, Loyal (1999)   *Everybody's Story.* SUNY Press.

Ruhlen, Merritt (1987)   *A Guide to the World's Languages.* Vol. 1, *Classification.* Stanford University Press.

Sabloff, Jeremy A., and William L. Rathje (1975)   "The Rise of a Maya Merchant Class." *Scientific American,* October, pp. 72–82.

Saggs, H. W. F. (1989)   *Civilization Before Greece and Rome.* Yale University Press.

Sahlins, Marshall (1961)   "The Segmentary Lineage: An Organization of Predatory Expansions." *American Anthropologist* 63:322–45.

—————— (1963)   "Poor Man, Rich Man, Big-man, Chief: Political Types in Melanesia and Polynesia." *Comparative Studies in Society and History* 5:285–303.

———— (1972)   *Stone Age Economics*. Aldine-Atherton.

Sanders, William T., and Barbara J. Price (1968)   *Mesoamerica: The Evolution of a Civilization*. Random House.

Sanderson, Stephen K. (1990)   *Social Evolutionism: A Critical History*. Blackwell.

Sapp, Jan (1994)   *Evolution by Association: A History of Symbiosis*. Oxford University Press.

Schelling, Thomas C. (1980)   *The Strategy of Conflict*. Harvard University Press.

Schmandt-Besserat, Denise (1989)   "Two Precursors of Writing in One Reckoning Device," in Senner, ed. (1989).

Schrödinger, Erwin (1967)   *What Is Life & Mind and Matter*. Cambridge University Press.

Schwarzenberger, Georg (1989)   "International Law," in *Encyclopaedia Britannica*, 15th ed., vol. 21, pp. 724–31.

Senner, Wayne M., ed. (1989)   *The Origins of Writing*. University of Nebraska Press.

Service, Elman R. (1962)   *Primitive Social Organization: An Evolutionary Perspective*. Random House.

———— (1975)   *Origins of the State and Civilization: The Process of Cultural Evolution*. W. W. Norton.

———— (1978)   *Profiles in Ethnology*. HarperCollins.

Seyfarth, Robert M. (1987)   "Vocal Communication and Its Relation to Language," in Barbara B. Smuts et al., eds. (1987), *Primate Societies*. University of Chicago Press.

Shelton, Jo-Ann, ed. (1988)   *As the Romans Did*. Oxford University Press.

Singer, Charles, et al., eds. (1954–56)   *A History of Technology*, 2vols. Oxford University Press.

Singer, Peter (1981)   *The Expanding Circle: Ethics and Sociobiology*. Farrar, Straus and Giroux.

Smith, Adam (1937)   *The Wealth of Nations*. Random House Modern Library. (Originally published, 1776.)

Smith, Bruce D. (1995)   *The Emergence of Agriculture*. Scientific American Library.

Smith, Merritt Roe, and Leo Marx, eds. (1994)   *Does Technology Drive History?* MIT Press.

Smith, Michael E. (1997)   "Life in the Provinces of the Aztec Empire." *Scientific American*, September, pp. 74–83.

Solow, Robert M. (1957)   "Technical Change and the Aggregate Production Function." *Review of Economics and Statistics* 39:312–20.

Soros, George (1999)   "Empower the IMF." *Civilization*, June/July.

Spence, Jonathan (1990)   *The Search for Modern China*. W. W. Norton.

Spencer, Herbert (1851)   *Social Statics*. John Chapman.

Spybey, Tony (1996)   *Globalization and World Society*. Polity Press.

Starr, Chester G. (1991)   *A History of the Ancient World*. Oxford University Press.

Stein, Gil (1994a)   "The Organizational Dynamics of Complexity in Greater Mesopotamia," in Stein and Rothman, eds. (1994).

———— (1994b)   "Economy, Ritual, and Power in 'Ubaid Mesopotamia," in Stein and Rothman, eds. (1994).

———— (1998)   "Heterogeneity, Power, and Political Economy: Some Current Research Issues in the Archaeology of Old World Complex Societies." *Journal of Archaeological Research* 6:1–44.

Stein, Gil, and Mitchell S. Rothman, eds. (1994)   *Chiefdoms and Early States in the Near East: The Organizational Dynamics of Complexity.* Prehistory Press.

Steward, Julian (1938)   *Basin-Plateau Aboriginal Sociopolitical Groups.* Smithsonian Institution, U.S. Government Printing Office.

——— (1955)   *Theory of Culture Change: The Methodology of Multilinear Evolution.* University of Illinois Press.

Stock, Gregory (1993)   *Metaman.* Simon and Schuster.

Strayer, Joseph (1955)   *Western Europe in the Middle Ages.* Appleton-Century-Crofts.

Swimme, Brian, and Thomas Berry (1994)   *The Universe Story.* HarperCollins.

Tainter, Joseph A. (1990)   *The Collapse of Complex Societies.* Cambridge University Press.

Teilhard de Chardin, Pierre (1969)   *The Future of Man.* Harper Torchbooks.

Thomas, David Hurst (1972)   "Western Shoshone Ecology: Settlement Patterns and Beyond," in Don D. Fowler, ed., *Great Basin Cultural Ecology: A Symposium.* Desert Research Institute.

Thomas, Hugh (1979)   *A History of the World.* Harper and Row.

Thompson, E. A. (1948)   *A History of Attila and the Huns.* Clarendon Press.

Tomkins, Gordon (1975)   "The Metabolic Code." *Science* 189: 760–63.

Tooby, John, and Leda Cosmides (1990)   "On the Universality of Human Nature and the Uniqueness of the Individual: The Role of Genetics and Adaptation." *Journal of Personality* 58:1:17–67.

Toynbee, Arnold J. (1947)   *A Study of History.* Abridged by D. C. Somervell, 2 vols. Oxford University Press.

Toyoda, Takeshi (1989)   "Ancient and Medieval Japan," in *Encyclopaedia Britannica,* 15th ed., vol. 22, pp. 303–16.

Treadgold, Warren (1997)   *A History of the Byzantine State and Society.* Stanford University Press.

Trigger, Bruce (1993)   *Early Civilizations: Ancient Egypt in Context.* The American University in Cairo Press.

Trivers, Robert (1971)   "The Evolution of Reciprocal Altruism." *Quarterly Review of Biology* 46:35–56.

Tucker, Robert C., ed. (1978)   *The Marx-Engels Reader.* W. W. Norton.

Udovitch, Abraham L. (1970)   "Commercial Techniques in Early Medieval Islamic Trade," in D. S. Richards, ed. (1970), *Islam and the Trade of Asia.* University of Pennsylvania Press.

Waldman, Marilyn R. (1989)   "The Islamic World," in *Encyclopaedia Britannica,* 15th ed., vol. 22.

Waldrop, M. Mitchell (1992)   *Complexity.* Simon and Schuster.

Waters, Malcolm (1995)   *Globalization.* Routledge.

Watson, Adam (1992)   *The Evolution of International Society.* Routledge.

Watson, James (1969)   *The Double Helix.* Mentor.

Watson, Patty Jo (1995)   "Explaining the Transition to Agriculture," in Price and Gebauer, eds. (1995).

Weber, Max (1961)   *General Economic History.* Collier Books.

Weinberg, Steven (1979)   *The First Three Minutes.* Bantam.

Wenke, Robert J. (1981)   "Explaining the Evolution of Cultural Complexity: A Review." *Advances in Archaeological Method and Theory* 4:79–127.

—— (1984) *Patterns in Prehistory: Humankind's First Three Million Years.* Oxford University Press.

White, Leslie (1940) "The Symbol." *Philosophy of Science* 7:451–63.

—— (1943) "Energy and the Evolution of Culture." *American Anthropologist* 45:335–56.

—— (1959) *The Evolution of Culture.* McGraw-Hill.

—— (1987) *Ethnological Essays,* edited and with an introduction by Beth Dillingham and Robert L. Carneiro. University of New Mexico Press.

White, Lynn, Jr. (1964) *Medieval Technology and Social Change.* Oxford University Press.

Whittaker, R. H. (1969) "New Concepts of Kingdoms and Organisms." *Science* 163: 150–59.

Widmer, Randolph J. (1988) *The Evolution of the Calusa.* University of Alabama Press.

—— (1994) "The Structure of Southeastern Chiefdoms," in Charles Hudson and Carmen Chaves Tesser, eds. (1994), *The Forgotten Centuries.* University of Georgia Press.

Wiessner, Polly, and Wulf Schiefenhövel (1996) *Food and the Status Quest: An Interdisciplinary Perspective.* Berghahn Books.

Williams, George (1966) *Adaptation and Natural Selection: A Critique of Some Current Evolutionary Thought.* Princeton University Press.

Williamson, Raymond K. (1984) *Introduction to Hegel's Philosophy of Religion.* SUNY Press.

Wills, Christopher (1993) *The Runaway Brain: The Evolution of Human Uniqueness.* Basic Books.

Wilson, A. N. (1997) *Paul: The Mind of the Apostle.* W. W. Norton.

Wilson, David Sloan (1997) "Altruism and Organism: Disentangling the Themes of Multilevel Selection Theory." *The American Naturalist* 150:S122–S34.

Wilson, Edward O. (1971) *The Insect Societies.* Harvard University Press.

—— (1975) *Sociobiology: The New Synthesis.* Harvard University Press.

—— (1978) *On Human Nature.* Harvard University Press.

—— (1998) *Consilience: The Unity of Knowledge.* Alfred A. Knopf.

Wolfson, Harry A. (1967) "Philo Judaeus," in Edwards, ed. (1967).

Wrangham, Richard, and Dale Peterson (1996) *Demonic Males.* Houghton Mifflin.

Wright, Henry (1994) "Prestate Political Formations," in Stein and Rothman, eds. (1994).

Wright, Robert (1985) "The Computer Behind the Curtain." *The Sciences,* July–August, pp. 4–6.

—— (1988) *Three Scientists and Their Gods: Looking for Meaning in an Age of Information.* Times Books.

—— (1989a) "One World, Max." *The New Republic,* November 6, pp. 68–75.

—— (1989b) "Tao Jones" (review of Gilder, 1989). *The New Republic,* November 20, pp. 38–42.

—— (1990) "The Intelligence Test" (review of Gould, 1989). *The New Republic,* January 29, pp. 28–36.

—— (1991) "Quest for the Mother Tongue." *The Atlantic,* pp. 39–68.

———— (1992)   "Why Is It Like Something to Be Alive?," in William H. Shore, ed., *Mysteries of Life and the Universe.* Harcourt Brace Jovanovich.

———— (1994)   *The Moral Animal: Evolutionary Psychology and Everyday Life.* Pantheon Books.

———— (1995a)   "Be Very Afraid: Nukes, Nerve Gas, and Anthrax Spores." *The New Republic,* May 1, pp. 19–27.

———— (1995b)   "Chaos Theory." *The New Republic,* July 10, 1995, p. 4.

———— (1995c)   "Microturfs." *The New Republic,* November 20, p. 4.

———— (1996)   "Highbrow Tribalism" (review of Huntington, 1996). *Slate,* November 1.

———— (1997)   "We're All One-Worlders Now." *Slate,* April 24.

Yoffee, Norman (1993)   "Too Many Chiefs?," in Norman Yoffee and A. Sherratt, eds., *Archaeological Theory: Who Sets the Agenda?* Cambridge University Press.

———— (1995)   "Political Economy in Early Mesopotamian States." *Annual Review of Anthropology* 24:281–311.

Yoffee, Norman, and George L. Cowgill, eds. (1988)   *The Collapse of Ancient States and Civilizations.* University of Arizona Press.

INDEX